Stewart's Calculus
Second Edition, Volume II

Student Solutions Manual

Barbara Frank
The Pennsylvania State University

Brooks/Cole Publishing Company
Pacific Grove, California

The trademark ITP is used under license.

Brooks/Cole Publishing Company
A Division of Wadsworth, Inc.

© 1991, 1987 by Wadsworth, Inc., Belmont, California 94002. All rights reserved. No part of this book may be reproduced, stored in a retrieval system, or transcribed, in any form or by any means—electronic, mechanical, photocopying, recording, or otherwise—without the prior written permission of the publisher, Brooks/Cole Publishing Company, Pacific Grove, California 93950, a division of Wadsworth, Inc.

Printed in the United States of America

10 9 8 7 6 5

ISBN 0-534-13214-6

Contents

Chapter 10 Infinite Sequences and Series   1

- 10.1 Sequences   1
- 10.2 Series   4
- 10.3 The Integral Test   7
- 10.4 The Comparison Tests   9
- 10.5 Alternating Series   12
- 10.6 Absolute Convergence and the Ratio and Root Tests   14
- 10.7 Strategy for Testing Series   17
- 10.8 Power Series   19
- 10.9 Taylor and Maclaurin Series   23
- 10.10 The Binomial Series   30
- 10.11 Approximation by Taylor Polynomials   32
   Review   37

Problems Plus   42

Chapter 11 Three-Dimensional Analytic Geometry and Vectors   47

- 11.1 Three-Dimensional Coordinate Systems   47
- 11.2 Vectors   50
- 11.3 The Dot Product   52
- 11.4 The Cross Product   56
- 11.5 Equations of Lines and Planes   59
- 11.6 Quadric Surfaces   64
- 11.7 Vector Functions and Space Curves   69
- 11.8 Arc Length and Curvature   75
- 11.9 Motion in Space: Velocity and Acceleration   79
- 11.10 Cylindrical and Spherical Coordinates   82
   Review   85

Applications Plus   91

Chapter 12 Partial Derivatives   97

- 12.1 Functions of Several Variables   97
- 12.2 Limits and Continuity   102
- 12.3 Partial Derivatives   106
- 12.4 Tangent Planes and Differentials   110
- 12.5 The Chain Rule   112
- 12.6 Directional Derivatives and the Gradient Vector   115
- 12.7 Maximum and Minimum Values   119
- 12.8 Lagrange Multipliers   123
   Review   126

Problems Plus   131

Chapter 13 Multiple Integrals   136

- 13.1 Double Integrals over Rectangles   136
- 13.2 Iterated Integrals   137
- 13.3 Double Integrals over General Regions   139
- 13.4 Double Integrals in Polar Coordinates   143
- 13.5 Applications of Double Integrals   146
- 13.6 Surface Area   148
- 13.7 Triple Integrals   149
- 13.8 Triple Integrals in Cylindrical and Spherical Coordinates   155
- 13.9 Change of Variables in Multiple Integrals   158
  Review   161

Applications Plus   166

Chapter 14 Vector Calculus   169

- 14.1 Vector Fields   169
- 14.2 Line Integrals   171
- 14.3 The Fundamental Theorem for Line Integrals   173
- 14.4 Green's Theorem   175
- 14.5 Curl and Divergence   178
- 14.6 Parametric Surfaces and Their Areas   182
- 14.7 Surface Integrals   185
- 14.8 Stokes' Theorem   189
- 14.9 The Divergence Theorem   191
  Review   193

Problems Plus   197

Chapter 15 Differential Equations   199

- 15.1 Basic Concepts; Separable and Homogeneous Equations   199
- 15.2 First Order Linear Equations   203
- 15.3 Exact Equations   205
- 15.4 Strategy for Solving First Order Equations   207
- 15.5 Second Order Linear Equations   209
- 15.6 Nonhomogeneous Linear Equations   211
- 15.7 Applications of Second Order Differential **Equations**   **213**
- 15.8 Series Solutions   214
  Review   217

Applications Plus   220

Appendices   224

- D   Lies My Calculator or Computer Told Me   224
- F   Complex Numbers   228

CHAPTER TEN

EXERCISES 10.1

1. $a_n = \dfrac{n}{2n+1} \Rightarrow \left\{\dfrac{1}{3}, \dfrac{2}{5}, \dfrac{3}{7}, \dfrac{4}{9}, \dfrac{5}{11}, \ldots\right\}$

3. $a_n = \dfrac{(-1)^{n-1}n}{2^n} \Rightarrow \left\{\dfrac{1}{2}, -\dfrac{1}{2}, \dfrac{3}{8}, -\dfrac{1}{4}, \dfrac{5}{32}, \ldots\right\}$

5. $a_n = \dfrac{1 \cdot 3 \cdot 5 \cdots (2n-1)}{n!} \Rightarrow \left\{1, \dfrac{3}{2}, \dfrac{5}{2}, \dfrac{35}{8}, \dfrac{63}{8}, \ldots\right\}$

7. $a_n = \sin\dfrac{n\pi}{2} \Rightarrow \{1, 0, -1, 0, 1, \ldots\}$

9. $a_1 = 1, a_{n+1} = \dfrac{1}{1+a_n} \Rightarrow \left\{1, \dfrac{1}{2}, \dfrac{2}{3}, \dfrac{3}{5}, \dfrac{5}{8}, \ldots\right\}$

11. $a_n = \dfrac{1}{2^n}$

13. $a_n = 3n - 2$

15. $a_n = (-1)^n\, n!$

17. $a_n = (-1)^{n+1}\left(\dfrac{n+1}{2n+1}\right)$

19.

21. $\lim\limits_{n \to \infty} \dfrac{1}{4n^2} = \dfrac{1}{4} \lim\limits_{n \to \infty} \dfrac{1}{n^2} = \dfrac{1}{4} \cdot 0 = 0$. Convergent.

23. $\lim\limits_{n \to \infty} \dfrac{n^2 - 1}{n^2 + 1} = \lim\limits_{n \to \infty} \dfrac{1 - 1/n^2}{1 + 1/n^2} = 1$. Convergent.

25. $\{a_n\}$ diverges since $\dfrac{n^2}{n+1} = \dfrac{n}{1 + 1/n} \to \infty$ as $n \to \infty$.

27. $0 < |a_n| = \dfrac{n^2}{1+n^3} = \dfrac{1}{1/n^2 + n} < \dfrac{1}{n}$ and $\lim\limits_{n \to \infty} \dfrac{1}{n} = 0$, so by the Squeeze Theorem, $\lim\limits_{n \to \infty} \dfrac{n^2}{1+n^3} = 0$, and, by Theorem 10.6, $\lim\limits_{n \to \infty} (-1)^n\left(\dfrac{n^2}{1+n^3}\right) = 0$.

29. $\lim\limits_{n \to \infty} \dfrac{1}{5^n} = \lim\limits_{n \to \infty} \left(\dfrac{1}{5}\right)^n = 0$, by (10.8) with $r = \dfrac{1}{5}$.

31. $\{a_n\} = \{0, -1, 0, 1, 0, -1, 0, 1, \ldots\}$. This sequence oscillates among $0, -1$, and 1 and so diverges.

Section 10.1

33. $a_n = (\pi/3)^n$ so $\{a_n\}$ diverges by (10.8) with $r = \pi/3 > 1$.

35. $\lim_{n \to \infty} \arctan 2n = \pi/2$ since $2n \to \infty$ as $n \to \infty$. Convergent.

37. $0 < \frac{3 + (-1)^n}{n^2} \leq \frac{4}{n^2}$ and $\lim_{n \to \infty} \frac{4}{n^2} = 0$, so $\left\{ \frac{3 + (-1)^n}{n^2} \right\}$ converges to 0 by the Squeeze Theorem.

39. $\lim_{x \to \infty} \frac{\ln(x^2)}{x} = \lim_{x \to \infty} \frac{2 \ln x}{x} \stackrel{H}{=} \lim_{x \to \infty} \frac{2/x}{1} = 0$, so by Theorem 10.2, $\left\{ \frac{\ln(n^2)}{n} \right\}$ converges to 0.

41. $\sqrt{n+2} - \sqrt{n} = (\sqrt{n+2} - \sqrt{n}) \frac{\sqrt{n+2} + \sqrt{n}}{\sqrt{n+2} + \sqrt{n}} = \frac{2}{\sqrt{n+2} + \sqrt{n}} < \frac{2}{2\sqrt{n}} = \frac{1}{\sqrt{n}} \to 0$ as $n \to \infty$. So by the Squeeze Theorem $\{\sqrt{n+2} - \sqrt{n}\}$ converges to 0.

43. $\lim_{x \to \infty} \frac{x}{2^x} \stackrel{H}{=} \lim_{x \to \infty} \frac{1}{(\ln 2)2^x} = 0$, so by Theorem 10.2 $\{a_n\}$ converges to 0.

45. Let $y = x^{-\frac{1}{x}}$. Then $\ln y = -(\ln x)/x$ and $\lim_{x \to \infty} (\ln y) \stackrel{H}{=} \lim_{x \to \infty} -(1/x)/1 = 0$, so $\lim_{x \to \infty} y = e^0 = 1$, and so $\{a_n\}$ converges to 1.

47. $0 \leq \frac{\cos^2 n}{2^n} \leq \frac{1}{2^n}$ [since $0 \leq \cos^2 n \leq 1$], so since $\lim_{n \to \infty} \frac{1}{2^n} = 0$, $\{a_n\}$ converges to 0 by the Squeeze Theorem.

49. The series converges, since $a_n = \frac{1 + 2 + 3 + \cdots + n}{n^2} = \frac{n(n+1)/2}{n^2}$ [Theorem 5.3]
$= \frac{n+1}{2n} = \frac{1 + 1/n}{2} \to \frac{1}{2}$ as $n \to \infty$.

51. $a_n = \frac{1}{2} \cdot \frac{2}{2} \cdot \frac{3}{2} \cdots \frac{(n-1)}{2} \cdot \frac{n}{2} \geq \frac{1}{2} \cdot \frac{n}{2} = \frac{n}{4} \to \infty$ as $n \to \infty$, so $\{a_n\}$ diverges.

53. $0 < a_n = \frac{n^3}{n!} = \frac{n}{n} \cdot \frac{n}{(n-1)} \cdot \frac{n}{(n-2)} \cdot \frac{1}{(n-3)} \cdots \frac{1}{3} \cdot \frac{1}{2} \cdot 1 \leq \frac{n^2}{(n-1)(n-2)(n-3)}$ [for
$n \geq 4] = \frac{1/n}{(1 - 1/n)(1 - 2/n)(1 - 3/n)} \to 0$ as $n \to \infty$, so by the Squeeze Theorem, $\{a_n\}$ converges to 0.

55. $0 < a_n = \frac{1 \cdot 3 \cdot 5 \cdots (2n-1)}{(2n)^n} = \frac{1}{2n} \cdot \frac{3}{2n} \cdot \frac{5}{2n} \cdots \frac{2n-1}{2n} \leq \frac{1}{2n} \cdot (1) \cdot (1) \cdots (1) = \frac{1}{2n} \to 0$ as $n \to \infty$, so by the Squeeze Theorem $\{a_n\}$ converges to 0.

57. If $|r| \geq 1$, then $\{r^n\}$ diverges (10.8), so $\{nr^n\}$ diverges also since $|nr^n| = n|r^n| \geq |r^n|$. If $|r| < 1$ then $\lim_{x \to \infty} xr^x = \lim_{x \to \infty} \frac{x}{r^{-x}} \stackrel{H}{=} \lim_{x \to \infty} \frac{1}{(-\ln r)r^{-x}} = \lim_{x \to \infty} \frac{r^x}{-\ln r} = 0$, so $\lim_{n \to \infty} nr^n = 0$, and hence $\{nr^n\}$ converges whenever $|r| < 1$.

59. $3(n+1) + 5 > 3n + 5$ so $\frac{1}{3(n+1)+5} < \frac{1}{3n+5} \Leftrightarrow a_{n+1} < a_n$ so $\{a_n\}$ is decreasing.

Section 10.1

61. $\left\{\frac{n-2}{n+2}\right\}$ is increasing since $a_n < a_{n+1} \Leftrightarrow \frac{n-2}{n+2} < \frac{(n+1)-2}{(n+1)+2} \Leftrightarrow (n-2)(n+3)$

   $< (n+2)(n-1) \Leftrightarrow n^2 + n - 6 < n^2 + n - 2 \Leftrightarrow -6 < -2$, which is of course true.

63. $a_1 = 0 > a_2 = -1 < a_3 = 0$, so the sequence is not monotonic.

65. $\left\{\frac{n}{n^2+n-1}\right\}$ is decreasing since $a_{n+1} < a_n \Leftrightarrow \frac{n+1}{(n+1)^2+(n+1)-1} < \frac{n}{n^2+n-1} \Leftrightarrow$

   $(n+1)(n^2+n-1) < n(n^2+3n+1) \Leftrightarrow n^3 + 2n^2 - 1 < n^3 + 3n^2 + n \Leftrightarrow$

   $0 < n^2 + n + 1 = (n + \frac{1}{2})^2 + \frac{3}{4}$, which is obviously true.

67. $a_1 = 2^{1/2}$, $a_2 = 2^{3/4}$, $a_3 = 2^{7/8}, \ldots, a_n = 2^{(2^n-1)/2^n} = 2^{(1-1/2^n)}$.

   $\lim_{n\to\infty} a_n = \lim_{n\to\infty} 2^{(1-1/2^n)} = 2^1 = 2$

   Alternate Solution: Let $L = \lim_{n\to\infty} a_n$ (we could show the limit exists by showing $\{a_n\}$ is bounded and increasing). So L must satisfy $L = \sqrt{2 \cdot L} \Rightarrow L^2 = 2L \Rightarrow L(L-2) = 0$ ($L \neq 0$ since the sequence increases) so $L = 2$.

69. (a) We show by induction that $\{a_n\}$ is increasing and bounded above by 3. Let $P(n)$ be the proposition that $a_{n+1} \geq a_n$ and $a_n \leq 3$. Clearly $P(1)$ is true. Assume $P(n)$ is true. Then

   $a_{n+1} \geq a_n \Rightarrow \frac{1}{a_{n+1}} \leq \frac{1}{a_n} \Rightarrow -\frac{1}{a_{n+1}} \leq -\frac{1}{a_n} \Rightarrow a_{n+2} = 3 - \frac{1}{a_{n+1}} \geq 3 - \frac{1}{a_n} = a_{n+1} \Leftrightarrow$

   $P(n+1)$. This proves that $\{a_n\}$ is increasing and bounded above by 3, so

   $1 = a_1 \leq a_n \leq 3$, i.e. $\{a_n\}$ is bounded, and hence convergent by Theorem 10.11.

   (b) If $L = \lim_{n\to\infty} a_n$, then $\lim_{n\to\infty} a_{n+1} = L$ also, so L must satisfy $L = 3 - 1/L$, so

   $L^2 - 3L + 1 = 0$ and the quadratic formula gives $L = \frac{3 \pm \sqrt{5}}{2}$. But $L > 1$, so

   $L = \frac{3 + \sqrt{5}}{2}$.

71. (a) Let a_n be the number of rabbit pairs in the nth month. Clearly $a_1 = 1 = a_2$. In the nth month, each pair that is 2 or more months old (i.e. a_{n-2} pairs) will have a pair of children to add to the a_{n-1} pairs already present. Thus $a_n = a_{n-1} + a_{n-2}$, so that $\{a_n\} = \{f_n\}$, the Fibonacci sequence.

   (b) $a_{n-1} = f_n/f_{n-1} = (f_{n-1} + f_{n-2})/f_{n-1} = 1 + f_{n-2}/f_{n-1} = 1 + 1/a_{n-2}$. If

   $L = \lim_{n\to\infty} a_n$, then L must satisfy $L = 1 + 1/L$ or $L^2 - L - 1 = 0$, so $L = \frac{1+\sqrt{5}}{2}$ (since

   L must be non-negative).

73. If $\lim_{n\to\infty} |a_n| = 0$ then $\lim_{n\to\infty} -|a_n| = 0$, and since $-|a_n| \leq a_n \leq |a_n|$, we have that

   $\lim_{n\to\infty} a_n = 0$ by the Squeeze Theorem.

3

75. (a) First we show that $a > a_1 > b_1 > b$.

$a_1 - b_1 = \frac{a+b}{2} - \sqrt{ab} = \frac{1}{2}(a - 2\sqrt{ab} + b) = \frac{1}{2}(\sqrt{a} - \sqrt{b})^2 > 0$ (since $a > b$) $\Rightarrow a_1 > b_1$.

Also $a - a_1 = a - \frac{a+b}{2} = \frac{a-b}{2} > 0$ and $b - b_1 = b - \sqrt{ab} = \sqrt{b}(\sqrt{b} - \sqrt{a}) < 0$,

so $a > a_1 > b_1 > b$. In the same way we can show that $a_1 > a_2 > b_2 > b_1$ and so the given assertion is true for $n = 1$. Suppose it is true for $n = k$, i.e., $a_k > a_{k+1} > b_{k+1} > b_k$.

Then $a_{k+2} - b_{k+2} = \frac{a_{k+1} + b_{k+1}}{2} - \sqrt{a_{k+1}b_{k+1}} = \frac{1}{2}(a_{k+1} - 2\sqrt{a_{k+1}b_{k+1}} + b_{k+1})$

$= \frac{1}{2}(\sqrt{a_{k+1}} - \sqrt{b_{k+1}})^2 > 0$ and

$a_{k+1} - a_{k+2} = a_{k+1} - \frac{a_{k+1} + b_{k+1}}{2} = \frac{a_{k+1} - b_{k+1}}{2} > 0$

$b_{k+1} - b_{k+2} = b_{k+1} - \sqrt{a_{k+1}b_{k+1}} = \sqrt{b_{k+1}}(\sqrt{b_{k+1}} - \sqrt{a_{k+1}}) < 0 \Rightarrow$

$a_{k+1} > a_{k+2} > b_{k+2} > b_{k+1}$ so the assertion is true for $n = k + 1$. Thus it is true for all n by mathematical induction.

(b) From part (a) we have $a > a_n > a_{n+1} > b_{n+1} > b_n > b$, which shows that both sequences are monotonic and bounded. So they are both convergent by Theorem 10.11.

(c) Let $\lim_{n \to \infty} a_n = \alpha$ and $\lim_{n \to \infty} b_n = \beta$. Then $\lim_{n \to \infty} a_{n+1} = \lim_{n \to \infty} \frac{a_n + b_n}{2} \Rightarrow$

$\alpha = \frac{\alpha + \beta}{2} \Rightarrow 2\alpha = \alpha + \beta \Rightarrow \alpha = \beta$.

EXERCISES 10.2

1. $\sum_{n=1}^{\infty} 4\left(\frac{2}{5}\right)^{n-1}$ is a geometric series with $a = 4$, $r = \frac{2}{5}$, so it converges to $\frac{4}{1 - 2/5} = \frac{20}{3}$.

3. $\sum_{n=1}^{\infty} \frac{2}{3}\left(-\frac{1}{3}\right)^{n-1}$ is geometric with $a = \frac{2}{3}$, $r = -\frac{1}{3}$, so it converges to $\frac{2/3}{1 - (-1/3)} = \frac{1}{2}$.

5. $\sum_{n=1}^{\infty} \frac{1}{36}\left(\frac{6}{5}\right)^{n-1}$ diverges since $r = \frac{6}{5} > 1$.

7. $a = 2$, $r = 3/4 < 1$, so series converges to $\frac{2}{1 - 3/4} = 8$.

9. $a = 5e/3$, $r = e/3 < 1$, so series converges to $\frac{5e/3}{1 - e/3} = \frac{5e}{3 - e}$.

11. $a = 1$, $r = 5/8 < 1$, so series converges to $\frac{1}{1 - 5/8} = \frac{8}{3}$.

13. $a = 64/3$, $r = 8/3 > 1$, so series diverges.

Section 10.2

15. $a = \dfrac{(-2)^4}{5^{-1}} = 80$, $|r| = \left|-\dfrac{2}{5}\right| < 1$, series converges to $\dfrac{80}{1-(-2/5)} = \dfrac{400}{7}$.

17. $\lim\limits_{n \to \infty} \dfrac{n}{n+1} = 1 \neq 0$ so series diverges by the Test for Divergence.

19. This series diverges, since if it converged, so would [by Theorem 10.19(a)] $2 \cdot \sum\limits_{n=1}^{\infty} \dfrac{1}{2n} = \sum\limits_{n=1}^{\infty} \dfrac{1}{n}$, which we know diverges (Example 7).

21. Converges. $s_n = \sum\limits_{i=1}^{n} \dfrac{1}{(3i-2)(3i+1)} = \sum\limits_{i=1}^{n}\left[\dfrac{1/3}{3i-2} - \dfrac{1/3}{3i+1}\right]$ $\{partial\ fractions\}$

$= \left[\dfrac{1}{3} \cdot 1 - \dfrac{1}{3} \cdot \dfrac{1}{4}\right] + \left[\dfrac{1}{3} \cdot \dfrac{1}{4} - \dfrac{1}{3} \cdot \dfrac{1}{7}\right] + \left[\dfrac{1}{3} \cdot \dfrac{1}{7} - \dfrac{1}{3} \cdot \dfrac{1}{10}\right] + \cdots$

$+ \left[\dfrac{1}{3} \cdot \dfrac{1}{3n-2} - \dfrac{1}{3} \cdot \dfrac{1}{3n+1}\right] = \dfrac{1}{3} - \dfrac{1}{3(3n+1)}$ [telescoping series]

$\lim\limits_{n \to \infty} s_n = \dfrac{1}{3} \Rightarrow \sum\limits_{n=1}^{\infty} \dfrac{1}{(3n-2)(3n+1)} = \dfrac{1}{3}$

23. Converges by Theorem 10.19.

$\sum\limits_{n=1}^{\infty}[2(0.1)^n + (0.2)^n] = 2\sum\limits_{n=1}^{\infty}(0.1)^n + \sum\limits_{n=1}^{\infty}(0.2)^n = 2\left(\dfrac{0.1}{1-0.1}\right) + \dfrac{0.2}{1-0.2} = \dfrac{2}{9} + \dfrac{1}{4} = \dfrac{17}{36}$

25. Diverges by the Test for Divergence. $\lim\limits_{n \to \infty} n/\sqrt{1+n^2} = \lim\limits_{n \to \infty} 1/\sqrt{1+1/n^2} = 1 \neq 0$.

27. Converges. $s_n = \sum\limits_{i=1}^{n} \dfrac{1}{i(i+2)} = \sum\limits_{i=1}^{n}\left[\dfrac{1/2}{i} - \dfrac{1/2}{i+2}\right]$ $\{partial\ fractions\}$

$= \left[\dfrac{1}{2} - \dfrac{1}{6}\right] + \left[\dfrac{1}{4} - \dfrac{1}{8}\right] + \left[\dfrac{1}{6} - \dfrac{1}{10}\right] + \cdots + \left[\dfrac{1}{2n-2} - \dfrac{1}{2n+2}\right] + \left[\dfrac{1}{2n} - \dfrac{1}{2n+4}\right]$

$= \dfrac{1}{2} + \dfrac{1}{4} - \dfrac{1}{2n+2} - \dfrac{1}{2n+4}$ [telescoping series]

Thus $\sum\limits_{n=1}^{\infty} \dfrac{1}{n(n+2)} = \lim\limits_{n \to \infty}\left[\dfrac{1}{2} + \dfrac{1}{4} - \dfrac{1}{2n+2} - \dfrac{1}{2n+4}\right] = \dfrac{3}{4}$.

29. Converges. $\sum\limits_{n=1}^{\infty} \dfrac{3^n + 2^n}{6^n} = \sum\limits_{n=1}^{\infty}\left[\left(\dfrac{1}{2}\right)^n + \left(\dfrac{1}{3}\right)^n\right] = \dfrac{1/2}{1-1/2} + \dfrac{1/3}{1-1/3} = \dfrac{3}{2}$

31. Converges. $s_n = (\sin 1 - \sin 1/2) + (\sin 1/2 - \sin 1/3) + \cdots$

$+ [\sin(1/n) - \sin(1/(n+1))] = \sin 1 - \sin[1/(n+1)]$

$\sum\limits_{n=1}^{\infty}[\sin(1/n) - \sin(1/(n+1))] = \lim\limits_{n \to \infty} s_n = \sin 1 - \sin 0 = \sin 1$

33. Diverges since $\lim\limits_{n \to \infty} \arctan n = \dfrac{\pi}{2} \neq 0$.

35. $s_n = (\ln 1 - \ln 2) + (\ln 2 - \ln 3) + (\ln 3 - \ln 4) + \cdots + [\ln n - \ln(n+1)] = \ln 1 - \ln(n+1)$
$= -\ln(n+1)$ [telescoping series]. Thus $\lim\limits_{n \to \infty} s_n = -\infty$, so the series is divergent.

Section 10.2

37. $0.\bar{5} = .5 + .05 + .005 + \cdots = \frac{.5}{1-.1} = \frac{5}{9}$

39. $0.\overline{307} = .307 + .000307 + .000000307 + \cdots = \frac{.307}{1-.001} = \frac{307}{999}$

41. $0.123\overline{456} = \frac{123}{1000} + \frac{.000456}{1-.001} = \frac{123}{1000} + \frac{456}{999000} = \frac{123333}{999000} = \frac{41111}{333000}$

43. $\sum_{n=0}^{\infty}(x-3)^n$ is a geometric series with $r = x-3$, so converges whenever $|x-3| < 1 \Rightarrow$
    $-1 < x - 3 < 1 \Leftrightarrow 2 < x < 4$. The sum is $\frac{1}{1-(x-3)} = \frac{1}{4-x}$.

45. $\sum_{n=2}^{\infty}\left(\frac{x}{5}\right)^n$ is a geometric series with $r = \frac{x}{5}$, so converges whenever $\left|\frac{x}{5}\right| < 1 \Leftrightarrow -5 < x < 5$.
    The sum is $\frac{(x/5)^2}{1-x/5} = \frac{x^2}{25-5x}$.

47. $\sum_{n=0}^{\infty}(2\sin x)^n$ is geometric so converges whenever $|2\sin x| < 1 \Leftrightarrow -\frac{1}{2} < \sin x < \frac{1}{2} \Leftrightarrow$
    $n\pi - \frac{\pi}{6} < x < n\pi + \frac{\pi}{6}$, where the sum is $\frac{1}{1-2\sin x}$.

49. Total distance $= 4 + 2 + 2 + 1 + 1 + \frac{1}{2} + \frac{1}{2} + \frac{1}{4} + \frac{1}{4} + \cdots$
    $= 4 + 4 + 2 + 1 + \frac{1}{2} + \frac{1}{4} + \cdots = 10 + \sum_{n=0}^{\infty}\left(\frac{1}{2}\right)^n = 10 + \frac{1}{1-1/2} = 12$ m.

51. The series $1 - 1 + 1 - 1 + 1 - 1 + \cdots$ diverges (geometric series with $r = -1$) so we cannot say $0 = 1 - 1 + 1 - 1 + 1 - 1 + \cdots$.

53. $\sum_{n=1}^{\infty} ca_n = \lim_{n\to\infty}\sum_{i=1}^{n} ca_i = \lim_{n\to\infty} c\sum_{i=1}^{n} a_i = c\lim_{n\to\infty}\sum_{i=1}^{n} a_i = c\sum_{n=1}^{\infty} a_n$, which exists by hypothesis.

55. Suppose on the contrary that $\sum(a_n + b_n)$ converges. Then by Theorem 10.19 (c), so would $\sum[(a_n + b_n) - a_n] = \sum b_n$, a contradiction.

57. (a) $s_1 = \text{lm} = 10\frac{1}{1\cdot 2} = \frac{1}{2}$, $s_2 = \frac{1}{2} + \frac{1}{1\cdot 2\cdot 3} = \frac{5}{6}$, $s_3 = \frac{5}{6} + \frac{3}{1\cdot 2\cdot 3\cdot 4} = \frac{23}{24}$,
    $s_4 = \frac{23}{24} + \frac{4}{1\cdot 2\cdot 3\cdot 4\cdot 5} = \frac{119}{120}$.
    The denominators are $(n+1)!$ so a guess would be $s_n = \frac{(n+1)!-1}{(n+1)!}$.

    (b) For $n = 1$, $s_1 = \frac{1}{2} = \frac{2!-1}{2!}$, so the formula holds for $n = 1$. Assume $s_k = \frac{(k+1)!-1}{(k+1)!}$.
    Then $s_{k+1} = \frac{(k+1)!-1}{(k+1)!} + \frac{k+1}{(k+2)!} = \frac{(k+1)!-1}{(k+1)!} + \frac{k+1}{(k+1)!(k+2)}$

$$= \frac{(k+2)! - (k+2) + k + 1}{(k+2)!} = \frac{(k+2)! - 1}{(k+2)!}.$$ Thus the formula is true for $n = k+1$.

So by induction the guess is correct.

(c) $\lim_{n \to \infty} s_n = \lim_{n \to \infty} \frac{(n+1)! - 1}{(n+1)!} = \lim_{n \to \infty} \left[1 - \frac{1}{(n+1)!}\right] = 1$ and so $\sum_{n=0}^{\infty} \frac{n}{(n+1)!} = 1$.

EXERCISES 10.3

1. $\sum_{n=1}^{\infty} 2/\sqrt[3]{n} = 2 \sum_{n=1}^{\infty} \frac{1}{n^{1/3}}$, which is a p-series, $p = \frac{1}{3} < 1$, so it diverges.

3. $\sum_{n=5}^{\infty} \frac{1}{n^{1.0001}}$ is a p-series, $p = 1.0001 > 1$, so it converges.

5. $\sum_{n=5}^{\infty} \frac{1}{(n-4)^2} = \sum_{n=1}^{\infty} \frac{1}{n^2}$ is a p-series, $p = 2 > 1$, so it converges.

7. Since $\frac{1}{\sqrt{x}+1}$ is continuous, positive, and decreasing on $[0,\infty)$ we can apply the Integral Test.

    $\int_1^{\infty} \frac{1}{\sqrt{x}+1} dx = \lim_{t \to \infty} [2\sqrt{x} - 2\ln(\sqrt{x}+1)]\Big|_1^t$ [using the substitution $u = \sqrt{x}+1$, so $x = (u-1)^2$

    and $dx = 2(u-1)du$] $= \lim_{t \to \infty} [(2\sqrt{t} - 2\ln(\sqrt{t}+1)) - (2 - 2\ln 2)]$. Now $2\sqrt{t} - 2\ln(\sqrt{t}+1)$

    $= 2\ln\left(\frac{e^{\sqrt{t}}}{\sqrt{t}+1}\right)$ and so $\lim_{t \to \infty} (2\sqrt{t} - 2\ln(\sqrt{t}+1)) = \infty$ [using L'Hospital's Rule] so both the

    integral and the original series diverge.

9. $f(x) = xe^{-x^2}$ is continuous and positive on $[1,\infty)$, and since $f'(x) = e^{-x^2}(1 - 2x^2) < 0$ for

    $x > 1$, f is decreasing as well. We can use the Integral Test.

    $\int_1^{\infty} xe^{-x^2} dx = \lim_{t \to \infty} [-\frac{1}{2} e^{-x^2}]\Big|_1^t = 0 - \left(-\frac{e^{-1}}{2}\right) = \frac{1}{2e}$ so the series converges.

11. $f(x) = \frac{x}{x^2+1}$ is continuous and positive on $[1,\infty)$, and since $f'(x) = \frac{1-x^2}{(x^2+1)^2} < 0$ for $x > 1$,

    f is also decreasing. Using the Integral Test,

    $\int_1^{\infty} \frac{x}{x^2+1} dx = \lim_{t \to \infty} \left[\frac{\ln(x^2+1)}{2}\right]_1^t = \infty$, so the series diverges.

13. $f(x) = \frac{1}{x \ln x}$ is continuous and positive on $[2,\infty)$, and also decreasing since

    $f'(x) = -\frac{1 + \ln x}{x^2 (\ln x)^2} < 0$ for $x > 2$, so we can use the Integral Test.

Section 10.3

$\int_2^\infty \frac{1}{x \ln x} dx = \lim_{t \to \infty} [\ln(\ln x)]_2^t = \lim_{t \to \infty} [\ln(\ln t) - \ln(\ln 2)] = \infty$, so the series diverges.

15. $f(x) = \frac{\arctan x}{1+x^2}$ is continuous and positive on $[1, \infty)$. $f'(x) = \frac{1 - 2x \arctan x}{(1+x^2)^2} < 0$ for $x > 1$,

since $2x \arctan x \geq \frac{\pi}{2} > 1$ for $x \geq 1$. So f is decreasing and we can use the Integral Test.

$\int_1^\infty \frac{\arctan x}{1+x^2} dx = \lim_{t \to \infty} [\frac{1}{2} \arctan^2 x]_1^t = \frac{(\pi/2)^2}{2} - \frac{(\pi/4)^2}{2} = \frac{3\pi^2}{32}$, so the series converges.

17. $f(x) = \frac{\ln x}{x^2}$ is continuous and positive for $x \geq 2$, and $f'(x) = \frac{1 - 2\ln x}{x^3} < 0$ for $x \geq 2$ so f is

decreasing. $\int_2^\infty \frac{\ln x}{x^2} dx = \lim_{t \to \infty} [-\frac{\ln x}{x} - \frac{1}{x}]_1^t$ [using integration by parts] $= 1$ [by

L'Hospital's Rule]. Thus $\sum_{n=1}^\infty \frac{\ln n}{n^2} = \sum_{n=2}^\infty \frac{\ln n}{n^2}$ converges by the Integral Test.

19. $f(x) = \frac{\sin(1/x)}{x^2}$ is continuous and positive for $x \geq 1$, and

$f'(x) = -\frac{\cos(1/x) + 2x \sin(1/x)}{x^4} < 0$ if $x \geq 1$ [because then $0 < \frac{1}{x} \leq 1 < \frac{\pi}{2}$ so that both

$\cos(1/x)$ and $\sin(1/x)$ are positive] and so f is decreasing. Using the Integral Test,

$\int_1^\infty \frac{\sin(1/x)}{x^2} dx = \lim_{t \to \infty} [\cos(1/x)]_1^t = 1 - \cos 1$, and so the series converges.

21. $f(x) = \frac{1}{x^2 + 2x + 2}$ is continuous and positive on $[1, \infty)$, and $f'(x) = -\frac{2x+2}{(x^2+2x+2)^2} < 0$ if

$x \geq 1$, so f is decreasing and we can use the Integral Test. $\int_1^\infty \frac{1}{x^2 + 2x + 2} dx$

$= \int_1^\infty \frac{1}{(x+1)^2 + 1} dx = \lim_{t \to \infty} [\arctan(x+1)]_1^t = \frac{\pi}{2} - \arctan 2$,

so the series converges also.

23. We have already shown that when $p = 1$ the series diverges (in Exercise 13 above), so

assume $p \neq 1$. $f(x) = \frac{1}{x(\ln x)^p}$ is continuous and positive on $[2, \infty)$, and

$f'(x) = -\frac{p + \ln x}{x^2 (\ln x)^{p+1}} < 0$ if $x > e^{-p}$, so that f is eventually decreasing and we can use the

Integral Test. $\int_2^\infty \frac{1}{x(\ln x)^p} dx = \lim_{t \to \infty} [\frac{(\ln x)^{1-p}}{1-p}]_2^t$ [for $p \neq 1$]

$= \lim_{t \to \infty} (\frac{(\ln t)^{1-p}}{1-p}) - \frac{(\ln 2)^{1-p}}{1-p}$. This limit exists whenever $1 - p < 0 \Leftrightarrow p > 1$,

so the series converges for $p > 1$.

8

25. Clearly the series cannot converge if $p \geq -\frac{1}{2}$, because then $\lim_{n \to \infty} n(1+n^2)^p \neq 0$. Also, if $p = -1$ the series diverges (see Exercise 11 above). So assume $p < -\frac{1}{2}$, $p \neq -1$. Then $f(x) = x(1+x^2)^p$ is continuous, positive, and eventually decreasing on $[1, \infty)$, and we can use the Integral Test. $\int_1^\infty x(1+x^2)^p \, dx = \lim_{t \to \infty} \left[\frac{1}{2} \cdot \frac{(1+x^2)^{p+1}}{p+1} \right]_1^t =$

$\lim_{t \to \infty} \frac{1}{2} \cdot \frac{(1+x^2)^{p+1}}{p+1} - \frac{2^p}{p+1}$. This limit exists and is finite $\Leftrightarrow p+1 < 0 \Leftrightarrow p < -1$, so the series converges whenever $p < -1$.

27. Since this is a p-series with $p = x$, $\zeta(x)$ is defined when $x > 1$.

29. Using the proof of the Integral Test, in particular 10.21 and 10.22, we have $\int_1^n f \leq s_{n-1}$ and $s_n \leq a_1 + \int_1^n f$. Letting $n \to \infty$ in both inequalities, we get

$$\int_1^\infty \frac{1}{x^{1.001}} \, dx \leq s \leq 1 + \int_1^\infty \frac{1}{x^{1.001}} \, dx, \text{ where } s = \sum_{n=1}^\infty \frac{1}{n^{1.001}}$$

But $\int_1^\infty \frac{1}{x^{1.001}} \, dx = \lim_{t \to \infty} \int_1^t \frac{1}{x^{1.001}} \, dx = \lim_{t \to \infty} \left[\frac{1}{-0.001} \left(\frac{1}{t^{.001}} - 1 \right) \right] = \frac{1}{0.001} = 1000$.

So $1000 \leq s \leq 1001$.

EXERCISES 10.4

1. $\frac{1}{n^3+n^2} < \frac{1}{n^3}$ since $n^3 + n^2 > n^3$ for all n, and since $\sum_{n=1}^\infty \frac{1}{n^3}$ is a convergent p-series $(p = 3 > 1)$, $\sum_{n=1}^\infty \frac{1}{n^3+n^2}$ converges also by 10.24 (a).

3. $\frac{3}{n2^n} \leq \frac{3}{2^n}$. $\sum_{n=1}^\infty \frac{3}{2^n}$ is a geometric series with $|r| = \frac{1}{2} < 1$, and hence converges, so $\sum_{n=1}^\infty \frac{3}{n2^n}$ converges also by the Comparison Test.

5. $\frac{1+5^n}{4^n} > \frac{5^n}{4^n} = \left(\frac{5}{4}\right)^n$. $\sum_{n=0}^\infty \left(\frac{5}{4}\right)^n$ is a divergent geometric series $(|r| = \frac{5}{4} > 1)$ so $\sum_{n=0}^\infty \frac{1+5^n}{4^n}$ diverges by the Comparison Test.

7. $\frac{3}{n(n+3)} < \frac{3}{n^2}$. $\sum_{n=1}^\infty \frac{3}{n^2} = 3 \sum_{n=1}^\infty \frac{1}{n^2}$ is a convergent p-series $(p = 2 > 1)$ so $\sum_{n=1}^\infty \frac{3}{n(n+3)}$ converges by the Comparison Test.

Section 10.5

9. $\frac{1}{\sqrt{n^3+1}} < \frac{1}{\sqrt{n^3}} = \frac{1}{n^{3/2}}$. $\sum_{n=1}^{\infty} \frac{1}{n^{3/2}}$ is a convergent p-series ($p = \frac{3}{2} > 1$) so $\sum_{n=1}^{\infty} \frac{1}{\sqrt{n^3+1}}$ converges by the Comparison Test.

11. $\frac{\sqrt{n}}{n-1} > \frac{\sqrt{n}}{n} = \frac{1}{n^{1/2}}$. $\sum_{n=2}^{\infty} \frac{1}{n^{1/2}}$ is a divergent p-series ($p = \frac{1}{2} < 1$) so $\sum_{n=2}^{\infty} \frac{\sqrt{n}}{n-1}$ diverges by the Comparison Test.

13. $n^3 + 1 > n^3 \Rightarrow \frac{1}{n^3+1} < \frac{1}{n^3} \Rightarrow \frac{n}{n^3+1} < \frac{n}{n^3} \Rightarrow \frac{n-1}{n^3+1} < \frac{n}{n^3} = \frac{1}{n^2}$. $\sum_{n=1}^{\infty} \frac{1}{n^2}$ is a convergent p-series ($p = 2 > 1$) so $\sum_{n=1}^{\infty} \frac{n-1}{n^3+1}$ converges by the Comparison Test.

15. $\frac{3 + \cos n}{3^n} \le \frac{4}{3^n}$ since $\cos n \le 1$. $\sum_{n=1}^{\infty} \frac{4}{3^n}$ is a geometric series with $|r| = \frac{1}{3} < 1$ so it converges, and so $\sum_{n=1}^{\infty} \frac{3 + \cos n}{3^n}$ converges by the Comparison Test.

17. $(n+1)(2n+1) > n \cdot 2n = 2n^2$ so $\frac{4}{(n+1)(2n+1)} < \frac{4}{2n^2} = \frac{2}{n^2}$. $\sum_{n=1}^{\infty} \frac{2}{n^2} = 2\sum_{n=1}^{\infty} \frac{1}{n^2}$ is a convergent p-series ($p = 2 > 1$), so $\sum_{n=1}^{\infty} \frac{4}{(n+1)(2n+1)}$ converges by the Comparison Test.

19. $\frac{n}{\sqrt{n^5+4}} < \frac{n}{\sqrt{n^5}} = \frac{1}{n^{3/2}}$. $\sum_{n=1}^{\infty} \frac{1}{n^{3/2}}$ is a convergent p-series ($p = \frac{3}{2} > 1$) so $\sum_{n=1}^{\infty} \frac{n}{\sqrt{n^5+4}}$ converges by the Comparison Test.

21. $\frac{2^n}{1+3^n} < \frac{2^n}{3^n} = \left(\frac{2}{3}\right)^n$. $\sum_{n=1}^{\infty} \left(\frac{2}{3}\right)^n$ is a convergent geometric series ($|r| = \frac{2}{3} < 1$), so $\sum_{n=1}^{\infty} \frac{2^n}{1+3^n}$ converges by the Comparison Test.

23. Let $a_n = \frac{1}{1+\sqrt{n}}$ and $b_n = \frac{1}{\sqrt{n}}$. Then $\lim_{n \to \infty} \frac{a_n}{b_n} = \lim_{n \to \infty} \frac{\sqrt{n}}{1+\sqrt{n}} = 1 > 0$. Since $\sum_{n=1}^{\infty} \frac{1}{\sqrt{n}}$ is a divergent p-series ($p = \frac{1}{2} < 1$), $\sum_{n=1}^{\infty} \frac{1}{1+\sqrt{n}}$ also diverges by the Limit Comparison Test.

25. Let $a_n = \frac{n^2+1}{n^4+1}$ and $b_n = \frac{1}{n^2}$. Then $\lim_{n \to \infty} \frac{a_n}{b_n} = \lim_{n \to \infty} \frac{n^4+n^2}{n^4+1} = 1$. Since $\sum_{n=1}^{\infty} \frac{1}{n^2}$ is a convergent p-series ($p = 2 > 1$), so is $\sum_{n=1}^{\infty} \frac{n^2+1}{n^4+1}$ by the Limit Comparison Test.

27. Let $a_n = \frac{n^2-n+2}{\sqrt[4]{n^{10}+n^5+3}}$ and $b_n = \frac{1}{\sqrt{n}}$. Then $\lim_{n \to \infty} \frac{a_n}{b_n}$

$$= \lim_{n \to \infty} \frac{n^{5/2} - n^{3/2} + 2n^{1/2}}{4\sqrt{n^{10} + n^5 + 3}} = \lim_{n \to \infty} \frac{1 - n^{-1} + 2n^{-2}}{4\sqrt{1 + n^{-5} + 3n^{-10}}} = 1.$$ Since $\sum_{n=1}^{\infty} \frac{1}{\sqrt{n}}$ is a

divergent p–series ($p = \frac{1}{2} < 1$), $\sum_{n=1}^{\infty} \frac{n^2 - n + 2}{4\sqrt{n^{10} + n^5 + 3}}$ diverges by the Limit Comparison Test.

29. Let $a_n = \frac{n+1}{n2^n}$ and $b_n = \frac{1}{2^n}$. Then $\lim_{n \to \infty} \frac{a_n}{b_n} = \lim_{n \to \infty} \frac{n+1}{n} = 1.$ Since $\sum_{n=1}^{\infty} \frac{1}{2^n}$ is a convergent geometric series ($|r| = \frac{1}{2} < 1$), $\sum_{n=1}^{\infty} \frac{n+1}{n2^n}$ converges by the Limit Comparison Test.

31. Let $a_n = \frac{\ln n}{n^3}$ and $b_n = \frac{1}{n^2}$. Then $\lim_{n \to \infty} \frac{a_n}{b_n} = \lim_{n \to \infty} \frac{\ln n}{n} = \lim_{n \to \infty} \frac{1/n}{1} = 0.$ So since $\sum_{n=1}^{\infty} \frac{1}{n^2}$ converges (p–series, $p = 2 > 1$), so does $\sum_{n=1}^{\infty} \frac{\ln n}{n^3}$ by part (b) of the Limit Comparison Test.

33. Clearly $n! = n(n-1)(n-2) \cdots (3)(2) \geq 2 \cdot 2 \cdot 2 \cdots 2 \cdot 2 = 2^{n-1}$, so $\frac{1}{n!} \leq \frac{1}{2^{n-1}}$. $\sum_{n=1}^{\infty} \frac{1}{2^{n-1}}$ is a convergent geometric series ($|r| = \frac{1}{2} < 1$) so $\sum_{n=1}^{\infty} \frac{1}{n!}$ converges by the Comparison Test.

35. $\frac{n!}{n^2} = \frac{n(n-1)(n-2) \cdots 1}{n^2} = \frac{(n-1)(n-2) \cdots 1}{n} \geq \frac{1}{n}$ and $\sum_{n=1}^{\infty} \frac{1}{n}$ diverges (harmonic series), so $\sum_{n=1}^{\infty} \frac{n!}{n^2}$ diverges by the Comparison Test.

[OR: $\lim_{n \to \infty} \frac{n!}{n^2} = \lim_{n \to \infty} \frac{n(n-1)!}{n \cdot n} = \lim_{n \to \infty} \frac{(n-1)!}{n} = \lim_{n \to \infty} \left(1 - \frac{1}{n}\right)(n-2)! = \infty$, so the series $\sum_{n=1}^{\infty} \frac{n!}{n^2}$ diverges by the Test for Divergence (10.18).]

37. Let $a_n = \sin\left(\frac{1}{n}\right)$ and $b_n = \frac{1}{n}$. Then $\lim_{n \to \infty} \frac{a_n}{b_n} = \lim_{n \to \infty} \frac{\sin(1/n)}{1/n} = \lim_{\theta \to 0} \frac{\sin \theta}{\theta} = 1$, so since $\sum_{n=1}^{\infty} b_n$ is the harmonic series (which diverges), $\sum_{n=1}^{\infty} \sin\left(\frac{1}{n}\right)$ diverges as well by the Limit Comparison Test.

39. Since $\frac{d_n}{10^n} \leq \frac{9}{10^n}$ for each n, and since $\sum_{n=1}^{\infty} \frac{9}{10^n}$ is a convergent geometric series ($|r| = \frac{1}{10} < 1$), $0.d_1 d_2 d_3 \cdots = \sum_{n=1}^{\infty} \frac{d_n}{10^n}$ will always converge by the Comparison Test.

41. Since $\sum a_n$ converges, $\lim_{n \to \infty} a_n = 0$, so there exists N such that $|a_n - 0| < 1$ for all $n > N$ $\Rightarrow 0 \leq a_n < 1$ for all $n > N \Rightarrow 0 \leq a_n^2 \leq a_n$. Since $\sum a_n$ converges, so will $\sum a_n^2$ by the Comparison Test.

43. We wish to prove that if $\lim_{n \to \infty} \frac{a_n}{b_n} = \infty$ and $\sum b_n$ diverges, then so does $\sum a_n$. So suppose on

11

the contrary that $\sum a_n$ converges. Since $\lim_{n\to\infty} \frac{a_n}{b_n} = \infty$, we have that $\lim_{n\to\infty} \frac{b_n}{a_n} = 0$, so by part (b) of the Limit Comparison Test (proved in Exercise 42), if $\sum a_n$ converges, so must $\sum b_n$. But this contradicts our hypothesis, so $\sum a_n$ must diverge.

45. $\lim_{n\to\infty} na_n = \lim_{n\to\infty} \frac{a_n}{1/n}$, so we apply the Limit Comparison Test with $b_n = \frac{1}{n}$. Since $\lim_{n\to\infty} na_n > 0$ we know that either both series converge or both series diverge, and we also know that $\sum_{n=0}^{\infty} \frac{1}{n}$ diverges (p-series with $p = 1$). Therefore $\sum a_n$ must be divergent.

EXERCISES 10.5

1. $\sum_{n=1}^{\infty} (-1)^{n-1} \frac{3}{n+4}$. $a_n = \frac{3}{n+4} > 0$ and $a_{n+1} < a_n$ for all n; $\lim_{n\to\infty} a_n = 0$ so the series converges by the Alternating Series Test.

3. $\sum_{n=1}^{\infty} (-1)^n \frac{n}{n+1}$. $\lim_{n\to\infty} \frac{n}{n+1} = 1$ so $\lim_{n\to\infty} (-1)^n \frac{n}{n+1}$ does not exist and the series diverges by the Test for Divergence.

5. $\sum_{n=1}^{\infty} (-1)^{n-1} \frac{1}{n^2}$. $a_n = \frac{1}{n^2} > 0$ and $a_{n+1} < a_n$ for all n, and $\lim_{n\to\infty} \frac{1}{n^2} = 0$, so the series converges by the Alternating Series Test.

7. $\sum_{n=1}^{\infty} (-1)^{n+1} \frac{n}{5n+1}$. $\lim_{n\to\infty} \frac{n}{5n+1} = \frac{1}{5}$ so $\lim_{n\to\infty} (-1)^{n+1} \frac{n}{5n+1}$ does not exist and the series diverges by the Test for Divergence.

9. $\sum_{n=1}^{\infty} (-1)^n \frac{n}{n^2+1}$. $a_n = \frac{n}{n^2+1} > 0$ for all n. $a_{n+1} < a_n \Leftrightarrow \frac{n+1}{(n+1)^2+1} < \frac{n}{n^2+1}$ $\Leftrightarrow (n+1)(n^2+1) < [(n+1)^2+1]n \Leftrightarrow n^3+n^2+n+1 < n^3+2n^2+2n \Leftrightarrow 0 < n^2+n-1$, which is true for all $n \geq 1$. Also $\lim_{n\to\infty} \frac{n}{n^2+1} = \lim_{n\to\infty} \frac{1/n}{1+1/n^2} = 0$. Therefore the series converges by the Alternating Series Test.

11. $\sum_{n=1}^{\infty} (-1)^{n-1} \frac{\sqrt{n}}{n+4}$. $a_n = \frac{\sqrt{n}}{n+4} > 0$ for all n. Let $f(x) = \frac{\sqrt{x}}{x+4}$. Then $f'(x) = \frac{4-x}{2\sqrt{x}(x+4)^2} < 0$ if $x > 4$, so $\{a_n\}$ is decreasing after $n = 4$.

$\lim_{n\to\infty} \frac{\sqrt{n}}{n+4} = \lim_{n\to\infty} \frac{1}{\sqrt{n}+4/\sqrt{n}} = 0$. So the series converges by the Alternating Series Test.

Section 10.5

13. $\sum_{n=2}^{\infty} (-1)^n \frac{n}{\ln n}$. $\lim_{n \to \infty} \frac{n}{\ln n} = \lim_{n \to \infty} \frac{1}{1/n} = \infty$ so the series diverges by the Test for Divergence.

15. $\sum_{n=1}^{\infty} (-1)^{n+1} \frac{n+10}{n(n+1)}$. $a_n = \frac{n+10}{n(n+1)} > 0$ for all n. Let $f(x) = \frac{x+10}{x(x+1)}$. Then

$f'(x) = -\frac{x^2 + 20x + 10}{(x^2+x)^2} < 0$ for $x \geq 1$, so $\{a_n\}$ is decreasing.

$\lim_{n \to \infty} \frac{n+10}{n(n+1)} = \lim_{n \to \infty} \frac{1/n + 10/n^2}{1 + 1/n} = 0$, so the series converges by the Alternating Series Test.

17. $\sum_{n=1}^{\infty} \frac{\cos n\pi}{n^{3/4}} = \sum_{n=1}^{\infty} \frac{(-1)^n}{n^{3/4}}$. $a_n = \frac{1}{n^{3/4}}$ is decreasing and positive, and $\lim_{n \to \infty} \frac{1}{n^{3/4}} = 0$ so the series converges by the Alternating Series Test.

19. $\sum_{n=1}^{\infty} (-1)^n \sin\left(\frac{\pi}{n}\right)$. $a_n = \sin\left(\frac{\pi}{n}\right) > 0$ for $n \geq 2$ and $\sin\left(\frac{\pi}{n}\right) \geq \sin\left(\frac{\pi}{n+1}\right)$, and

$\lim_{n \to \infty} \sin\left(\frac{\pi}{n}\right) = \sin 0 = 0$, so the series converges by the Alternating Series Test.

21. $\sum_{n=1}^{\infty} (-1)^n \frac{1}{\sqrt[n]{n}}$. Let $L = \lim_{n \to \infty} \frac{1}{\sqrt[n]{n}}$ (if it exists). Then $\ln L = \lim_{n \to \infty} \ln(n^{-1/n})$

$= \lim_{n \to \infty} -\frac{\ln n}{n} = \lim_{n \to \infty} -\frac{1/n}{1} = 0$, so $L = e^0 = 1$, so $\lim_{n \to \infty} (-1)^n \frac{1}{\sqrt[n]{n}}$ does not exist

and the series diverges by the Test for Divergence.

23. First consider $f(x) = \frac{x^2 + x}{(x+2)^3} \Rightarrow f'(x) = \frac{(2x+1)(x+2)^3 - (x^2+x)3(x+2)^2}{(x+2)^6} = \frac{-x^2 + 2x + 2}{(x+2)^4}$

$\Rightarrow f'(x) < 0$ for $x > 1 + \sqrt{3}$, so f is decreasing on $(1 + \sqrt{3}, \infty) \Rightarrow \{a_n\}$ is decreasing for $n \geq 3$.

Also $\lim_{n \to \infty} \frac{n(n+1)}{(n+2)^3} = \lim_{n \to \infty} \frac{1 + 1/n}{n(1 + 2/n)^3} = 0$, so by the Alternating Series test the series is convergent.

25. $\sum_{n=1}^{\infty} \frac{(-3)^n}{n^2}$. $\lim_{n \to \infty} \frac{3^n}{n^2} = \infty$, so $\lim_{n \to \infty} \frac{(-3)^n}{n^2}$ does not exist. Thus the given series is divergent by the Test for Divergence.

27. Let $\sum b_n$ be the series for which $b_n = 0$ if n is odd and $b_n = \frac{1}{n^2}$ if n is even. Then

$\sum b_n = \sum \frac{1}{(2n)^2}$ clearly converges (by comparison with the p–series for $p = 2$). So suppose

that $\sum (-1)^{n-1} a_n$ converges. Then by Theorem 10.19 (b) so does $\sum [(-1)^{n-1} a_n + b_n]$

$= 1 + \frac{1}{3} + \frac{1}{5} + \cdots = \sum \frac{1}{2n-1}$. But this diverges (comparison with harmonic series), a

contradiction. $\sum (-1)^{n-1} a_n$ must diverge. The Alternating Series Test does not apply since $\{a_n\}$ is not decreasing.

13

Section 10.6

29. Clearly $a_n = \frac{1}{n+p}$ is decreasing and eventually positive and $\lim_{n \to \infty} a_n = 0$ for any p. So the series will converge (by the Alternating Series Test) for any p for which all the a_n's are defined, i.e., $n + p \neq 0$ for $n \geq 1$, or p is not a negative integer.

31. If $a_n = \frac{1}{n^2}$, then $a_{11} = \frac{1}{121} < 0.01$, so by Theorem 10.28, $\sum_{n=1}^{\infty} \frac{1}{n^2} \approx \sum_{n=1}^{10} \frac{1}{n^2} \approx 0.82$.

33. $\sum_{n=0}^{\infty} (-1)^n \frac{2^n}{n!}$ Since $\frac{2}{n} < \frac{2}{3}$ for $n \geq 4$, $0 < \frac{2^n}{n!} < \frac{2}{1} \cdot \frac{2}{2} \cdot \frac{2}{3} \cdot \left(\frac{2}{3}\right)^{n-3} \to 0$ as $n \to \infty$, so by the Squeeze Theorem, $\lim_{n \to \infty} \frac{2^n}{n!} = 0$, and hence $\sum_{n=0}^{\infty} (-1)^n \frac{2^n}{n!}$ is a convergent alternating series.

$\frac{2^8}{8!} = \frac{256}{40320} < 0.01$, so $\sum_{n=0}^{\infty} (-1)^n \frac{2^n}{n!} \approx \sum_{n=0}^{7} (-1)^n \frac{2^n}{n!} \approx 0.13$.

35. $\sum_{n=1}^{\infty} \frac{(-1)^{n-1}}{(2n-1)!}$ $a_5 = \frac{1}{(2 \cdot 5 - 1)!} = \frac{1}{362880} < 0.00001$, so $\sum_{n=1}^{\infty} \frac{(-1)^{n-1}}{(2n-1)!}$

$\approx \sum_{n=1}^{4} \frac{(-1)^{n-1}}{(2n-1)!} \approx 0.8415$.

37. $\sum_{n=0}^{\infty} \frac{(-1)^n}{2^n n!}$ $a_6 = \frac{1}{2^6 6!} = \frac{1}{46080} < 0.000022$, so $\sum_{n=0}^{\infty} \frac{(-1)^n}{2^n n!} \approx \sum_{n=0}^{5} \frac{(-1)^n}{2^n n!} \approx 0.6065$.

39. (a) We will prove this by induction. Let P(n) be the proposition that $s_{2n} = h_{2n} - h_n$. P(1) is true by an easy calculation. So suppose that P(n) is true. We will show that P(n+1) must be true as a consequence.

$h_{2n+2} - h_{n+1} = \left(h_{2n} + \frac{1}{2n+1} + \frac{1}{2n+2}\right) - \left(h_n + \frac{1}{n+1}\right) = (h_{2n} - h_n) + \frac{1}{2n+1} - \frac{1}{2n+2} = s_{2n}$

$+ \frac{1}{2n+1} - \frac{1}{2n+2} = s_{2n+2}$, which is P(n+1), and proves that $s_{2n} = h_{2n} - h_n$ for all n.

(b) We know that $h_{2n} - \ln(2n) \to \gamma$ and $h_n - \ln(n) \to \gamma$ as $n \to \infty$. So

$s_{2n} = h_{2n} - h_n = [h_{2n} - \ln(2n)] - [h_n - \ln(n)] + [\ln(2n) - \ln n]$, and

$\lim_{n \to \infty} s_{2n} = \gamma - \gamma + \lim_{n \to \infty} [\ln(2n) - \ln n] = \lim_{n \to \infty} (\ln 2 + \ln n - \ln n) = \ln 2$.

EXERCISES 10.6

1. $\sum_{n=1}^{\infty} \frac{1}{n\sqrt{n}} = \sum_{n=1}^{\infty} \frac{1}{n^{3/2}}$ is a convergent p-series ($p = \frac{3}{2} > 1$), so the given series is absolutely convergent.

Section 10.6

3. $\lim_{n\to\infty} \left|\frac{a_{n+1}}{a_n}\right| = \lim_{n\to\infty} \left|\frac{(-3)^{n+1}/(n+1)^3}{(-3)^n/n^3}\right| = 3\lim_{n\to\infty} \left(\frac{n}{n+1}\right)^3 = 3 > 1$, so the series diverges by the Ratio Test.

5. $\sum_{n=1}^{\infty} \frac{1}{2n+1}$ diverges (use the Integral Test or the Limit Comparison Test with $b_n = \frac{1}{n}$), but since $\lim_{n\to\infty} \frac{1}{2n+1} = 0$, $\sum_{n=1}^{\infty} \frac{(-1)^{n+1}}{2n+1}$ converges by the Alternating Series Test, and so is conditionally convergent.

7. $\lim_{n\to\infty} \left|\frac{a_{n+1}}{a_n}\right| = \lim_{n\to\infty} \frac{1/(2n+1)!}{1/(2n-1)!} = \lim_{n\to\infty} \frac{1}{(2n+1)2n} = 0$, so by the Ratio Test the series is absolutely convergent.

9. $\sum_{n=1}^{\infty} \frac{n}{n^2+4}$ diverges (use the Limit Comparison Test with $b_n = \frac{1}{n}$). But since
$0 \leq \frac{n+1}{(n+1)^2+4} < \frac{n}{n^2+4} \Leftrightarrow n^3 + n^2 + 4n + 4 < n^3 + 2n^2 + 5n \Leftrightarrow 0 < n^2 + n - 4$ (which is true for $n \geq 2$), and since $\lim_{n\to\infty} \frac{n}{n^2+4} = 0$, $\sum_{n=1}^{\infty} (-1)^n \frac{n}{n^2+4}$ converges by the Alternating Series Test, and so converges conditionally.

11. $\lim_{n\to\infty} \frac{2n}{3n-4} = \frac{2}{3}$ so $\sum_{n=1}^{\infty} (-1)^n \frac{2n}{3n-4}$ diverges by the Test for Divergence.

13. $\left|\frac{\sin 2n}{n^2}\right| \leq \frac{1}{n^2}$ and $\sum_{n=1}^{\infty} \frac{1}{n^2}$ converges (p-series, $p = 2 > 1$), so $\sum_{n=1}^{\infty} \frac{\sin 2n}{n^2}$ converges absolutely (by the Comparison Test).

15. $\lim_{n\to\infty} \left|\frac{a_{n+1}}{a_n}\right| = \lim_{n\to\infty} \left|\frac{2^{n+1}/[(n+1)3^{n+2}]}{2^n/(n3^{n+1})}\right| = \frac{2}{3} \lim_{n\to\infty} \frac{n}{n+1} = \frac{2}{3} < 1$ so the series converges absolutely by the Ratio Test.

17. $\lim_{n\to\infty} \left|\frac{a_{n+1}}{a_n}\right| = \lim_{n\to\infty} \frac{(n+2)5^{n+1}/[(n+1)3^{2(n+1)}]}{(n+1)5^n/(n3^{2n})} = \lim_{n\to\infty} \frac{5n(n+2)}{9(n+1)^2} = \frac{5}{9} < 1$ so the series converges absolutely by the Ratio Test.

19. $\sum_{n=2}^{\infty} \frac{1}{\ln n}$ diverges (since $\frac{1}{\ln n} > \frac{1}{n}$ and $\sum_{n=1}^{\infty} \frac{1}{n}$ diverges), but $\sum_{n=2}^{\infty} \frac{(-1)^n}{\ln n}$ converges by the Alternating Series Test (since $\frac{1}{\ln n}$ decreases to 0), so the series converges conditionally.

21. $\lim_{n\to\infty} \left|\frac{a_{n+1}}{a_n}\right| = \lim_{n\to\infty} \frac{(n+1)!/10^{n+1}}{n!/10^n} = \lim_{n\to\infty} \frac{n+1}{10} = \infty$ so the series diverges by the Ratio Test.

23. $\dfrac{|\cos(n\pi/3)|}{n!} \leq \dfrac{1}{n!}$ and $\sum_{n=1}^{\infty} \dfrac{1}{n!}$ converges (Exercise 33, Section 10.4), so the given series converges absolutely by the Comparison Test.

25. $\lim_{n \to \infty} \left|\dfrac{a_{n+1}}{a_n}\right| = \lim_{n \to \infty} \dfrac{(n+1)^{n+1}/5^{2n+5}}{n^n/5^{2n+3}} = \lim_{n \to \infty} \dfrac{1}{25}\left(\dfrac{n+1}{n}\right)^n (n+1) = \infty$, so the series diverges by the Ratio Test.

27. $\lim_{n \to \infty} \sqrt[n]{|a_n|} = \lim_{n \to \infty} \left|\dfrac{1-3n}{3+4n}\right| = \dfrac{3}{4} < 1$, so the series converges absolutely by the Root Test.

29. $\lim_{n \to \infty} \left|\dfrac{a_{n+1}}{a_n}\right| = \lim_{n \to \infty} \dfrac{(n+1)!/(1 \cdot 3 \cdot 5 \cdots (2n+1))}{n!/(1 \cdot 3 \cdot 5 \cdots (2n-1))} = \lim_{n \to \infty} \dfrac{n+1}{2n+1} = \dfrac{1}{2} < 1$, so the series converges absolutely by the Ratio Test.

31. $\sum_{n=1}^{\infty} \dfrac{2 \cdot 4 \cdot 6 \cdots (2n)}{n!} = \sum_{n=1}^{\infty} \dfrac{2^n n!}{n!} = \sum_{n=1}^{\infty} 2^n$ which diverges by the Test for Divergence since $\lim_{n \to \infty} 2^n = \infty$.

33. $\lim_{n \to \infty} \left|\dfrac{a_{n+1}}{a_n}\right| = \lim_{n \to \infty} \dfrac{(n+3)!/[(n+1)! \, 10^{n+1}]}{(n+2)!/(n! \, 10^n)} = \dfrac{1}{10} \lim_{n \to \infty} \dfrac{n+3}{n+1} = \dfrac{1}{10} < 1$ so the series converges absolutely by the Ratio Test.

35. $\dfrac{|\sin 3n| \, n^2}{(1.1)^n} \leq \dfrac{1 \cdot n^2}{(1.1)^n}$ for all n, so we test the series $\sum_{n=1}^{\infty} \dfrac{n^2}{(1.1)^n}$ using the Ratio Test.

$\lim_{n \to \infty} \left|\dfrac{a_{n+1}}{a_n}\right| = \dfrac{(n+1)^2/(1.1)^{n+1}}{n^2/(1.1)^n} = \dfrac{1}{1.1} \lim_{n \to \infty} \left(\dfrac{n+1}{n}\right)^2 = \dfrac{1}{1.1} < 1$ so $\sum_{n=1}^{\infty} \dfrac{n^2}{(1.1)^n}$ converges absolutely, and by the Comparison Test, so does $\sum_{n=1}^{\infty} \dfrac{(\sin 3n) \, n^2}{(1.1)^n}$.

37. (a) $\lim_{n \to \infty} \left|\dfrac{1/(n+1)^3}{1/n^3}\right| = \lim_{n \to \infty} \dfrac{n^3}{(n+1)^3} = \lim_{n \to \infty} \dfrac{1}{(1+1/n)^3} = 1$. So inconclusive.

(b) $\lim_{n \to \infty} \left|\dfrac{(n+1) \cdot 2^n}{2^{n+1} \cdot n}\right| = \lim_{n \to \infty} \dfrac{n+1}{2n} = \lim_{n \to \infty} \left(\dfrac{1}{2} + \dfrac{1}{2n}\right) = \dfrac{1}{2}$. So conclusive (convergent).

(c) $\lim_{n \to \infty} \left|\dfrac{(-3)^n}{\sqrt{n+1}} \cdot \dfrac{\sqrt{n}}{(-3)^{n-1}}\right| = 3 \lim_{n \to \infty} \sqrt{\dfrac{n}{n+1}} = 3 \lim_{n \to \infty} \sqrt{\dfrac{1}{1+1/n}} = 3$. So conclusive.

(Divergent)

(d) $\lim_{n \to \infty} \left|\dfrac{\sqrt{n+1}}{1+(n+1)^2} \cdot \dfrac{1+n^2}{\sqrt{n}}\right| = \lim_{n \to \infty} \left[\sqrt{1+\dfrac{1}{n}} \cdot \dfrac{(1/n^2)+1}{(1/n^2)+(1+1/n)^2}\right] = 1$. So inconclusive.

39. By the Triangle Inequality [see Exercise 4.1.46] we have

$\left|\sum_{i=1}^{n} a_i\right| \leq \sum_{i=1}^{n} |a_i| \Rightarrow -\sum_{i=1}^{n} |a_i| \leq \sum_{i=1}^{n} a_i \leq \sum_{i=1}^{n} |a_i| \Rightarrow$

Section 10.7

$$-\lim_{n\to\infty}\sum_{i=1}^{n}|a_i| \leq \lim_{n\to\infty}\sum_{i=1}^{n}a_i \leq \lim_{n\to\infty}\sum_{i=1}^{n}|a_i| \Rightarrow -\sum_{n=1}^{\infty}|a_n| \leq \sum_{n=1}^{\infty}a_n \leq \sum_{n=1}^{\infty}|a_n|$$

$$\Rightarrow \left|\sum_{n=1}^{\infty}a_n\right| \leq \sum_{n=1}^{\infty}|a_n|.$$

41. (a) Since $\sum a_n$ is absolutely convergent, and since $|a_n^+| \leq |a_n|$ and $|a_n^-| \leq |a_n|$ (because a_n^+ and a_n^- each equal either a_n or 0), we conclude by the Comparison Test that both $\sum a_n^+$ and $\sum a_n^-$ must be absolutely convergent. [Or use Theorem 10.19.]

(b) We will show by contradiction that both $\sum a_n^+$ and $\sum a_n^-$ must diverge. For suppose that $\sum a_n^+$ converged. Then so would $\sum (a_n^+ - \frac{1}{2}a_n)$ by Theorem 10.19. But $\sum (a_n^+ - \frac{1}{2}a_n) = \sum \left(\frac{a_n + |a_n|}{2} - \frac{a_n}{2}\right) = \frac{1}{2}\sum |a_n|$ which diverges because $\sum a_n$ is only conditionally convergent. Hence $\sum a_n^+$ can't converge. Similarly, neither can $\sum a_n^-$.

EXERCISES 10.7

1. Use the Comparison Test, with $a_n = \frac{\sqrt{n}}{n^2+1}$ and $b_n = \frac{1}{n^{3/2}}$: $\frac{\sqrt{n}}{n^2+1} < \frac{\sqrt{n}}{n^2} = \frac{1}{n^{3/2}}$, and $\sum_{n=1}^{\infty}\frac{1}{n^{3/2}}$ is a convergent p–series ($p = \frac{3}{2} > 1$), so $\sum_{n=1}^{\infty} a_n = \sum_{n=1}^{\infty}\frac{\sqrt{n}}{n^2+1}$ converges as well.

3. $\sum_{n=1}^{\infty}\frac{4^n}{3^{2n-1}} = 3\sum_{n=1}^{\infty}\left(\frac{4}{9}\right)^n$ which is a convergent geometric series ($|r| = \frac{4}{9} < 1$).

5. Converges by the Alternating Series Test, since $a_n = \frac{1}{(\ln n)^2}$ is decreasing ($\ln x$ is an increasing function) and $\lim_{n\to\infty} a_n = 0$.

7. $\sum_{k=1}^{\infty}\frac{1}{k^{1.7}}$ is a convergent p–series ($p = 1.7 > 1$).

9. $\lim_{n\to\infty}\left|\frac{a_{n+1}}{a_n}\right| = \lim_{n\to\infty}\frac{(n+1)/e^{n+1}}{n/e^n} = \frac{1}{e}\lim_{n\to\infty}\frac{n+1}{n} = \frac{1}{e} < 1$, so the series converges by the Ratio Test.

11. Use the Limit Comparison Test with $a_n = \frac{n^3+1}{n^4-1}$ and $b_n = \frac{1}{n}$. $\lim_{n\to\infty}\frac{a_n}{b_n}$
$= \lim_{n\to\infty}\frac{n^4+n}{n^4-1} = \lim_{n\to\infty}\frac{1+1/n^3}{1-1/n^4} = 1$, and since $\sum_{n=2}^{\infty} b_n$ diverges (harmonic series) so does $\sum_{n=2}^{\infty}\frac{n^3+1}{n^4-1}$.

13. Let $f(x) = \dfrac{2}{x(\ln x)^3}$. $f(x)$ is clearly positive and decreasing for $x \geq 2$, so we apply the Integral Test. $\displaystyle\int_2^\infty \dfrac{2}{x(\ln x)^3}\, dx = \lim_{t\to\infty}\left[\dfrac{-1}{(\ln x)^2}\right]_2^t = 0 - \dfrac{-1}{(\ln 2)^2}$ which is finite. So $\displaystyle\sum_{n=2}^\infty \dfrac{2}{n(\ln n)^3}$ converges.

15. $\displaystyle\lim_{n\to\infty}\left|\dfrac{a_{n+1}}{a_n}\right| = \lim_{n\to\infty} \dfrac{3^{n+1}(n+1)^2/(n+1)!}{3^n n^2/n!} = 3\lim_{n\to\infty}\dfrac{n+1}{n^2} = 0$, so the series converges by the Ratio Test.

17. $\dfrac{3^n}{5^n + n} \leq \dfrac{3^n}{5^n} = \left(\dfrac{3}{5}\right)^n$. Since $\displaystyle\sum_{n=1}^\infty \left(\dfrac{3}{5}\right)^n$ is a convergent geometric series ($|r| = \tfrac{3}{5} < 1$), $\displaystyle\sum_{n=1}^\infty \dfrac{3^n}{5^n + n}$ will converge by the Comparison Test.

19. $\displaystyle\lim_{n\to\infty}\left|\dfrac{a_{n+1}}{a_n}\right| = \lim_{n\to\infty}\dfrac{(n+1)!/(2\cdot 5\cdot 8 \cdots (3n+5))}{n!/(2\cdot 5\cdot 8 \cdots (3n+2))} = \lim_{n\to\infty}\dfrac{n+1}{3n+5} = \tfrac{1}{3} < 1$, so the series converges by the Ratio Test.

21. Use the Limit Comparison Test with $a_i = \dfrac{1}{\sqrt{i(i+1)}}$ and $b_i = \dfrac{1}{i}$. $\displaystyle\lim_{i\to\infty}\dfrac{a_i}{b_i} = \lim_{i\to\infty}\dfrac{i}{\sqrt{i(i+1)}} = \lim_{i\to\infty}\dfrac{1}{\sqrt{1+1/i}} = 1$. Since $\displaystyle\sum_{i=1}^\infty b_i$ diverges (harmonic series) so does $\displaystyle\sum_{i=1}^\infty \dfrac{1}{\sqrt{i(i+1)}}$.

23. $\displaystyle\lim_{n\to\infty} 2^{1/n} = 2^0 = 1$, so $\displaystyle\lim_{n\to\infty}(-1)^n 2^{1/n}$ does not exist and the series diverges by the Test for Divergence.

25. Let $f(x) = \dfrac{\ln x}{\sqrt{x}}$. Then $f'(x) = \dfrac{2 - \ln x}{2x^{3/2}} < 0$ when $\ln x > 2$ or $x > e^2$ so $\dfrac{\ln n}{\sqrt{n}}$ is decreasing for $n > e^2$. By l'Hospital's Rule, $\displaystyle\lim_{n\to\infty}\dfrac{\ln n}{\sqrt{n}} = \lim_{n\to\infty}\dfrac{1/n}{1/(2\sqrt{n})} = \lim_{n\to\infty}\dfrac{2}{\sqrt{n}} = 0$, so the series converges by the Alternating Series Test.

27. The series diverges since it is a geometric series with $r = -\pi$ and $|r| = \pi > 1$. [Or use the Test for Divergence.]

29. $\displaystyle\sum_{n=1}^\infty \dfrac{(-2)^{2n}}{n^n} = \sum_{n=1}^\infty \left(\dfrac{4}{n}\right)^n$. $\displaystyle\lim_{n\to\infty}\sqrt[n]{|a_n|} = \lim_{n\to\infty}\dfrac{4}{n} = 0$, so the series converges by the Root Test.

31. Since $\dfrac{k\ln k}{(k+1)^3} < \dfrac{k\ln k}{k^3} = \dfrac{\ln k}{k^2}$, and since $\displaystyle\sum_{n=1}^\infty \dfrac{\ln n}{n^2}$ converges (Exercise 17, Section 10.3),

Section 10.8

the given series converges by the Comparison Test.

33. $\lim_{n \to \infty} \left| \frac{a_{n+1}}{a_n} \right| = \lim_{n \to \infty} \frac{2^{n+1}/(2n+3)!}{2^n/(2n+1)!} = 2 \lim_{n \to \infty} \frac{1}{(2n+3)(2n+2)} = 0$, so the series converges by the Ratio Test.

35. $0 < \frac{\tan^{-1} n}{n^{3/2}} < \frac{\pi/2}{n^{3/2}}$. $\sum_{n=1}^{\infty} \frac{\pi/2}{n^{3/2}} = \frac{\pi}{2} \sum_{n=1}^{\infty} \frac{1}{n^{3/2}}$ which is a convergent p-series (p = $\frac{3}{2}$ > 1), so $\sum_{n=1}^{\infty} \frac{\tan^{-1} n}{n^{3/2}}$ converges by the Comparison Test.

37. Since $\frac{3}{\pi} < 1$, $\lim_{n \to \infty} \frac{1}{1 + (3/\pi)^n} = \frac{1}{1+0} = 1 \neq 0$, so the series diverges by the Test for Divergence.

39. $\lim_{n \to \infty} \sqrt[n]{|a_n|} = \lim_{n \to \infty} (2^{1/n} - 1) = 1 - 1 = 0$, so the series converges by the Root Test.

EXERCISES 10.8

NOTE: "R" stands for "radius of convergence" and "I" stands for "interval of convergence" in this section.

1. (a) We are given that the power series $\sum_{n=0}^{\infty} a_n x^n$ is convergent for $x = 4$. So by Theorem 10.38 it must converge at least for $-4 < x \leq 4$. In particular it converges when $x = -2$, that is, $\sum_{n=0}^{\infty} a_n (-2)^n$ is convergent.

   (b) But it does not follow that $\sum_{n=0}^{\infty} a_n (-4)^n$ is necessarily convergent. [See the comments after Theorem 10.38. An example is $a_n = (-1)^n / (n 4^n)$.]

3. If $u_n = \frac{x^n}{n+2}$, then $\lim_{n \to \infty} \left| \frac{u_{n+1}}{u_n} \right| = \lim_{n \to \infty} \left| \frac{x^{n+1}}{n+3} \cdot \frac{n+2}{x^n} \right| = |x| \lim_{n \to \infty} \frac{n+2}{n+3} = |x| < 1$ for convergence (by the Ratio Test). So R = 1. When $x = 1$, the series is $\sum_{n=0}^{\infty} \frac{1}{n+2}$ which diverges (by the Integral Test or Comparison Test), and when $x = -1$, it is $\sum_{n=0}^{\infty} \frac{(-1)^n}{n+2}$ which converges (by the Alternating Series Test), so I = $[-1, 1)$.

5. If $u_n = nx^n$, then $\lim_{n \to \infty} \left| \frac{u_{n+1}}{u_n} \right| = \lim_{n \to \infty} \left| \frac{(n+1)x^{n+1}}{nx^n} \right| = |x| \lim_{n \to \infty} \frac{n+1}{n} = |x| < 1$ for convergence (by the Ratio Test). So R = 1. When $x = 1$ or -1, $\lim_{n \to \infty} nx^n$ does not exist,

Section 10.8

so $\sum\limits_{n=0}^{\infty} nx^n$ diverges for these values. So $I = (-1, 1)$.

7. If $u_n = \frac{x^n}{n!}$, then $\lim\limits_{n \to \infty} \left|\frac{u_{n+1}}{u_n}\right| = \lim\limits_{n \to \infty} \left|\frac{x^{n+1}/(n+1)!}{x^n/n!}\right| = |x| \lim\limits_{n \to \infty} \frac{1}{n+1} = 0 < 1$ for all x.

So, by the Ratio Test, $R = \infty$, and $I = (-\infty, \infty)$.

9. If $u_n = \frac{(-1)^n x^n}{n2^n}$, then $\lim\limits_{n \to \infty} \left|\frac{u_{n+1}}{u_n}\right| = \lim\limits_{n \to \infty} \left|\frac{x^{n+1}/[(n+1)2^{n+1}]}{x^n/(n2^n)}\right| = \left|\frac{x}{2}\right| \lim\limits_{n \to \infty} \frac{n}{n+1}$

$= \left|\frac{x}{2}\right| < 1$ for convergence, so $|x| < 2$ and $R = 2$. When $x = 2$, $\sum\limits_{n=1}^{\infty} \frac{(-1)^n x^n}{n2^n}$

$= \sum\limits_{n=1}^{\infty} \frac{(-1)^n}{n}$ which converges by the Alternating Series Test. When $x = -2$

$\sum\limits_{n=1}^{\infty} \frac{(-1)^n x^n}{n2^n} = \sum\limits_{n=1}^{\infty} \frac{1}{n}$ which diverges (harmonic series), so $I = (-2, 2]$.

11. If $u_n = \frac{3^n x^n}{(n+1)^2}$, then $\lim\limits_{n \to \infty} \left|\frac{u_{n+1}}{u_n}\right| = \lim\limits_{n \to \infty} \left|\frac{3^{n+1} x^{n+1}}{(n+2)^2} \cdot \frac{(n+1)^2}{3^n x^n}\right|$

$= 3|x| \lim\limits_{n \to \infty} \left(\frac{n+1}{n+2}\right)^2 = 3|x| < 1$ for convergence, so $|x| < \frac{1}{3}$ and $R = \frac{1}{3}$. When $x = \frac{1}{3}$,

$\sum\limits_{n=0}^{\infty} \frac{3^n x^n}{(n+1)^2} = \sum\limits_{n=0}^{\infty} \frac{1}{(n+1)^2} = \sum\limits_{n=1}^{\infty} \frac{1}{n^2}$ which is a convergent p–series $(p = 2 > 1)$. When

$x = -\frac{1}{3}$, $\sum\limits_{n=0}^{\infty} \frac{3^n x^n}{(n+1)^2} = \sum\limits_{n=0}^{\infty} \frac{(-1)^n}{(n+1)^2}$ which converges by the Alternating Series Test, so

$I = [-1/3, 1/3]$.

13. If $u_n = \frac{x^n}{\ln n}$, then $\lim\limits_{n \to \infty} \left|\frac{u_{n+1}}{u_n}\right| = \lim\limits_{n \to \infty} \left|\frac{x^{n+1}}{\ln(n+1)} \cdot \frac{\ln n}{x^n}\right| = |x| \lim\limits_{n \to \infty} \frac{\ln n}{\ln(n+1)} = |x|$ (using

l'Hospital's Rule), so $R = 1$. When $x = 1$, $\sum\limits_{n=2}^{\infty} \frac{x^n}{\ln n} = \sum\limits_{n=2}^{\infty} \frac{1}{\ln n}$ which diverges because

$\frac{1}{\ln n} > \frac{1}{n}$ and $\sum\limits_{n=2}^{\infty} \frac{1}{n}$ is the divergent harmonic series. When $x = -1$, $\sum\limits_{n=2}^{\infty} \frac{x^n}{\ln n} = \sum\limits_{n=2}^{\infty} \frac{(-1)^n}{\ln n}$

which converges by the Alternating Series Test. So $I = [-1, 1)$.

15. If $u_n = \frac{n}{4^n}(2x-1)^n$, then $\left|\frac{u_{n+1}}{u_n}\right| = \left|\frac{(n+1)(2x-1)^{n+1}}{4^{n+1}} \cdot \frac{4^n}{n(2x-1)^n}\right| = \left|\frac{2x-1}{4}\left(1+\frac{1}{n}\right)\right|$

$\to \frac{1}{2}\left|x - \frac{1}{2}\right|$ as $n \to \infty$. For convergence, $\frac{1}{2}\left|x - \frac{1}{2}\right| < 1 \Rightarrow \left|x - \frac{1}{2}\right| < 2 \Rightarrow R = 2$ and

$-2 < x - \frac{1}{2} < 2 \Rightarrow -\frac{3}{2} < x < \frac{5}{2}$. If $x = -\frac{3}{2}$, the series becomes $\sum\limits_{n=0}^{\infty} \frac{n}{4^n}(-4)^n = \sum\limits_{n=0}^{\infty} (-1)^n n$

which is divergent by the Test for Divergence. If $x = \frac{5}{2}$, the series is $\sum\limits_{n=0}^{\infty} \frac{n}{4^n} 4^n = \sum\limits_{n=0}^{\infty} n$ which is

20

also divergent by the Test for Divergence. So $I = (-\frac{3}{2}, \frac{5}{2})$.

17. If $u_n = \frac{(-1)^n(x-1)^n}{\sqrt{n}}$, then $\lim_{n \to \infty} \left|\frac{u_{n+1}}{u_n}\right| = \lim_{n \to \infty} \left|\frac{(x-1)^{n+1}}{\sqrt{n+1}} \cdot \frac{\sqrt{n}}{(x-1)^n}\right|$

$= |x-1| \lim_{n \to \infty} \sqrt{\frac{n}{n+1}} = |x-1| < 1$ for convergence, or $0 < x < 2$, and $R = 1$. When

$x = 0$, $\sum_{n=1}^{\infty} \frac{(-1)^n(x-1)^n}{\sqrt{n}} = \sum_{n=1}^{\infty} \frac{1}{\sqrt{n}}$ which is a divergent p-series ($p = \frac{1}{2} < 1$). When

$x = 2$, the series is $\sum_{n=1}^{\infty} \frac{(-1)^n}{\sqrt{n}}$ which converges by the Alternating Series Test. So $I = (0, 2]$.

19. If $u_n = \frac{(x-2)^n}{n^n}$, then $\lim_{n \to \infty} \sqrt[n]{|u_n|} = \lim_{n \to \infty} \frac{x-2}{n} = 0$, so the series converges for all x (by the Root Test). $R = \infty$ and $I = (-\infty, \infty)$.

21. If $u_n = \frac{2^n(x-3)^n}{n+3}$, then $\lim_{n \to \infty} \left|\frac{u_{n+1}}{u_n}\right| = \lim_{n \to \infty} \left|\frac{2^{n+1}(x-3)^{n+1}}{n+4} \cdot \frac{n+3}{2^n(x-3)^n}\right|$

$= 2|x-3| \lim_{n \to \infty} \frac{n+3}{n+4} = 2|x-3| < 1$ for convergence, or $|x-3| < \frac{1}{2} \Leftrightarrow \frac{5}{2} < x < \frac{7}{2}$, and

$R = \frac{1}{2}$. When $x = \frac{5}{2}$, $\sum_{n=0}^{\infty} \frac{2^n(x-3)^n}{n+3} = \sum_{n=0}^{\infty} \frac{(-1)^n}{n+3}$ which converges by the Alternating

Series Test. When $x = \frac{7}{2}$, $\sum_{n=0}^{\infty} \frac{2^n(x-3)^n}{n+3} = \sum_{n=0}^{\infty} \frac{1}{n+3} = \sum_{n=3}^{\infty} \frac{1}{n}$, the harmonic series, which

diverges. So $I = [5/2, 7/2)$.

23. If $u_n = \frac{n(x+10)^n}{(n^2+1)4^n}$, then $\lim_{n \to \infty} \left|\frac{u_{n+1}}{u_n}\right|$

$= \lim_{n \to \infty} \left|\frac{(n+1)(x+10)^{n+1}}{[(n+1)^2+1]4^{n+1}} \cdot \frac{(n^2+1)4^n}{n(x+10)^n}\right|$

$= \frac{|x+10|}{4} \lim_{n \to \infty} \frac{n^3+n^2+n+1}{n^3+2n^2+2n} = \frac{|x+10|}{4} < 1$ for convergence, so $|x+10| < 4$,

$-14 < x < -6$, and $R = 4$. When $x = -14$, $\sum_{n=0}^{\infty} \frac{n(x+10)^n}{(n^2+1)4^n} = \sum_{n=0}^{\infty} \frac{(-1)^n n}{n^2+1}$ which

converges by the Alternating Series Test. When $x = -6$,

$\sum_{n=0}^{\infty} \frac{n(x+10)^n}{(n^2+1)4^n} = \sum_{n=0}^{\infty} \frac{n}{n^2+1}$ which diverges (by the Integral Test or the Limit

Comparison Test with $b_n = 1/n$). So $I = [-14, -6)$.

25. If $u_n = \left(\frac{n}{2}\right)^n (x+6)^n$, then $\lim_{n \to \infty} \sqrt[n]{|u_n|} = \lim_{n \to \infty} \frac{n(x+6)}{2} = \infty$ unless $x = -6$, in which case

21

the limit is 0. So by the Root Test, the series converges only for $x = -6$. $R = 0$ and $I = \{-6\}$.

27. If $u_n = \dfrac{(2x-1)^n}{n^3}$, then $\lim\limits_{n \to \infty} \left|\dfrac{u_{n+1}}{u_n}\right| = |2x-1| \lim\limits_{n \to \infty} \left(\dfrac{n}{n+1}\right)^3 = |2x-1| < 1$ for convergence, so $|x - 1/2| < 1/2 \Leftrightarrow 0 < x < 1$, and $R = 1/2$. The series $\sum\limits_{n=1}^{\infty} \dfrac{(2x-1)^n}{n^3}$ converges both for $x = 0$ and $x = 1$ (in the first case because of the Alternating Series Test and in the last case because we get a p-series with $p = 3 > 1$). So $I = [0, 1]$.

29. If $u_n = \dfrac{n}{\sqrt{n+1}}(x-e)^n$, then $\lim\limits_{n \to \infty} \left|\dfrac{u_{n+1}}{u_n}\right| = \lim\limits_{n \to \infty} \left|\dfrac{(n+1)(x-e)^{n+1}/\sqrt{n+2}}{n(x-e)^n/\sqrt{n+1}}\right|$

$= |x-e| \lim\limits_{n \to \infty} \left(\dfrac{n^3 + 3n^2 + 3n + 1}{n^3 + 2n^2}\right)^{1/2} = |x-e| < 1$ for convergence, so $e - 1 < x < e + 1$ and $R = 1$. When $x = e \pm 1$, $\sum\limits_{n=0}^{\infty} \dfrac{n}{\sqrt{n+1}}(x-e)^n$ will diverge by the Test for Divergence since $\lim\limits_{n \to \infty} \dfrac{n}{\sqrt{n+1}} = \infty$. So $I = (e-1, e+1)$.

31. If $u_n = \dfrac{n! x^n}{(2n)!}$, then $\lim\limits_{n \to \infty} \left|\dfrac{u_{n+1}}{u_n}\right| = \lim\limits_{n \to \infty} \left|\dfrac{(n+1)! x^{n+1}}{(2n+2)!} \cdot \dfrac{(2n)!}{n! x^n}\right|$

$= \lim\limits_{n \to \infty} \dfrac{n+1}{(2n+1)(2n+2)} |x| = 0 < 1$ for all x, so $R = \infty$ and $I = (-\infty, \infty)$.

33. If $u_n = \dfrac{(-1)^n x^{2n+1}}{n!(n+1)! 2^{2n+1}}$, then $\lim\limits_{n \to \infty} \left|\dfrac{u_{n+1}}{u_n}\right| = \left(\dfrac{x}{2}\right)^2 \lim\limits_{n \to \infty} \dfrac{1}{(n+1)(n+2)} = 0$ for all x.

So $J_1(x)$ converges for all x; the domain is $(-\infty, \infty)$.

35. $s_{2n-1} = 1 + 2x + x^2 + 2x^3 + \cdots + x^{2n-2} + 2x^{2n-1} = (1 + 2x)(1 + x^2 + x^4 + \cdots + x^{2n-2})$

$= (1 + 2x)\dfrac{1 - x^{2n}}{1 - x^2} \to \dfrac{1 + 2x}{1 - x^2}$ as $n \to \infty$, when $|x| < 1$. Also $s_{2n} = s_{2n-1} + x^{2n} \to \dfrac{1 + 2x}{1 - x^2}$

since $x^{2n} \to 0$ for $|x| < 1$. Therefore $s_n \to \dfrac{1 + 2x}{1 - x^2}$ by Exercise 10.1.76(a).

Thus the interval of convergence is $(-1, 1)$ and $f(x) = \dfrac{1 + 2x}{1 - x^2}$.

37. We use the Root Test on the series $\sum a_n x^n$. $\lim\limits_{n \to \infty} \sqrt[n]{|a_n x^n|} = |x| \lim\limits_{n \to \infty} \sqrt[n]{|a_n|} = a|x| < 1$ for convergence, or $|x| < \dfrac{1}{a}$, so $R = \dfrac{1}{a}$.

Section 10.9

EXERCISES 10.9

1. $f(x) = \cos x$ $f(0) = 1$
    $f'(x) = -\sin x$ $f'(0) = 0$
    $f''(x) = -\cos x$ $f''(0) = -1$
    $f^{(3)}(x) = \sin x$ $f^{(3)}(0) = 0$
    $f^{(4)}(x) = \cos x$ $f^{(4)}(0) = 1$
    ...   ...

    $\cos x = f(0) + f'(0)\,x + \dfrac{f''(0)}{2!}x^2 + \dfrac{f^{(3)}(0)}{3!}x^3 + \dfrac{f^{(4)}(0)}{4!}x^4 + \cdots = 1 - \dfrac{x^2}{2!} + \dfrac{x^4}{4!} - \cdots$

    $= \displaystyle\sum_{n=0}^{\infty} \dfrac{(-1)^n x^{2n}}{(2n)!}.$ If $u_n = \dfrac{(-1)^n x^{2n}}{(2n)!}$, then $\displaystyle\lim_{n\to\infty} \left|\dfrac{u_{n+1}}{u_n}\right| = x^2 \lim_{n\to\infty} \dfrac{1}{(2n+2)(2n+1)} = 0$

    < 1 for all x. So $R = \infty$.

3. $f(x) = \sin x$ $f(\pi/4) = \sqrt{2}/2$
    $f'(x) = \cos x$ $f'(\pi/4) = \sqrt{2}/2$
    $f''(x) = -\sin x$ $f''(\pi/4) = -\sqrt{2}/2$
    $f^{(3)}(x) = -\cos x$ $f^{(3)}(\pi/4) = -\sqrt{2}/2$
    $f^{(4)}(x) = \sin x$ $f^{(4)}(\pi/4) = \sqrt{2}/2$

    $\sin x = f(\pi/4) + f'(\pi/4)(x - \tfrac{\pi}{4}) + \dfrac{f''(\pi/4)}{2!}(x-\tfrac{\pi}{4})^2 + \dfrac{f^{(3)}(\pi/4)}{3!}(x-\tfrac{\pi}{4})^3 + \dfrac{f^{(4)}(\pi/4)}{4!}(x-\tfrac{\pi}{4})^4$

    $+ \cdots = \dfrac{\sqrt{2}}{2}\left[1 + (x-\tfrac{\pi}{4}) - \tfrac{1}{2!}(x-\tfrac{\pi}{4})^2 - \tfrac{1}{3!}(x-\tfrac{\pi}{4})^3 + \tfrac{1}{4!}(x-\tfrac{\pi}{4})^4 + \cdots\right]$

    $= \dfrac{\sqrt{2}}{2} \displaystyle\sum_{n=0}^{\infty} \dfrac{(-1)^{n(n-1)/2}(x-\pi/4)^n}{n!}.$ If $u_n = \dfrac{(-1)^{n(n-1)/2}(x-\pi/4)^n}{n!}$, then $\displaystyle\lim_{n\to\infty} \left|\dfrac{u_{n+1}}{u_n}\right|$

    $= \displaystyle\lim_{n\to\infty} \dfrac{|x-\pi/4|}{n+1} = 0 < 1$ for all x, so $R = \infty$.

5. $f(x) = (1+x)^{-2}$ $f(0) = 1$
    $f'(x) = -2(1+x)^{-3}$ $f'(0) = -2$
    $f''(x) = 2 \cdot 3(1+x)^{-4}$ $f''(0) = 2 \cdot 3$
    $f^{(3)}(x) = -2 \cdot 3 \cdot 4(1+x)^{-5}$ $f^{(3)}(0) = -2 \cdot 3 \cdot 4$
    $f^{(4)}(x) = 2 \cdot 3 \cdot 4 \cdot 5(1+x)^{-6}$ $f^{(4)}(0) = 2 \cdot 3 \cdot 4 \cdot 5$
    ...   ...

    So $f^{(n)}(0) = (-1)^n(n+1)!$, and

23

$$\frac{1}{(1+x)^2} = \sum_{n=0}^{\infty} \frac{(-1)^n (n+1)!}{n!} x^n = \sum_{n=0}^{\infty} (-1)^n (n+1) x^n.$$

If $u_n = (-1)^n (n+1) x^n$, then $\lim_{n \to \infty} \left| \frac{u_{n+1}}{u_n} \right| = |x|$ so $R = 1$.

7. $f(x) = x^{-1}$ $\qquad\qquad\qquad\qquad$ $f(1) = 1$
    $f'(x) = -x^{-2}$ $\qquad\qquad\qquad\quad$ $f'(1) = -1$
    $f''(x) = 2x^{-3}$ $\qquad\qquad\qquad\quad$ $f''(1) = 2$
    $f^{(3)}(x) = -3 \cdot 2 x^{-4}$ $\qquad\qquad\;$ $f^{(3)}(1) = -3 \cdot 2$
    $f^{(4)}(x) = 4 \cdot 3 \cdot 2 x^{-5}$ $\qquad\qquad$ $f^{(4)}(1) = 4 \cdot 3 \cdot 2$
    $\ldots$ $\qquad\qquad\qquad\qquad\qquad\quad$ $\ldots$

So $f^{(n)}(1) = (-1)^n n!$, and $\frac{1}{x} = \sum_{n=0}^{\infty} \frac{(-1)^n n!}{n!}(x-1)^n = \sum_{n=0}^{\infty} (-1)^n (x-1)^n$. If $u_n = (-1)^n (x-1)^n$ then $\lim_{n \to \infty} \left| \frac{u_{n+1}}{u_n} \right| = |x-1| < 1$ for convergence, so $0 < x < 2$ and $R = 1$.

9. Clearly $f^{(n)}(x) = e^x$, so $f^{(n)}(3) = e^3$ and $e^x = \sum_{n=0}^{\infty} \frac{e^3}{n!} (x-3)^n$. If $u_n = \frac{e^3}{n!}(x-3)^n$ then $\lim_{n \to \infty} \left| \frac{u_{n+1}}{u_n} \right| = \lim_{n \to \infty} \frac{|x-3|}{n+1} = 0$ for all x, so $R = \infty$.

11. $f(x) = \sinh x$ $\qquad\qquad\qquad\quad$ $f(0) = 0$
    $f'(x) = \cosh x$ $\qquad\qquad\qquad\quad$ $f'(0) = 1$
    $f''(x) = \sinh x$ $\qquad\qquad\qquad\quad$ $f''(0) = 0$
    $f^{(3)}(x) = \cosh x$ $\qquad\qquad\qquad$ $f^{(3)}(0) = 1$
    $f^{(4)}(x) = \sinh x$ $\qquad\qquad\qquad$ $f^{(4)}(0) = 0$
    $\ldots$ $\qquad\qquad\qquad\qquad\qquad\quad$ $\ldots$

So $f^{(n)}(0) = \begin{cases} 0 & \text{if } n \text{ is even} \\ 1 & \text{if } n \text{ is odd} \end{cases}$, and $\sinh x = \sum_{n=0}^{\infty} \frac{x^{2n+1}}{(2n+1)!}$. If $u_n = \frac{x^{2n+1}}{(2n+1)!}$ then $\lim_{n \to \infty} \left| \frac{u_{n+1}}{u_n} \right| = x^2 \lim_{n \to \infty} \frac{1}{(2n+3)(2n+2)} = 0 < 1$ for all x, so $R = \infty$.

13. $f(x) = \frac{1}{1+x} = \frac{1}{1-(-x)} = \sum_{n=0}^{\infty} (-1)^n x^n$ with $|-x| < 1 \Leftrightarrow |x| < 1$ so $R = 1$.

15. $f(x) = \frac{1}{(1+x)^2} = -\frac{d}{dx}\left(\frac{1}{1+x}\right) = -\frac{d}{dx}\left(\sum_{n=0}^{\infty} (-1)^n x^n\right)$ [from Exercise 13]

$= \sum_{n=1}^{\infty} (-1)^{n+1} n x^{n-1} = \sum_{n=0}^{\infty} (-1)^n (n+1) x^n$ with $R = 1$.

Section 10.9

17. $f(x) = \dfrac{1}{1+4x^2} = \sum\limits_{n=0}^{\infty} (-1)^n (4x^2)^n$ [substituting $4x^2$ for x in the series from Exercise 13

    above] $= \sum\limits_{n=0}^{\infty} (-1)^n 4^n x^{2n}$, with $|4x^2| < 1$ so $x^2 < \frac{1}{4}$, $|x| < \frac{1}{2}$ and $R = \frac{1}{2}$.

19. $f(x) = \dfrac{1}{4+x^2} = \dfrac{1}{4}\left(\dfrac{1}{1+x^2/4}\right) = \dfrac{1}{4} \sum\limits_{n=0}^{\infty} (-1)^n \left(\dfrac{x^2}{4}\right)^n$ [using Exercise 13]

    $= \sum\limits_{n=0}^{\infty} \dfrac{(-1)^n x^{2n}}{4^{n+1}}$, with $\left|\dfrac{x^2}{4}\right| < 1 \Leftrightarrow x^2 < 4 \Leftrightarrow |x| < 2$, so $R = 2$.

21. $f(x) = \dfrac{1}{1-x^2} = \sum\limits_{n=0}^{\infty} (x^2)^n = \sum\limits_{n=0}^{\infty} x^{2n}$, with $|x^2| < 1 \Leftrightarrow |x| < 1$ so $R = 1$.

23. $f(x) = \ln(1+x) - \ln(1-x) = \int \dfrac{dx}{1+x} + \int \dfrac{dx}{1-x} = \int \left[\sum\limits_{n=0}^{\infty} (-1)^n x^n + \sum\limits_{n=0}^{\infty} x^n\right] dx$

    $= \int \sum\limits_{n=0}^{\infty} 2x^{2n}\, dx = \sum\limits_{n=0}^{\infty} \dfrac{2x^{2n+1}}{2n+1} + C$. But $f(0) = \ln 1 - \ln 1 = 0$, so $C = 0$ and we have

    $f(x) = \sum\limits_{n=0}^{\infty} \dfrac{2x^{2n+1}}{2n+1}$ with $R = 1$.

25. $e^{3x} = \sum\limits_{n=0}^{\infty} \dfrac{(3x)^n}{n!} = \sum\limits_{n=0}^{\infty} \dfrac{3^n x^n}{n!}$, with $R = \infty$.

27. $x^2 \cos x = x^2 \sum\limits_{n=0}^{\infty} \dfrac{(-1)^n x^{2n}}{(2n)!} = \sum\limits_{n=0}^{\infty} \dfrac{(-1)^n x^{2n+2}}{(2n)!}$, with $R = \infty$.

29. $x \sin\left(\dfrac{x}{2}\right) = x \sum\limits_{n=0}^{\infty} \dfrac{(-1)^n (x/2)^{2n+1}}{(2n+1)!} = \sum\limits_{n=0}^{\infty} \dfrac{(-1)^n x^{2n+2}}{(2n+1)! 2^{2n+1}}$, with $R = \infty$.

31. $\sin^2 x = \dfrac{1}{2}[1 - \cos 2x] = \dfrac{1}{2}\left[1 - \sum\limits_{n=0}^{\infty} \dfrac{(-1)^n (2x)^{2n}}{(2n)!}\right] = \dfrac{1}{2}\left[1 - 1 - \sum\limits_{n=1}^{\infty} \dfrac{(-1)^n (2x)^{2n}}{(2n)!}\right]$

    $= \sum\limits_{n=1}^{\infty} \dfrac{(-1)^{n+1} 2^{2n-1} x^{2n}}{(2n)!}$, with $R = \infty$.

33. $\dfrac{\sin x}{x} = \dfrac{1}{x} \sum\limits_{n=0}^{\infty} \dfrac{(-1)^n x^{2n+1}}{(2n+1)!} = \sum\limits_{n=0}^{\infty} \dfrac{(-1)^n x^{2n}}{(2n+1)!}$ and this series also gives the required

    value at $x = 0$, so $R = \infty$.

35. $f(x) = (1+x)^{1/2}$ $f(0) = 1$

    $f'(x) = \frac{1}{2}(1+x)^{-1/2}$ $f'(0) = \frac{1}{2}$

    $f''(x) = -\frac{1}{4}(1+x)^{-3/2}$ $f''(0) = -\frac{1}{4}$

25

$f^{(3)}(x) = \frac{3}{8}(1+x)^{-5/2}$ $\quad\quad f^{(3)}(0) = \frac{3}{8}$

$f^{(4)}(x) = -\frac{15}{16}(1+x)^{-7/2}$ $\quad\quad f^{(4)}(0) = -\frac{15}{16}$

... ...

So $f^{(n)}(0) = \frac{(-1)^{n-1} 1 \cdot 3 \cdot 5 \cdots (2n-3)}{2^n}$ for $n \geq 2$, and $\sqrt{1+x} = 1 + \frac{x}{2}$

$+ \sum_{n=2}^{\infty} \frac{(-1)^{n-1} 1 \cdot 3 \cdot 5 \cdots (2n-3)}{2^n n!} x^n$. If $u_n = \frac{(-1)^{n+1} 1 \cdot 3 \cdot 5 \cdots (2n-3)}{2^n n!} x^n$ then

$\lim_{n \to \infty} \left| \frac{u_{n+1}}{u_n} \right| = \frac{|x|}{2} \lim_{n \to \infty} \frac{2n-1}{n+1} = |x| < 1$ for convergence, so $R = 1$.

37. $f(x) = (1-x)^{-1/3}$ $\quad\quad f(0) = 1$

$f'(x) = \frac{1}{3}(1-x)^{-4/3}$ $\quad\quad f'(0) = \frac{1}{3}$

$f''(x) = \frac{4}{9}(1-x)^{-7/3}$ $\quad\quad f''(0) = \frac{4}{9}$

$f^{(3)}(x) = \frac{28}{27}(1-x)^{-10/3}$ $\quad\quad f^{(3)}(0) = \frac{28}{27}$

$f^{(4)}(x) = \frac{280}{81}(1-x)^{-13/3}$ $\quad\quad f^{(4)}(0) = \frac{280}{81}$

... ...

So $f^{(n)}(0) = \frac{1 \cdot 4 \cdot 7 \cdot 10 \cdot 13 \cdots (3n-2)}{3^n}$ for $n \geq 1$, and

$f(x) = 1 + \sum_{n=1}^{\infty} \frac{1 \cdot 4 \cdot 7 \cdot 10 \cdot 13 \cdots (3n-2)}{3^n n!} x^n$. If $u_n = \frac{1 \cdot 4 \cdot 7 \cdot 10 \cdot 13 \cdots (3n-2)}{3^n n!} x^n$

then $\lim_{n \to \infty} \left| \frac{u_{n+1}}{u_n} \right| = \frac{|x|}{3} \lim_{n \to \infty} \frac{3n+1}{n+1} = |x| < 1$ for convergence, so $R = 1$.

39. $f(x) = (1+x)^{-3} = -\frac{1}{2} \frac{d}{dx}\left[\frac{1}{(1+x)^2}\right] = -\frac{1}{2} \frac{d}{dx}\left[\sum_{n=0}^{\infty}(-1)^n(n+1)x^n\right]$ [from Exercise 15

above] $= -\frac{1}{2} \sum_{n=1}^{\infty}(-1)^n n(n+1)x^{n-1} = \sum_{n=0}^{\infty} \frac{(-1)^n(n+1)(n+2)x^n}{2}$, with $R = 1$ (since

that is the R in Exercise 15).

41. $\ln(5+x) = \ln[5(1+x/5)] = \ln(5) + \ln(1+x/5) = \ln(5) + \frac{1}{5} \int \frac{dx}{1+x/5}$

$= \ln(5) + \frac{1}{5} \int \sum_{n=0}^{\infty} (-1)^n \left(\frac{x}{5}\right)^n dx = \ln(5) + \sum_{n=0}^{\infty} \frac{(-1)^n x^{n+1}}{(n+1)5^{n+1}} = \ln(5) + \sum_{n=1}^{\infty} \frac{(-1)^{n-1} x^n}{n 5^n}$,

with $R = 5$.

43. $\ln(1+x) = \int \frac{dx}{1+x} = \int \sum_{n=0}^{\infty} (-1)^n x^n \, dx = \sum_{n=1}^{\infty} \frac{(-1)^{n-1} x^n}{n}$ with $R = 1$, so

$\ln(1.1) = \sum_{n=1}^{\infty} \frac{(-1)^{n-1}(0.1)^n}{n}$. This is an alternating series with $a_5 = \frac{(0.1)^5}{5} = .000002$, so to 5 decimals, $\ln(1.1) \approx \sum_{n=1}^{4} \frac{(-1)^{n-1}(0.1)^n}{n} \approx 0.09531$.

45. $\int \sin(x^2) \, dx = \int \sum_{n=0}^{\infty} (-1)^n \frac{(x^2)^{2n+1}}{(2n+1)!} \, dx = \int \sum_{n=0}^{\infty} \frac{(-1)^n x^{4n+2}}{(2n+1)!} \, dx$

$= C + \sum_{n=0}^{\infty} \frac{(-1)^n x^{4n+3}}{(4n+3)(2n+1)!}$.

47. $\int \frac{dx}{1+x^4} = \int \sum_{n=0}^{\infty} (-1)^n x^{4n} \, dx = C + \sum_{n=0}^{\infty} \frac{(-1)^n x^{4n+1}}{4n+1}$.

49. Using the series we obtained in Exercise 35, we get

$\sqrt{x^3+1} = 1 + \frac{x^3}{2} + \sum_{n=2}^{\infty} \frac{(-1)^{n-1} 1 \cdot 3 \cdot 5 \cdots (2n-3)}{2^n n!} x^{3n}$ so

$\int \sqrt{x^3+1} \, dx = \int \left(1 + \frac{x^3}{2} + \sum_{n=2}^{\infty} \frac{(-1)^{n-1} 1 \cdot 3 \cdot 5 \cdots (2n-3)}{2^n n!} x^{3n}\right) dx$

$= C + x + \frac{x^4}{8} + \sum_{n=2}^{\infty} \frac{(-1)^{n-1} 1 \cdot 3 \cdot 5 \cdots (2n-3)}{2^n n! (3n+1)} x^{3n+1}$.

51. Using our series from Exercise 45, we get $\int_0^1 \sin(x^2) \, dx = \sum_{n=0}^{\infty} \left[\frac{(-1)^n x^{4n+3}}{(4n+3)(2n+1)!}\right]_0^1$

$= \sum_{n=0}^{\infty} \frac{(-1)^n}{(4n+3)(2n+1)!}$ and $|a_3| = \frac{1}{75600} < 0.000014$ so by Theorem 10.28, we have

$\sum_{n=0}^{2} \frac{(-1)^n}{(4n+3)(2n+1)!} \approx \frac{1}{3} - \frac{1}{42} + \frac{1}{1320} \approx 0.310$.

53. $\int_0^{0.5} \frac{dx}{1+x^6} = \int_0^{0.5} \sum_{n=0}^{\infty} (-1)^n x^{6n} \, dx = \sum_{n=0}^{\infty} \left[\frac{(-1)^n x^{6n+1}}{6n+1}\right]_0^{1/2} = \sum_{n=0}^{\infty} \frac{(-1)^n}{(6n+1)2^{6n+1}}$ and

$a_2 = \frac{1}{106496} < 0.00001$, so by Theorem 10.28, we use $\sum_{n=0}^{1} \frac{(-1)^n}{(6n+1)2^{6n+1}} = \frac{1}{2} - \frac{1}{896} \approx 0.4989$.

55. $\int_0^{0.5} x^2 e^{-x^2} \, dx = \int_0^{0.5} \sum_{n=0}^{\infty} \frac{(-1)^n x^{2n+2}}{n!} \, dx = \sum_{n=0}^{\infty} \left[\frac{(-1)^n x^{2n+3}}{n!(2n+3)}\right]_0^{1/2}$

$= \sum_{n=0}^{\infty} \frac{(-1)^n}{n!(2n+3)2^{2n+3}}$ and since $a_2 = \frac{1}{1792} < 0.001$ we use

$\sum_{n=0}^{1} \frac{(-1)^n}{n!(2n+3)2^{2n+3}} = \frac{1}{24} - \frac{1}{160} \approx 0.0354$.

57. (a) $J_0(x) = \sum_{n=0}^{\infty} \dfrac{(-1)^n x^{2n}}{2^{2n}(n!)^2}$, $J_0'(x) = \sum_{n=1}^{\infty} \dfrac{(-1)^n 2n x^{2n-1}}{2^{2n}(n!)^2}$, and

$J_0''(x) = \sum_{n=1}^{\infty} \dfrac{(-1)^n 2n(2n-1) x^{2n-2}}{2^{2n}(n!)^2}$, so $x^2 J_0''(x) + x J_0'(x) + x^2 J_0(x)$

$= \sum_{n=1}^{\infty} \dfrac{(-1)^n 2n(2n-1) x^{2n}}{2^{2n}(n!)^2} + \sum_{n=1}^{\infty} \dfrac{(-1)^n 2n x^{2n}}{2^{2n}(n!)^2} + \sum_{n=0}^{\infty} \dfrac{(-1)^n x^{2n+2}}{2^{2n}(n!)^2}$

$= \sum_{n=1}^{\infty} \dfrac{(-1)^n 2n(2n-1) x^{2n}}{2^{2n}(n!)^2} + \sum_{n=1}^{\infty} \dfrac{(-1)^n 2n x^{2n}}{2^{2n}(n!)^2} + \sum_{n=1}^{\infty} \dfrac{(-1)^{n-1} x^{2n}}{2^{2n-2}((n-1)!)^2}$

$= \sum_{n=1}^{\infty} (-1)^n \left[\dfrac{2n(2n-1) + 2n - 2^2 n^2}{2^{2n}(n!)^2} \right] x^{2n} = \sum_{n=1}^{\infty} (-1)^n \left[\dfrac{4n^2 - 2n + 2n - 4n^2}{2^{2n}(n!)^2} \right] x^{2n}$

$= 0$.

(b) $\int_0^1 J_0(x)\,dx = \int_0^1 \left[\sum_{n=0}^{\infty} \dfrac{(-1)^n x^{2n}}{2^{2n}(n!)^2} \right] dx$

$= \int_0^1 dx + \int_0^1 -\dfrac{x^2}{4}\,dx + \int_0^1 \dfrac{x^4}{64}\,dx + \int_0^1 -\dfrac{x^6}{2304}\,dx + \cdots = \left[x - \dfrac{x^3}{3\cdot 4} + \dfrac{x^5}{5\cdot 64} - \dfrac{x^7}{7\cdot 2304} + \cdots \right]_0^1$

$= 1 - \dfrac{1}{12} + \dfrac{1}{320} - \dfrac{1}{16128} + \cdots$. Since $\dfrac{1}{16128} \approx 0.000062$, it follows from Theorem 10.28

that, correct to 3 decimal places, $\int_0^1 J_0(x)\,dx \approx 1 - \dfrac{1}{12} + \dfrac{1}{320} \approx 0.920$.

59. $\sum_{n=0}^{\infty} (-1)^n \dfrac{x^{4n}}{n!} = \sum_{n=0}^{\infty} \dfrac{(-x^4)^n}{n!} = e^{-x^4}$ by (10.47).

61. $\sum_{n=0}^{\infty} \dfrac{(-1)^n \pi^{2n+1}}{4^{2n+1}(2n+1)!} = \sum_{n=0}^{\infty} \dfrac{(-1)^n (\pi/4)^{2n+1}}{(2n+1)!} = \sin \dfrac{\pi}{4} = \dfrac{1}{\sqrt{2}}$ by (10.49).

63. $\sum_{n=0}^{\infty} \dfrac{x^{n+1}}{(n+1)!} = \dfrac{x}{1!} + \dfrac{x^2}{2!} + \dfrac{x^3}{3!} + \cdots = \left(1 + \dfrac{x}{1!} + \dfrac{x^2}{2!} + \dfrac{x^3}{3!} + \cdots \right) - 1 = e^x - 1$ by (10.47).

65. As in Example 10(a), we have $e^{-x^2} = 1 - \dfrac{x^2}{1!} + \dfrac{x^4}{2!} + \dfrac{x^6}{3!} + \cdots$ and we know that

$\cos x = 1 - \dfrac{x^2}{2!} + \dfrac{x^4}{4!} - \cdots$ from Equation 10.50. Therefore

$e^{-x^2} \cos x = (1 - x^2 + \tfrac{1}{2}x^4 - \cdots)(1 - \tfrac{1}{2}x^2 + \tfrac{1}{24}x^4 - \cdots)$

$= 1 - \tfrac{1}{2}x^2 + \tfrac{1}{24}x^4 - x^2 + \tfrac{1}{2}x^4 + \tfrac{1}{2}x^4 + \cdots = 1 - \tfrac{3}{2}x^2 + \tfrac{25}{24}x^4 + \cdots$

67. From Example 2 we have $\ln(1-x) = -x - \dfrac{x^2}{2} - \dfrac{x^3}{3} - \cdots$, $|x| < 1$. Therefore

$$y = \dfrac{\ln(1-x)}{e^x} = \dfrac{-x - \tfrac{1}{2}x^2 - \tfrac{1}{3}x^3 - \cdots}{1 + x + \tfrac{1}{2}x^2 + \tfrac{1}{6}x^3 + \cdots}$$

Long division gives

Section 10.9

$$
\begin{array}{r}
1+x+\tfrac{1}{2}x^2+\tfrac{1}{6}x^3-\cdots \overline{\Big)\begin{array}{l} -x+\tfrac{1}{2}x^2-\tfrac{1}{3}x^3+\cdots \\ -x-\tfrac{1}{2}x^2-\tfrac{1}{3}x^3-\cdots \\ \overline{-x-\ x^2-\tfrac{1}{2}x^3-\cdots} \\ \quad\ \ \tfrac{1}{2}x^2+\tfrac{1}{6}x^3+\cdots \\ \quad\ \ \tfrac{1}{2}x^2+\tfrac{1}{2}x^3+\cdots \\ \overline{\qquad\qquad -\tfrac{1}{3}x^3+\cdots} \\ \qquad\qquad -\tfrac{1}{3}x^3+\cdots \end{array}}
\end{array}
$$

So $\dfrac{\ln(1-x)}{e^x} = -x + \tfrac{1}{2}x^2 - \tfrac{1}{3}x^3 + \cdots$, $|x| < 1$.

69. If $u_n = \dfrac{x^n}{n^2}$, then $\lim\limits_{n \to \infty} \left|\dfrac{u_{n+1}}{u_n}\right| = |x| \lim\limits_{n \to \infty} \left(\dfrac{n}{n+1}\right)^2 = |x| < 1$ for convergence so $R = 1$.

When $x = \pm 1$, $\sum\limits_{n=1}^{\infty} \left|\dfrac{x^n}{n^2}\right| = \sum\limits_{n=1}^{\infty} \dfrac{1}{n^2}$ which is a convergent p-series ($p = 2 > 1$), so the interval of convergence for f is $[-1, 1]$. By Theorem 10.39, the radii of convergence of f' and f'' are both 1, so we need only check the endpoints.

$f'(x) = \sum\limits_{n=1}^{\infty} \dfrac{nx^{n-1}}{n^2} = \sum\limits_{n=0}^{\infty} \dfrac{x^n}{n+1}$, and this series diverges for $x = 1$ (harmonic series) and converges for $x = -1$ (Alternating Series Test), so the interval of convergence is $[-1, 1)$. $f''(x) = \sum\limits_{n=1}^{\infty} \dfrac{nx^{n-1}}{n+1}$ diverges at both 1 and -1 (Test for Divergence) since $\lim\limits_{n \to \infty} \dfrac{n}{n+1} = 1 \neq 0$, so its interval of convergence is $(-1, 1)$.

71. $f(x) = \begin{cases} e^{-1/x^2} & \text{if } x \neq 0 \\ 0 & \text{if } x = 0 \end{cases}$, so $f'(0) = \lim\limits_{x \to 0} \dfrac{f(x) - f(0)}{x - 0} = \lim\limits_{x \to 0} \dfrac{e^{-1/x^2}}{x} = \lim\limits_{x \to 0} \dfrac{1/x}{e^{1/x^2}}$

$= \lim\limits_{x \to 0} \dfrac{x}{2e^{1/x^2}}$ [using l'Hospital's Rule and simplifying] $= 0$. Similarly, we can use the definition of the derivative and l'Hospital's Rule to show that $f''(0) = 0$, $f^{(3)}(0) = 0, \ldots$ $f^{(n)}(0) = 0$, so that the Maclaurin series for f consists entirely of zero terms. But since $f(x) \neq 0$ except for $x = 0$, we see that f cannot equal its Maclaurin series except at $x = 0$.

Section 10.10

EXERCISES 10.10

1. $(1+x)^{1/2} = \sum_{n=0}^{\infty} \binom{1/2}{n} x^n = 1 + \left(\frac{1}{2}\right)x + \frac{\left(\frac{1}{2}\right)\left(-\frac{1}{2}\right)}{2!} x^2 + \frac{\left(\frac{1}{2}\right)\left(-\frac{1}{2}\right)\left(-\frac{3}{2}\right)}{3!} x^3 + \cdots$

$= 1 + \frac{x}{2} - \frac{x^2}{2^2 \cdot 2!} + \frac{1 \cdot 3 \, x^3}{2^3 \cdot 3!} - \frac{1 \cdot 3 \cdot 5 \, x^4}{2^4 \cdot 4!} + \cdots$

$= 1 + \frac{x}{2} + \sum_{n=2}^{\infty} \frac{(-1)^{n-1} 1 \cdot 3 \cdot 5 \cdots (2n-3) \, x^n}{2^n \cdot n!}. \quad R = 1.$

3. $[1 + (2x)]^{-4} = 1 + (-4)(2x) + \frac{(-4)(-5)}{2!}(2x)^2 + \frac{(-4)(-5)(-6)}{3!}(2x)^3 + \cdots$

$= 1 + \sum_{n=1}^{\infty} \frac{(-1)^n 2^n 4 \cdot 5 \cdot 6 \cdots (n+3)}{n!} x^n = \sum_{n=0}^{\infty} (-1)^n \frac{2^n (n+1)(n+2)(n+3)}{6} x^n.$

$|2x| < 1 \Leftrightarrow |x| < \frac{1}{2}$ so $R = \frac{1}{2}$.

5. $[1 + (-x)]^{-1/2} = \sum_{n=0}^{\infty} \binom{-1/2}{n} (-x)^n = 1 + \left(-\frac{1}{2}\right)(-x) + \frac{\left(-\frac{1}{2}\right)\left(-\frac{3}{2}\right)}{2!}(-x)^2 + \cdots$

$= 1 + \frac{x}{2} + \frac{1 \cdot 3}{2^2 2!} x^2 + \frac{1 \cdot 3 \cdot 5}{2^3 3!} x^3 + \frac{1 \cdot 3 \cdot 5 \cdot 7}{2^4 4!} x^4 + \cdots = 1 + \sum_{n=1}^{\infty} \frac{1 \cdot 3 \cdot 5 \cdots (2n-1)}{2^n \cdot n!} x^n$

so $\frac{x}{\sqrt{1-x}} = x + \sum_{n=1}^{\infty} \frac{1 \cdot 3 \cdot 5 \cdots (2n-1)}{2^n \cdot n!} x^{n+1}$ with $R = 1$.

7. $(8+x)^{-1/3} = \frac{1}{2}\left(1 + \frac{x}{8}\right)^{-1/3} = \frac{1}{2}\left[1 + \left(-\frac{1}{3}\right)\left(\frac{x}{8}\right) + \frac{\left(-\frac{1}{3}\right)\left(-\frac{4}{3}\right)}{2!}\left(\frac{x}{8}\right)^2 + \cdots\right]$

$= \frac{1}{2}\left[1 + \sum_{n=1}^{\infty} \frac{(-1)^n 1 \cdot 4 \cdot 7 \cdots (3n-2)}{3^n \, n! \, 8^n} x^n\right]$ and $\left|\frac{x}{8}\right| < 1 \Leftrightarrow |x| < 8$, so $R = 8$.

9. $(1 - x^4)^{1/4} = 1 + \left(\frac{1}{4}\right)(-x^4) + \frac{\left(\frac{1}{4}\right)\left(-\frac{3}{4}\right)}{2!}(-x^4)^2 + \frac{\left(\frac{1}{4}\right)\left(-\frac{3}{4}\right)\left(-\frac{7}{4}\right)}{3!}(-x^4)^3 + \cdots$

$= 1 - \frac{x^4}{4} - \sum_{n=2}^{\infty} \frac{3 \cdot 7 \cdot 11 \cdots (4n-5)}{4^n \, n!} x^{4n}$ with $R = 1$.

11. $(1-x)^{-5} = 1 + (-5)(-x) + \frac{(-5)(-6)}{2!}(-x)^2 + \frac{(-5)(-6)(-7)}{3!}(-x)^3 + \cdots$

$= 1 + \sum_{n=1}^{\infty} \frac{5 \cdot 6 \cdot 7 \cdots (n+4)}{n!} x^n = \sum_{n=0}^{\infty} \frac{(n+4)!}{4! \, n!} x^n \Rightarrow \frac{x^5}{(1-x)^5} = \sum_{n=0}^{\infty} \frac{(n+4)!}{4! \, n!} x^{n+5}$

$\left(\text{or } \sum_{n=0}^{\infty} \frac{(n+1)(n+2)(n+3)(n+4)}{24} x^{n+5}\right)$ with $R = 1$.

13. (a) $(1 - x^2)^{-1/2} = 1 + \left(-\frac{1}{2}\right)(-x^2) + \frac{\left(-\frac{1}{2}\right)\left(-\frac{3}{2}\right)}{2!}(-x^2)^2 + \frac{\left(-\frac{1}{2}\right)\left(-\frac{3}{2}\right)\left(-\frac{5}{2}\right)}{3!}(-x^2)^3 + \cdots$

Section 10.10

$$= 1 + \sum_{n=1}^{\infty} \frac{1 \cdot 3 \cdot 5 \cdots (2n-1)}{2^n \, n!} x^{2n}$$

(b) $\sin^{-1} x = \int \frac{1}{\sqrt{1-x^2}} dx = C + x + \sum_{n=1}^{\infty} \frac{1 \cdot 3 \cdot 5 \cdots (2n-1)}{(2n+1) \, 2^n \, n!} x^{2n+1}$

$$= x + \sum_{n=1}^{\infty} \frac{1 \cdot 3 \cdot 5 \cdots (2n-1)}{(2n+1) \, 2^n \, n!} x^{2n+1} \text{ since } 0 = \sin^{-1} 0 = C.$$

15. (a) $(1+x)^{-1/2} = 1 + \left(-\frac{1}{2}\right)x + \frac{\left(-\frac{1}{2}\right)\left(-\frac{3}{2}\right)}{2!} x^2 + \frac{\left(-\frac{1}{2}\right)\left(-\frac{3}{2}\right)\left(-\frac{5}{2}\right)}{3!} x^3 + \cdots$

$$= 1 + \sum_{n=1}^{\infty} \frac{(-1)^n 1 \cdot 3 \cdot 5 \cdots (2n-1)}{2^n \, n!} x^n$$

(b) Take $x = 0.1$ in the above series. $\frac{1 \cdot 3 \cdot 5 \cdot 7}{2^4 \, 4!} (0.1)^4 < 0.00003$, so

$$\frac{1}{\sqrt{1.1}} \approx 1 - \frac{0.1}{2} + \frac{1 \cdot 3}{2^2 \cdot 2!} (0.1)^2 - \frac{1 \cdot 3 \cdot 5}{2^3 \, 3!} (0.1)^3 \approx 0.953.$$

17. (a) $(1-x)^{-2} = 1 + (-2)(-x) + \frac{(-2)(-3)}{2!}(-x)^2 + \cdots = \sum_{n=0}^{\infty} (n+1) x^n$, so

$$\frac{x}{(1-x)^2} = \sum_{n=0}^{\infty} (n+1) x^{n+1} = \sum_{n=1}^{\infty} n x^n.$$

(b) With $x = \frac{1}{2}$ in part (a), we have $\sum_{n=1}^{\infty} \frac{n}{2^n} = \frac{1/2}{(1-1/2)^2} = 2$.

19. (a) $(1+x^2)^{1/2} = 1 + \left(\frac{1}{2}\right) x^2 + \frac{\left(\frac{1}{2}\right)\left(-\frac{1}{2}\right)}{2!}(x^2)^2 + \frac{\left(\frac{1}{2}\right)\left(-\frac{1}{2}\right)\left(-\frac{3}{2}\right)}{3!}(x^2)^3 + \cdots$

$$= 1 + \frac{x^2}{2} + \sum_{n=2}^{\infty} \frac{(-1)^{n-1} 1 \cdot 3 \cdot 5 \cdots (2n-3)}{2^n \, n!} x^{2n}$$

(b) The coefficient of x^{10} in the above Maclaurin series will be $\frac{f^{(10)}(0)}{10!}$, so

$$f^{(10)}(0) = 10! \left(\frac{1 \cdot 3 \cdot 5 \cdot 7}{2^5 \, 5!}\right) = 99{,}225.$$

21. (a) $g'(x) = \sum_{n=1}^{\infty} \binom{k}{n} n x^{n-1}$. $(1+x) g'(x) = (1+x) \sum_{n=1}^{\infty} \binom{k}{n} n x^{n-1}$

$$= \sum_{n=1}^{\infty} \binom{k}{n} n x^{n-1} + \sum_{n=1}^{\infty} \binom{k}{n} n x^n = \sum_{n=0}^{\infty} \binom{k}{n+1}(n+1) x^n + \sum_{n=0}^{\infty} \binom{k}{n} n x^n$$

$$= \sum_{n=0}^{\infty} \left((n+1) \frac{k(k-1)(k-2) \cdots (k-n)}{(n+1)!}\right) x^n + \sum_{n=0}^{\infty} \left((n) \frac{k(k-1)(k-2) \cdots (k-n+1)}{n!}\right) x^n$$

$$= \sum_{n=0}^{\infty} \frac{(n+1) k(k-1)(k-2) \cdots (k-n+1)}{(n+1)!} [(k-n) + n] x^n$$

$$= \sum_{n=0}^{\infty} \frac{k^2 (k-1)(k-2) \cdots (k-n+1)}{n!} x^n = k \sum_{n=0}^{\infty} \binom{k}{n} x^n = k \, g(x).$$

31

Section 10.11

So $g'(x) = \dfrac{kg(x)}{1+x}$.

(b) $h'(x) = -k(1+x)^{-k-1}g(x) + (1+x)^{-k}g'(x)$

$= -k(1+x)^{-k-1}g(x) + (1+x)^{-k}\left(\dfrac{kg(x)}{1+x}\right) = -k(1+x)^{-k-1}g(x) + k(1+x)^{-k-1}g(x)$

$= 0$

(c) From part (b) we see that h must be constant for $x \in (-1,1)$, so $h(x) = h(0) = 1$ for $x \in (-1,1)$. Thus $h(x) = 1 = (1+x)^{-k}g(x) \Leftrightarrow g(x) = (1+x)^k$ for $x \in (-1,1)$.

EXERCISES 10.11

1. $f(x) = 1 + 2x + 3x^2 + 4x^3$ $f(-1) = -2$
$f'(x) = 2 + 6x + 12x^2$ $f'(-1) = 8$
$f''(x) = 6 + 24x$ $f''(-1) = -18$
$f^{(3)}(x) = 24$ $f^{(3)}(-1) = 24$
$f^{(4)}(x) = 0$ $f^{(4)}(-1) = 0$

$T_4(x) = \sum\limits_{n=0}^{4} \dfrac{f^{(n)}(-1)}{n!}(x+1)^n = -2 + 8(x+1) - 9(x+1)^2 + 4(x+1)^3$

3. $f(x) = \sin x$ $f(\tfrac{\pi}{6}) = \tfrac{1}{2}$
$f'(x) = \cos x$ $f'(\tfrac{\pi}{6}) = \tfrac{\sqrt{3}}{2}$
$f''(x) = -\sin x$ $f''(\tfrac{\pi}{6}) = -\tfrac{1}{2}$
$f^{(3)}(x) = -\cos x$ $f^{(3)}(\tfrac{\pi}{6}) = -\tfrac{\sqrt{3}}{2}$

$T_3(x) = \sum\limits_{n=0}^{3} \dfrac{f^{(n)}(\pi/6)}{n!}(x - \tfrac{\pi}{6})^n = \tfrac{1}{2} + \tfrac{\sqrt{3}}{2}(x - \tfrac{\pi}{6}) - \tfrac{1}{4}(x - \tfrac{\pi}{6})^2 - \tfrac{\sqrt{3}}{12}(x - \tfrac{\pi}{6})^3$

$= -\tfrac{1}{2} - \tfrac{\sqrt{3}}{2}(x - \tfrac{2\pi}{3}) + \tfrac{1}{4}(x - \tfrac{2\pi}{3})^2 + \tfrac{\sqrt{3}}{12}(x - \tfrac{2\pi}{3})^3 - \tfrac{1}{48}(x - \tfrac{2\pi}{3})^4$

5. $f(x) = \tan x$ $f(0) = 0$
$f'(x) = \sec^2 x$ $f'(0) = 1$
$f''(x) = 2\sec^2 x \tan x$ $f''(0) = 0$
$f^{(3)}(x) = 4\sec^2 x \tan^2 x + 2\sec^4 x$ $f^{(3)}(0) = 2$
$f^{(4)}(x) = 8\sec^2 x \tan^3 x + 16\sec^4 x \tan x$ $f^{(4)}(0) = 0$

$T_4(x) = \sum\limits_{n=0}^{4} \dfrac{f^{(n)}(0)}{n!}x^n = x + \tfrac{2}{3!}x^3 = x + \tfrac{x^3}{3}$

Section 10.11

7. $f(x) = e^x \sin x$ $\qquad f(0) = 0$
    $f'(x) = e^x(\sin x + \cos x)$ $\qquad f'(0) = 1$
    $f''(x) = 2e^x \cos x$ $\qquad f''(0) = 2$
    $f^{(3)}(x) = 2e^x(\cos x - \sin x)$ $\qquad f^{(3)}(0) = 2$
    $$T_3(x) = \sum_{n=0}^{3} \frac{f^{(n)}(0)}{n!} x^n = x + x^2 + \frac{x^3}{3}$$

9. $f(x) = x^{1/2}$ $\qquad f(9) = 3$
    $f'(x) = \frac{1}{2}x^{-1/2}$ $\qquad f'(9) = \frac{1}{6}$
    $f''(x) = -\frac{1}{4}x^{-3/2}$ $\qquad f''(9) = -\frac{1}{108}$
    $f^{(3)}(x) = \frac{3}{8}x^{-5/2}$ $\qquad f^{(3)}(9) = \frac{1}{648}$
    $$T_3(x) = \sum_{n=0}^{3} \frac{f^{(n)}(9)}{n!}(x-9)^n = 3 + \tfrac{1}{6}(x-9) - \tfrac{1}{216}(x-9)^2 + \tfrac{1}{3888}(x-9)^3.$$

11. $f(x) = \ln(\sin x)$ $\qquad f(\tfrac{\pi}{2}) = 0$
    $f'(x) = \cot x$ $\qquad f'(\tfrac{\pi}{2}) = 0$
    $f''(x) = -\csc^2 x$ $\qquad f''(\tfrac{\pi}{2}) = -1$
    $f^{(3)}(x) = 2\csc^2 x \cot x$ $\qquad f^{(3)}(\tfrac{\pi}{2}) = 0$
    $$T_3(x) = \sum_{n=0}^{3} \frac{f^{(n)}(\pi/2)}{n!}(x-\pi/2)^n = -\tfrac{1}{2}\left(x-\tfrac{\pi}{2}\right)^2$$

13. $f(x) = \cos x$ $\qquad f(0) = 1$
    $f'(x) = -\sin x$ $\qquad f'(0) = 0$
    $f''(x) = -\cos x$ $\qquad f''(0) = -1$
    $f^{(3)}(x) = \sin x$ $\qquad f^{(3)}(0) = 0$
    $f^{(4)}(x) = \cos x$ $\qquad f^{(4)}(0) = 1$

    $T_1(x) = 1$

    $T_2(x) = 1 - \frac{x^2}{2}$

    $T_3(x) = 1 - \frac{x^2}{2}$

    $T_4(x) = 1 - \frac{x^2}{2} + \frac{x^4}{24}$

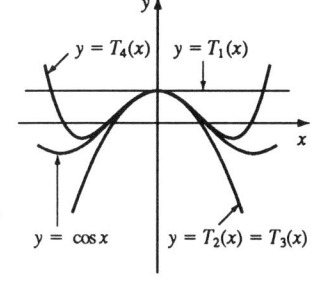

15. $f(x) = (1+x)^{1/2}$ $\qquad f(0) = 1$
    $f'(x) = \tfrac{1}{2}(1+x)^{-1/2}$ $\qquad f'(0) = \tfrac{1}{2}$
    $f''(x) = -\tfrac{1}{4}(1+x)^{-3/2}$

    (a) $(1+x)^{1/2} = 1 + \tfrac{1}{2}x + R_1(x)$ where $R_1(x) = \dfrac{f''(z)}{2!}x^2 = -\dfrac{1}{8(1+z)^{3/2}}x^2$ and z lies

33

between 0 and x.

(b) $0 \leq x \leq 0.1 \Rightarrow 0 \leq x^2 \leq 0.01$, and $0 < z < 0.1 \Rightarrow 1 < 1+z < 1.1$ so $|R_1(x)| < \frac{0.01}{8.1} = 0.00125$.

17. $f(x) = \sin x$ $\quad\quad f(\frac{\pi}{4}) = \frac{\sqrt{2}}{2}$

$f'(x) = \cos x$ $\quad\quad f'(\frac{\pi}{4}) = \frac{\sqrt{2}}{2}$

$f''(x) = -\sin x$ $\quad\quad f''(\frac{\pi}{4}) = -\frac{\sqrt{2}}{2}$

$f^{(3)}(x) = -\cos x$ $\quad\quad f^{(3)}(\frac{\pi}{4}) = -\frac{\sqrt{2}}{2}$

$f^{(4)}(x) = \sin x$ $\quad\quad f^{(4)}(\frac{\pi}{4}) = \frac{\sqrt{2}}{2}$

$f^{(5)}(x) = \cos x$ $\quad\quad f^{(5)}(\frac{\pi}{4}) = \frac{\sqrt{2}}{2}$

$f^{(6)}(x) = -\sin x$

(a) $\sin x = \frac{\sqrt{2}}{2} + \frac{\sqrt{2}}{2}(x - \frac{\pi}{4}) - \frac{\sqrt{2}}{4}(x - \frac{\pi}{4})^2 - \frac{\sqrt{2}}{12}(x - \frac{\pi}{4})^3 + \frac{\sqrt{2}}{48}(x - \frac{\pi}{4})^4$
$+ \frac{\sqrt{2}}{240}(x - \frac{\pi}{4})^5 + R_5(x)$ where $R_5(x) = \frac{f^{(6)}(z)}{6!}(x - \frac{\pi}{4})^6 = \frac{-\sin z}{720}(x - \frac{\pi}{4})^6$ and z lies between $\frac{\pi}{4}$ and x.

(b) Since $0 \leq x \leq \frac{\pi}{2}$, $-\frac{\pi}{4} \leq x - \frac{\pi}{4} \leq \frac{\pi}{4} \Rightarrow 0 \leq (x - \frac{\pi}{4})^6 \leq (\frac{\pi}{4})^6$, and since $0 < z < \frac{\pi}{2}$,
$0 < \sin z < 1$, so $|R_5(x)| < \frac{(\pi/4)^6}{720} \approx 0.00033$.

19. $f(x) = (1 + 2x)^{-4}$ $\quad\quad f(0) = 1$

$f'(x) = -8(1 + 2x)^{-5}$ $\quad\quad f'(0) = -8$

$f''(x) = 80(1 + 2x)^{-6}$ $\quad\quad f''(0) = 80$

$f^{(3)}(x) = -960(1 + 2x)^{-7}$ $\quad\quad f^{(3)}(0) = -960$

$f^{(4)}(x) = 13440(1 + 2x)^{-8}$

(a) $(1 + 2x)^{-4} = 1 - 8x + 40x^2 - 160x^3 + R_3(x)$ where
$R_3(x) = \frac{f^{(4)}(z)}{4!}x^4 = \frac{13440(1 + 2z)^{-8}}{4!}x^4 = \frac{560\,x^4}{(1 + 2z)^8}$.

(b) $|x| \leq 0.1 \Rightarrow 0 \leq x^4 \leq 0.0001$ and $|z| < 0.1 \Rightarrow 0.8 < 1 + 2z < 1.2$, so
$|R_3(x)| < \frac{560 \cdot (0.0001)}{(0.8)^8} < 0.34$.

21. $f(x) = \tan x$ $\quad\quad f(0) = 0$

$f'(x) = \sec^2 x$ $\quad\quad f'(0) = 1$

$f''(x) = 2\sec^2 x \tan x$ $\quad\quad f''(0) = 0$

$f^{(3)}(x) = 4\sec^2 x \tan^2 x + 2\sec^4 x$ $\quad\quad f^{(3)}(0) = 2$

Section 10.11

$f^{(4)}(x) = 8\sec^2 x \tan^3 x + 16 \sec^4 x \tan x$

(a) $\tan x = x + \frac{x^3}{3} + R_3(x)$ where $R_3(x) = \frac{f^{(4)}(z)}{4!}x^4$

$= \frac{8\sec^2 z \tan^3 z + 16\sec^4 z \tan z}{4!}x^4 = \frac{\sec^2 z \tan^3 z + 2\sec^4 z \tan z}{3}x^4$ where z lies between 0 and x.

(b) $0 \le x^4 \le \left(\frac{\pi}{6}\right)^4$ and $0 < z < \frac{\pi}{6} \Rightarrow \sec^2 z < \frac{4}{3}$ and $\tan z < \frac{\sqrt{3}}{3}$ so

$|R_3(x)| < \frac{(4/3)[1/(3\sqrt{3})] + 2(16/9)(1/\sqrt{3})}{3}\left(\frac{\pi}{6}\right)^4 = \frac{4\sqrt{3}}{9}\left(\frac{\pi}{6}\right)^4 < .06$

23. $f(x) = e^{x^2}$ $f(0) = 1$
$f'(x) = e^{x^2}(2x)$ $f'(0) = 0$
$f''(x) = e^{x^2}(2 + 4x^2)$ $f''(0) = 2$
$f^{(3)}(x) = e^{x^2}(12x + 8x^3)$ $f^{(3)}(0) = 0$
$f^{(4)}(x) = e^{x^2}(12 + 48x^2 + 16x^4)$

(a) $e^{x^2} = 1 + x^2 + R_3(x)$ where $R_3(x) = \frac{f^{(4)}(z)}{4!}x^4 = \frac{e^{z^2}(3 + 12z^2 + 4z^4)}{6}x^4$ and z lies between 0 and x.

(b) $0 \le x \le 0.1 \Rightarrow |R_3(x)| < \frac{e^{0.01}(3 + 0.12 + 0.0004)}{6}(0.0001) < 0.00006.$

25. $f(x) = x^{3/4}$ $f(16) = 8$
$f'(x) = \frac{3}{4}x^{-1/4}$ $f'(16) = \frac{3}{8}$
$f''(x) = \frac{-3}{16}x^{-5/4}$ $f''(16) = \frac{-3}{512}$
$f^{(3)}(x) = \frac{15}{64}x^{-9/4}$ $f^{(3)}(16) = \frac{15}{32768}$
$f^{(4)}(x) = -\frac{135}{256}x^{-13/4}$

(a) $x^{3/4} = 8 + \frac{3}{8}(x-16) - \frac{3}{1024}(x-16)^2 + \frac{5}{65536}(x-16)^3 + R_3(x)$ where

$R_3(x) = \frac{f^{(4)}(z)}{4!}(x-16)^4 = -\frac{135(x-16)^4}{256 \cdot 4! z^{13/4}}$ and z lies between 16 and x.

(b) $|x - 16| \le 1$ and $z > 15 \Rightarrow |R_3(x)| < \frac{135}{256 \cdot 24 \cdot 15^{13/4}} < 0.0000034.$

27. $e^x = 1 + x + \frac{x^2}{2!} + \cdots + \frac{x^n}{n!} + R_n(x)$ where $R_n(x) = \frac{e^z}{(n+1)!}x^{n+1}$ and z lies between 0 and x.

So taking $x = 0.1$, we have $z < 0.1 \Rightarrow e^z < 3^{0.1} < 2 \Rightarrow |R_3(0.1)| < \frac{2}{4!}(0.0001) < 0.00001$

and $e^{0.1} \approx 1 + 0.1 + \frac{(0.1)^2}{2} + \frac{(0.1)^3}{6} \approx 1.10517.$

29. $f(x) = (1+x)^{1/5} = 1 + \frac{1}{5}x + \frac{\left(\frac{1}{5}\right)\left(-\frac{4}{5}\right)}{2!}x^2 + R_2(x) = 1 + \frac{1}{5}x - \frac{2}{25}x^2 + R_2(x)$ where

$|R_2(x)| = \frac{1 \cdot 4 \cdot 9}{5^3 \cdot 3!}|1+z|^{-14/5}|x|^3$ and z lies between 0 and x, so $|R_2(0.1)|$

$< \frac{1 \cdot 4 \cdot 9}{5^3 \cdot 3!}(0.001) = 0.000048 < 0.00005$. Thus $(1.1)^{1/5} \approx 1 + \frac{0.1}{5} - \frac{2}{25}(0.1)^2 = 1.0192$

(correct to four decimals).

31. If $f(x) = \ln(1+x)$ then $f^{(n)}(x) = (-1)^n(n-1)!(1+x)^{-n}$ so

$\ln(1+x) = x - \frac{x^2}{2} + \frac{x^3}{3} - \frac{x^4}{4} + \frac{x^5}{5} + R_5(x)$ where $|R_5(x)| = \frac{|x|^6}{6|1+z|^6}$ and z is between 0 and

x. So $|R_5(0.4)| < \frac{(0.4)^6}{6} \approx 0.00068 < 0.001$ and

$\ln(1.4) \approx 0.4 - \frac{(0.4)^2}{2} + \frac{(0.4)^3}{3} - \frac{(0.4)^4}{4} + \frac{(0.4)^5}{5} \approx 0.336.$

33. $\sin x = x - \frac{x^3}{3!} + \frac{x^5}{5!} + R_5(x)$, where $R_5(x) = \frac{-\sin z}{6!}x^6$ and z lies between 0 and x. So

$|R_5(0.5)| < \frac{1 \cdot (0.5)^6}{720} \approx 0.00002 < 0.0001$, and $\sin(0.5) \approx 0.5 - \frac{(0.5)^3}{3!} + \frac{(0.5)^5}{5!} \approx 0.4794.$

35. $\cos x = 1 - \frac{x^2}{2!} + \frac{x^4}{4!} + R_4(x)$ where $R_4(x) = \frac{\cos z}{5!}x^5$ and z is between 0 and x, so

$|R_4(10°)| = |R_4(\frac{\pi}{18})| < \frac{(\pi/18)^5}{5!} < 0.0000014$ so $\cos(10°) \approx 1 - \frac{(\pi/18)^2}{2!} + \frac{(\pi/18)^4}{4!} \approx 0.98481.$

37. Using the information of Exercise 3, above, we see that

$\sin x = \frac{1}{2} + \frac{\sqrt{3}}{2}(x - \frac{\pi}{6}) - \frac{1}{4}(x - \frac{\pi}{6})^2 - \frac{\sqrt{3}}{12}(x - \frac{\pi}{6})^3 + R_3(x)$, where $R_3(x) = \frac{\sin z}{4!}(x - \frac{\pi}{6})^4$ and z

lies between $\frac{\pi}{6}$ and x. Now 35° is $\frac{\pi}{6} + \frac{\pi}{36}$ in radian measure, so

$|R_3(\frac{\pi}{36})| < \frac{(\pi/36)^4}{4!} < 0.000003$ and $\sin 35° \approx \frac{1}{2} + \frac{\sqrt{3}}{2}(\frac{\pi}{36}) - \frac{1}{4}(\frac{\pi}{36})^2 - \frac{\sqrt{3}}{12}(\frac{\pi}{36})^3 \approx 0.57358.$

39. $\sin x = x - \frac{x^3}{6} + R_4(x)$ where $R_4(x) = \frac{\sin z}{5!}x^5$ and z is between 0 and x. So

$|R_4(x)| < \frac{|x|^5}{120} < 0.01 \Rightarrow |x|^5 < 1.2 \Rightarrow |x| < (1.2)^{1/5} \approx 1.037$. This is certainly true if $|x| \leq 1$.

41. We will use Theorem 10.56. $R_n(x) = \frac{f^{(n+1)}(z)}{(n+1)!}x^{n+1}$ with $f^{(n+1)}(z) = \pm \sin z$ or $\pm \cos z$, and

with z between 0 and x. In every case $|f^{(n+1)}(z)| \leq 1$. Thus $|R_n(x)| \leq \frac{|x|^{n+1}}{(n+1)!} \to 0$ as $n \to \infty$

by (10.57), so $\sin x$ is equal to its Taylor series by Theorem 10.56.

43. $R_n(x) = \frac{f^{(n+1)}(z)}{(n+1)!}x^{n+1}$ where $f^{(n+1)}(z) = \sinh z$ or $\cosh z$. Since z lies between 0 and x,

$|\sinh z| < |\sinh x|$ and $|\cosh z| < |\cosh x|$, so in the case $f^{(n+1)}(z) = \sinh z$ we have

Chapter 10 Review

$|R_n(x)| < |\sinh z| \frac{|x|^{n+1}}{(n+1)!} \to 0$ as $n \to \infty$ by (10.57) (and similarly if $f^{(n+1)}(z) = \cosh z$). So by Theorem 10.56, $\sinh x$ is equal to its Taylor series.

45. From Section 10.10 $(1+x)^n = 1 + nx + \frac{n(n-1)}{2!}x^2 + \cdots > 1 + nx$ for all $x > 0$ and $n > 1$.

47. $\lim_{x \to 0} \frac{\sin x - x + \frac{1}{6}x^3}{x^5} = \lim_{x \to 0} \frac{(x - \frac{1}{6}x^3 + \frac{1}{5!}x^5 - \frac{1}{7!}x^7 + \cdots) - x + \frac{1}{6}x^3}{x^5}$

$= \lim_{x \to 0} \frac{\frac{1}{5!}x^5 - \frac{1}{7!}x^7 + \cdots}{x^5} = \lim_{x \to 0} (\frac{1}{5!} - \frac{1}{7!}x^2 + \frac{1}{9!}x^4 - \cdots) = \frac{1}{5!} = \frac{1}{120}$

since power series are continuous functions.

49. $y = T_1(x) = f(c) + f'(c)(x-c) \Leftrightarrow y - f(c) = f'(c)(x-c)$ is the equation of the line passing through $(c, f(c))$ with slope $f'(c)$, and this describes the tangent line.

51. Using Taylor's Formula (10.54) with $n = 1$, $c = x_n$, $x = r$, we get

$f(r) = f(x_n) + f'(x_n)(r - x_n) + R_2(x)$, where $R_2(x) = \frac{1}{2}f''(z)(r - x_n)^2$ and z lies between x_n and r. But r is a root, so $f(r) = 0$ and Taylor's Formula becomes

$$0 = f(x_n) + f'(x_n)(r - x_n) + \frac{1}{2}f''(z)(r - x_n)^2$$

Taking the first two terms to the left side and dividing by $f'(x_n)$, we have

$$x_n - r - \frac{f(x_n)}{f'(x_n)} = \frac{1}{2}\frac{f''(z)}{f'(x_n)}|x_n - r|^2$$

So the formula for Newton's method $\Rightarrow$

$$|x_{n+1} - r| = \left|x_n - \frac{f(x_n)}{f'(x_n)} - r\right| = \frac{1}{2}\frac{|f''(z)|}{|f'(x_n)|}|x_n - r|^2 \leq \frac{M}{2K}|x_n - r|^2$$

since $|f''(z)| \leq M$ and $|f'(x_n)| \geq K$.

53. $q!(e - s_q) = q!\left(\frac{p}{q} - 1 - \frac{1}{1!} - \frac{1}{2!} - \cdots - \frac{1}{q!}\right) = p(q-1)! - q! - q! - \frac{q!}{2!} - \cdots - 1$, which is clearly an integer, and $q!(e - s_q) = q!\left[\frac{e^z}{(q+1)!}\right] = \frac{e^z}{q+1}$. We have

$0 < \frac{e^z}{q+1} < \frac{e}{q+1} < \frac{e}{3} < 1$ since $0 < z < 1$ and $q > 2$, and so $0 < q!(e - s_q) < 1$,

which is a contradiction since we have already shown $q!(e - s_q)$ must be an integer. So e cannot be rational.

REVIEW EXERCISES FOR CHAPTER 10

1. False. See the warning in Note 2 after Theorem 10.17.

3. False. For example, take $a_n = (-1)^n/(n6^n)$.

Chapter 10 Review

5. False, since $\lim_{n\to\infty} \left|\frac{a_{n+1}}{a_n}\right| = \lim_{n\to\infty} \left|\frac{n^3}{(n+1)^3}\right| = \lim_{n\to\infty} \frac{1}{(1+1/n)^3} = 1$.

7. False. See the remarks after Example 3 in Section 10.4.

9. False. A power series has the form $a_0 + a_1 x + a_2 x^2 + a_3 x^3 + \cdots$.

11. True. See Example 8 in Section 10.1.

13. True. By Theorem 10.44 the coefficient of x^3 is $\frac{f'''(0)}{3!} = \frac{1}{3} \Rightarrow f'''(0) = 2$.
    [Or use Theorem 10.39 to differentiate f three times.]

15. $\lim_{n\to\infty} \frac{n}{2n+5} = \lim_{n\to\infty} \frac{1}{2+5/n} = \frac{1}{2}$ and the sequence is convergent.

17. $\{2n+5\}$ is divergent since $2n+5 \to \infty$ as $n \to \infty$.

19. $\{\sin n\}$ is divergent since $\lim_{n\to\infty} \sin n$ does not exist.

21. $\left\{\left(1+\frac{3}{n}\right)^{4n}\right\}$ is convergent. Let $y = \left(1+\frac{3}{x}\right)^{4x}$. Then $\lim_{x\to\infty} \ln y$

    $= \lim_{x\to\infty} 4x \ln(1+3/x) = \lim_{x\to\infty} \frac{\ln(1+3/x)}{1/(4x)} \stackrel{H}{=} \lim_{x\to\infty} \frac{\frac{1}{1+3/x}\cdot(-3/x^2)}{-1/(4x^2)} = \lim_{x\to\infty} \frac{12}{1+3/x}$

    $= 12$, so $\lim_{x\to\infty} y = \lim_{n\to\infty}\left(1+\frac{3}{n}\right)^{4n} = e^{12}$.

23. Use the Limit Comparison Test with $a_n = \frac{n^2}{n^3+1}$ and $b_n = \frac{1}{n}$.

    $\lim_{n\to\infty} \frac{a_n}{b_n} = \lim_{n\to\infty} \frac{n^2/(n^3+1)}{1/n} = \lim_{n\to\infty} \frac{1}{1+1/n^3} = 1$. Since $\sum_{n=1}^{\infty} \frac{1}{n}$ (the harmonic series) diverges, $\sum_{n=1}^{\infty} \frac{n^2}{n^3+1}$ diverges also.

25. An alternating series with $a_n = \frac{1}{n^{1/4}}$, $a_n > 0$ for all n, and $a_n > a_{n+1}$.

    $\lim_{n\to\infty} a_n = \lim_{n\to\infty} \frac{1}{n^{1/4}} = 0$ so the series converges by the Alternating Series Test.

27. $\lim_{n\to\infty} \sqrt[n]{|a_n|} = \lim_{n\to\infty} \frac{n}{3n+1} = \frac{1}{3} < 1$, so series converges by the Root Test.

29. $\frac{|\sin n|}{1+n^2} \leq \frac{1}{1+n^2} < \frac{1}{n^2}$ and since $\sum_{n=1}^{\infty} \frac{1}{n^2}$ converges (p-series with $p = 2 > 1$), so does

    $\sum_{n=1}^{\infty} \frac{|\sin n|}{1+n^2}$ by the Comparison Test.

31. $\lim_{n\to\infty} \left|\frac{a_{n+1}}{a_n}\right| = \lim_{n\to\infty} \frac{1\cdot 3\cdot 5 \cdots (2n-1)(2n+1)}{5^{n+1}(n+1)!} \cdot \frac{5^n n!}{1\cdot 3\cdot 5 \cdots (2n-1)} = \lim_{n\to\infty} \frac{2n+1}{5(n+1)}$

Chapter 10 Review

$= \frac{2}{5} < 1$, so the series converges by the Ratio Test.

33. $\lim_{n\to\infty} \left|\frac{a_{n+1}}{a_n}\right| = \lim_{n\to\infty} \frac{4^{n+1}}{(n+1)3^{n+1}} \cdot \frac{n3^n}{4^n} = \frac{4}{3}\lim_{n\to\infty} \frac{n}{n+1} = \frac{4}{3} > 1$ so the series diverges by the Ratio Test.

35. Consider the series of absolute values: $\sum_{n=1}^{\infty} n^{-1/3}$ is a p-series with $p = \frac{1}{3} < 1$ and is therefore divergent. But if we apply the Alternating Series Test we see that $a_{n+1} < a_n$ and $\lim_{n\to\infty} n^{-1/3} = 0$. Therefore $\sum_{n=1}^{\infty} (-1)^{n-1} n^{-1/3}$ is conditionally convergent.

37. $\left|\frac{a_{n+1}}{a_n}\right| = \left|\frac{(-1)^{n+1}(n+2)3^{n+1}}{2^{2n+3}} \cdot \frac{2^{2n+1}}{(-1)^n(n+1)3^n}\right| = \frac{n+2}{n+1} \cdot \frac{3}{4} = \frac{1+(2/n)}{1+(1/n)} \cdot \frac{3}{4} \to \frac{3}{4} < 1$ as $n \to \infty$

so by the Ratio Test $\sum_{n=1}^{\infty} (-1)^n(n+1)3^n/2^{2n+1}$ is absolutely convergent.

39. Convergent geometric series. $\sum_{n=1}^{\infty} \frac{2^{2n+1}}{5^n} = 2\sum_{n=1}^{\infty} \frac{4^n}{5^n} = 2\left(\frac{4/5}{1-4/5}\right) = 8$.

41. $\sum_{n=1}^{\infty} \left[\tan^{-1}(n+1) - \tan^{-1} n\right]$

$= \lim_{n\to\infty} [(\tan^{-1} 2 - \tan^{-1} 1) + (\tan^{-1} 3 - \tan^{-1} 2) + \cdots + (\tan^{-1}(n+1) - \tan^{-1} n)]$

$= \lim_{n\to\infty} [\tan^{-1}(n+1) - \tan^{-1} 1] = \frac{\pi}{2} - \frac{\pi}{4} = \frac{\pi}{4}$

43. $1.2 + 0.0\overline{345} = \frac{12}{10} + \frac{345/10000}{1-1/1000} = \frac{12}{10} + \frac{345}{9990} = \frac{4111}{3330}$.

45. $\sum_{n=1}^{\infty} \frac{(-1)^{n+1}}{n^5} = 1 - \frac{1}{32} + \frac{1}{243} - \frac{1}{1024} + \frac{1}{3125} - \frac{1}{7776} + \frac{1}{16807} - \frac{1}{32768} + \cdots$ Since

$\frac{1}{32768} < 0.000031$, $\sum_{n=1}^{\infty} \frac{(-1)^{n+1}}{n^5} \approx \sum_{n=1}^{7} \frac{(-1)^{n+1}}{n^5} \approx 0.9721$.

47. Use the Limit Comparison Test. $\lim_{n\to\infty} \left|\frac{\left(\frac{n+1}{n}\right)a_n}{a_n}\right| = \lim_{n\to\infty} \frac{n+1}{n}$

$= \lim_{n\to\infty} \left(1 + \frac{1}{n}\right) = 1 > 0$. Since $\sum |a_n|$ is convergent, so is $\sum \left|\left(\frac{n+1}{n}\right)a_n\right|$ by the Limit Comparison Test.

49. $\lim_{n\to\infty} \left|\frac{u_{n+1}}{u_n}\right| = \lim_{n\to\infty} \left|\frac{x^{n+1}}{3^{n+1}(n+1)^3} \cdot \frac{3^n n^3}{x^n}\right| = \frac{|x|}{3}\lim_{n\to\infty} \left(\frac{n}{n+1}\right)^3 = \frac{|x|}{3} < 1$ for convergence

(Ratio Test) $\Rightarrow |x| < 3$ and the radius of convergence is 3. When $x = \pm 3$, $\sum_{n=1}^{\infty} |u_n| = \sum_{n=1}^{\infty} \frac{1}{n^3}$ which is a convergent p-series ($p = 3 > 1$), so the interval of convergence is $[-3, 3]$.

51. $\lim_{n\to\infty} \left|\frac{u_{n+1}}{u_n}\right| = \lim_{n\to\infty} \left|\frac{2^{n+1}(x-3)^{n+1}}{\sqrt{n+4}} \cdot \frac{\sqrt{n+3}}{2^n(x-3)^n}\right| = 2|x-3|\lim_{n\to\infty} \sqrt{\frac{n+3}{n+4}} = 2|x-3| < 1$

Chapter 10 Review

$\Leftrightarrow |x - 3| < \frac{1}{2}$ so the radius of convergence is 1/2. For $x = \frac{7}{2}$ the series becomes

$\sum_{n=0}^{\infty} \frac{1}{\sqrt{n+3}} = \sum_{n=3}^{\infty} \frac{1}{n^{1/2}}$ which diverges ($p = \frac{1}{2} < 1$), but for $x = \frac{5}{2}$ we get $\sum_{n=0}^{\infty} \frac{(-1)^n}{\sqrt{n+3}}$

which is a convergent alternating series, so the interval of convergence is $[5/2, 7/2)$.

53. $f(x) = \sin x$ $f(\frac{\pi}{6}) = \frac{1}{2}$
    $f'(x) = \cos x$ $f'(\frac{\pi}{6}) = \frac{\sqrt{3}}{2}$
    $f''(x) = -\sin x$ $f''(\frac{\pi}{6}) = -\frac{1}{2}$
    $f^{(3)}(x) = -\cos x$ $f^{(3)}(\frac{\pi}{6}) = -\frac{\sqrt{3}}{2}$
    $f^{(4)}(x) = \sin x$ $f^{(4)}(\frac{\pi}{6}) = \frac{1}{2}$
    ...   ...

$f^{(2n)}(\frac{\pi}{6}) = (-1)^n \cdot \frac{1}{2}$ and $f^{(2n+1)}(\frac{\pi}{6}) = (-1)^n \cdot \frac{\sqrt{3}}{2}$

$\sin x = \sum_{n=0}^{\infty} \frac{f^{(n)}(\pi/6)}{n!}(x - \frac{\pi}{6})^n = \sum_{n=0}^{\infty} \frac{(-1)^n}{2(2n)!}(x - \frac{\pi}{6})^{2n} + \sum_{n=0}^{\infty} \frac{(-1)^n \sqrt{3}}{2(2n+1)!}(x - \frac{\pi}{6})^{2n+1}$

55. $\frac{1}{1+x} = \frac{1}{1-(-x)} = \sum_{n=0}^{\infty} (-1)^n x^n$ for $|x| < 1 \Rightarrow \frac{x^2}{1+x} = \sum_{n=0}^{\infty} (-1)^n x^{n+2}$ with $R = 1$.

57. $\frac{1}{1-x} = \sum_{n=0}^{\infty} x^n$ for $|x| < 1 \Rightarrow \ln(1-x) = -\int \frac{dx}{1-x} = -\int \sum_{n=0}^{\infty} x^n dx = C - \sum_{n=0}^{\infty} \frac{x^{n+1}}{n+1}$.

$\ln(1 - 0) = C - 0 \Rightarrow C = 0 \Rightarrow \ln(1-x) = -\sum_{n=0}^{\infty} \frac{x^{n+1}}{n+1} = \sum_{n=1}^{\infty} \frac{-x^n}{n}$ with $R = 1$.

59. $\sin x = \sum_{n=0}^{\infty} \frac{(-1)^n x^{2n+1}}{(2n+1)!} \Rightarrow \sin(x^4) = \sum_{n=0}^{\infty} \frac{(-1)^n (x^4)^{2n+1}}{(2n+1)!} = \sum_{n=0}^{\infty} \frac{(-1)^n x^{8n+4}}{(2n+1)!}$ for all x,

so the radius of convergence is ∞.

61. $(16 - x)^{-1/4} = \frac{1}{2}(1 - \frac{x}{16})^{-1/4} = \frac{1}{2}\left[1 + (-\frac{1}{4})(-\frac{x}{16}) + \frac{(-\frac{1}{4})(-\frac{5}{4})}{2!}(-\frac{x}{16})^2 + \cdots\right]$

$= \sum_{n=0}^{\infty} \frac{1 \cdot 5 \cdot 9 \cdots (4n-3)}{2 \cdot 4^n \cdot n! \cdot 16^n} x^n = \sum_{n=0}^{\infty} \frac{1 \cdot 5 \cdot 9 \cdots (4n-3)}{2^{6n+1} n!} x^n$ for $\left|\frac{x}{16}\right| < 1 \Rightarrow R = 16$.

63. $e^x = \sum_{n=0}^{\infty} \frac{x^n}{n!}$ so $\frac{e^x}{x} = \frac{1}{x} + \sum_{n=1}^{\infty} \frac{x^{n-1}}{n!}$ and $\int \frac{e^x}{x} dx = C + \ln|x| + \sum_{n=1}^{\infty} \frac{x^n}{n \cdot n!}$

65. $e^{-1/4} = \sum_{k=0}^{n} \frac{(-1/4)^k}{k!} + R_n(-1/4)$ where $R_n(-1/4) = \frac{e^z}{(n+1)!}(-\frac{1}{4})^{n+1}$ and

$-\frac{1}{4} < z < 0 \Rightarrow |R_n(-1/4)| < \frac{e^0}{(n+1)!}(\frac{1}{4})^{n+1} = \frac{1}{(n+1)!}(\frac{1}{4})^{n+1} \Rightarrow |R_4(-1/4)| < \frac{1}{5! \cdot 4^5}$

$< 0.0000082 < 0.0001$ so $e^{-1/4} \approx 1 - \frac{1}{4} + \frac{1}{32} - \frac{1}{384} + \frac{1}{6144} \approx 0.7788$.

67. $f(x) = x^{1/2}$ $\qquad\qquad\qquad\qquad$ $f(1) = 1$
$f'(x) = \frac{1}{2}x^{-1/2}$ $\qquad\qquad\qquad$ $f'(1) = \frac{1}{2}$
$f''(x) = -\frac{1}{4}x^{-3/2}$ $\qquad\qquad\;\;$ $f''(1) = -\frac{1}{4}$
$f^{(3)}(x) = \frac{3}{8}x^{-5/2}$ $\qquad\qquad\quad$ $f^{(3)}(1) = \frac{3}{8}$
$f^{(4)}(x) = -\frac{15}{16}x^{-7/2}$

$\sqrt{x} = 1 + \frac{1}{2}(x-1) - \frac{1}{8}(x-1)^2 + \frac{1}{16}(x-1)^3 + R_3(x)$ where

$R_3(x) = \frac{f^{(4)}(z)}{4!}(x-1)^4 = -\frac{5(x-1)^4}{128z^{7/2}}$ with z between x and 1. If $0.9 \leq x \leq 1.1$ then

$0 \leq |x-1| \leq 0.1$ and $z^{7/2} > (0.9)^{7/2}$ so $|R_3(x)| < \frac{5(0.1)^4}{128(0.9)^{7/2}} < 0.000006.$

69. $e^x = \sum\limits_{n=0}^{\infty} \frac{x^n}{n!} \Rightarrow e^{-1/x^2} = \sum\limits_{n=0}^{\infty} \frac{(-1/x^2)^n}{n!} = 1 - \frac{1}{x^2} + \frac{1}{2x^4} - \cdots \Rightarrow$

$x^2(1 - e^{-1/x^2}) = x^2(\frac{1}{x^2} - \frac{1}{2x^4} + \cdots) = 1 - \frac{1}{2x^2} + \cdots \to 1$ as $x \to \infty$.

71. $e^x = \sum\limits_{n=0}^{\infty} \frac{x^n}{n!} \Rightarrow e^{x^2} = \sum\limits_{n=0}^{\infty} \frac{(x^2)^n}{n!} = \sum\limits_{n=0}^{\infty} \frac{x^{2n}}{n!} = \sum\limits_{k=0}^{\infty} \frac{f^{(k)}(0)}{k!}x^k \Rightarrow$

$\frac{f^{(2n)}(0)}{(2n)!} = \frac{1}{n!} \Rightarrow f^{(2n)}(0) = \frac{(2n)!}{n!}.$

PROBLEMS PLUS (after Chapter 10)

1. It would be far too much work to compute 15 derivatives of f. The key idea is to remember that $f^{(n)}(0)$ occurs in the coefficient of x^n in the Maclaurin series of f. We start with the Maclaurin series for sin: $\sin x = x - \frac{x^3}{3!} + \frac{x^5}{5!} - \cdots$. Then $\sin(x^3) = x^3 - \frac{x^9}{3!} + \frac{x^{15}}{5!} - \cdots$ and so the coefficient of x^{15} is $\frac{f^{(15)}(0)}{15!} = \frac{1}{5!}$. Therefore,

$$f^{(15)}(0) = \frac{15!}{5!} = 6 \cdot 7 \cdot 8 \cdot 9 \cdot 10 \cdot 11 \cdot 12 \cdot 13 \cdot 14 \cdot 15 = 10{,}897{,}286{,}400.$$

3. (a) At each stage, each side is replaced by 4 shorter sides, each of length $\frac{1}{3}$ of the side length at the preceding stage. Writing s_0 and ℓ_0 for the number of sides and the length of the side of the initial triangle, we have

$$s_0 = 3 \qquad \ell_0 = 1$$
$$s_1 = 3 \cdot 4 \qquad \ell_1 = \tfrac{1}{3}$$
$$s_2 = 3 \cdot 4^2 \qquad \ell_2 = \tfrac{1}{3^2}$$
$$s_3 = 3 \cdot 4^3 \qquad \ell_3 = \tfrac{1}{3^3}$$
$$\cdots \qquad \cdots$$

In general, we have $s_n = 3 \cdot 4^n$ and $\ell_n = 1/3^n$. Thus the length of the perimeter at the nth stage of construction is $p_n = s_n \ell_n = 3 \cdot 4^n \cdot (1/3^n) = 4^n/3^{n-1}$.

(b) $p_n = \frac{4^n}{3^{n-1}} = 4(\frac{4}{3})^{n-1}$. Since $\frac{4}{3} > 1$, $p_n \to \infty$ as $n \to \infty$.

(c) The area of each of the small triangles added at a given stage is $\frac{1}{9}$th of the area of the triangle added at the preceding stage. Let a be the area of the original triangle. Then the area a_n of each of the small triangles added at stage n is $a_n = a \cdot \frac{1}{9^n} = \frac{a}{9^n}$. Since a small triangle is added to each side at every stage, it follows that the total area A_n added to the figure at the nth stage is $A_n = s_{n-1} \cdot a_n = 3 \cdot 4^{n-1} \cdot \frac{a}{9^n} = a \cdot \frac{4^{n-1}}{3^{2n-1}}$. Then the total area enclosed by the snowflake curve is $A = a + A_1 + A_2 + A_3 + \cdots$

$= a + a \cdot \frac{1}{3} + a \cdot \frac{4}{3^3} + a \cdot \frac{4^2}{3^5} + a \cdot \frac{4^3}{3^7} + \cdots$. After the first term, this is a geometric series with common ratio $4/9$, so $A = a + \frac{a/3}{1-(4/9)} = a + \frac{a}{3} \cdot \frac{9}{5} = \frac{8a}{5}$. But the area of the original equilateral triangle with side 1 is $a = \frac{1}{2} \cdot 1 \cdot \sin(\pi/3) = \frac{\sqrt{3}}{4}$. So the area enclosed by the snowflake curve is $\frac{8}{5} \cdot \frac{\sqrt{3}}{4} = \frac{2\sqrt{3}}{5}$.

5. (a) Using Equation 14a in Appendix B with $x = y = \theta$, we get $\tan 2\theta = \frac{2\tan\theta}{1-\tan^2\theta}$, so $\cot 2\theta = \frac{1-\tan^2\theta}{2\tan\theta} \Rightarrow 2\cot 2\theta = \frac{1-\tan^2\theta}{\tan\theta} = \cot\theta - \tan\theta$. Replacing θ by $\frac{1}{2}x$, we get $2\cot x = \cot\tfrac{1}{2}x - \tan\tfrac{1}{2}x$, or $\tan\tfrac{1}{2}x = \cot\tfrac{1}{2}x - 2\cot x$.

(b) From part (a) we have $\tan\frac{x}{2^n} = \cot\frac{x}{2^n} - 2\cot\frac{x}{2^{n-1}}$. So the partial sum of the given series

42

Problems Plus

is $s_n = \frac{1}{2}\tan\frac{x}{2} + \frac{1}{4}\tan\frac{x}{4} + \frac{1}{8}\tan\frac{x}{8} + \cdots + \frac{1}{2^n}\tan\frac{x}{2^n}$

$= (\frac{1}{2}\cot\frac{x}{2} - \cot x) + (\frac{1}{4}\cot\frac{x}{4} - \frac{1}{2}\cot\frac{x}{2}) + (\frac{1}{8}\cot\frac{x}{8} - \frac{1}{4}\cot\frac{x}{4}) + \cdots + (\frac{1}{2^n}\cot\frac{x}{2^n} - \frac{1}{2^{n-1}}\cot\frac{x}{2^{n-1}})$

$= -\cot x + \frac{1}{2^n}\cot\frac{x}{2^n}$ [telescoping series]

Now $\frac{1}{2^n}\cot\frac{x}{2^n} = \frac{\cos(x/2^n)}{2^n\sin(x/2^n)} = \frac{1}{x} \cdot \frac{\cos(x/2^n)}{\sin(x/2^n)/(x/2^n)} \to \frac{1}{x} \cdot \frac{1}{1} = \frac{1}{x}$ as $n \to \infty$ since $\frac{x}{2^n} \to 0$ for $x \neq 0$. Therefore, if $x \neq 0$ and $x \neq n\pi$, we have

$\sum_{n=1}^{\infty} \frac{1}{2^n}\tan\frac{x}{2^n} = \lim_{n \to \infty} \left(-\cot x + \frac{1}{2^n}\cot\frac{x}{2^n}\right) = -\cot x + \frac{1}{x}$.

If $x = 0$, then all terms in the series are 0, so the sum is 0.

7. $a_{n+1} = \frac{1}{2}(a_n + b_n)$, $b_{n+1} = \sqrt{b_n a_{n+1}}$. So $a_1 = \cos\theta$, $b_1 = 1 \Rightarrow a_2 = \frac{1}{2}(1 + \cos\theta) = \cos^2\frac{\theta}{2}$,
$b_2 = \sqrt{b_1 a_2} = \sqrt{\cos^2(\theta/2)} = \cos\frac{\theta}{2}$ since $-\frac{\pi}{2} \leq \theta \leq \frac{\pi}{2}$. Then $a_3 = \frac{1}{2}(\cos\frac{\theta}{2} + \cos^2\frac{\theta}{2})$
$= \cos\frac{\theta}{2} \cdot \frac{1}{2}(1 + \cos\frac{\theta}{2}) = \cos\frac{\theta}{2}\cos^2\frac{\theta}{4} \Rightarrow b_3 = \sqrt{b_2 a_3} = \sqrt{\cos\frac{\theta}{2}\cos\frac{\theta}{2}\cos^2\frac{\theta}{4}} = \cos\frac{\theta}{2}\cos\frac{\theta}{4}$
$\Rightarrow a_4 = \frac{1}{2}(\cos\frac{\theta}{2}\cos^2\frac{\theta}{4} + \cos\frac{\theta}{2}\cos\frac{\theta}{4}) = \cos\frac{\theta}{2}\cos\frac{\theta}{4} \cdot \frac{1}{2}(1 + \cos\frac{\theta}{4}) = \cos\frac{\theta}{2}\cos\frac{\theta}{4}\cos^2\frac{\theta}{8}$
$\Rightarrow b_4 = \sqrt{\cos\frac{\theta}{2}\cos\frac{\theta}{4}\cos\frac{\theta}{2}\cos\frac{\theta}{4}\cos^2\frac{\theta}{8}} = \cos\frac{\theta}{2}\cos\frac{\theta}{4}\cos\frac{\theta}{8}$. By now we see the pattern:
$b_n = \cos\frac{\theta}{2}\cos\frac{\theta}{2^2}\cos\frac{\theta}{2^3} \cdots \cos\frac{\theta}{2^{n-1}}$ and $a_n = b_n\cos\frac{\theta}{2^{n-1}}$. (This could be proved by mathematical induction.) Note that $\sin\theta = 2\cos\frac{\theta}{2}\sin\frac{\theta}{2} = 2\cos\frac{\theta}{2}(2\cos\frac{\theta}{4}\sin\frac{\theta}{4})$
$= 4\cos\frac{\theta}{2}\cos\frac{\theta}{4}\sin\frac{\theta}{4} = 8\cos\frac{\theta}{2}\cos\frac{\theta}{4}\cos\frac{\theta}{8}\sin\frac{\theta}{8} = \cdots = 2^{n-1}\cos\frac{\theta}{2}\cos\frac{\theta}{4}\cdots\cos\frac{\theta}{2^{n-1}}\sin\frac{\theta}{2^{n-1}}$.
So $b_n = \cos\frac{\theta}{2}\cos\frac{\theta}{2^2}\cos\frac{\theta}{2^3}\cdots\cos\frac{\theta}{2^{n-1}} = \frac{\sin\theta}{2^{n-1}\sin(\theta/2^{n-1})}$. But $2^{n-1}\sin(\theta/2^{n-1})$
$= \theta\frac{\sin(\theta/2^{n-1})}{\theta/2^{n-1}} \to \theta$ as $n \to \infty$, so $b_n \to \frac{\sin\theta}{\theta}$ and $a_n = b_n\cos\frac{\theta}{2^{n-1}} \to \frac{\sin\theta}{\theta} \cdot 1 = \frac{\sin\theta}{\theta}$.

9. $u = 1 + \frac{x^3}{3!} + \frac{x^6}{6!} + \frac{x^9}{9!} + \cdots$, $v = x + \frac{x^4}{4!} + \frac{x^7}{7!} + \frac{x^{10}}{10!} + \cdots$, $w = \frac{x^2}{2!} + \frac{x^5}{5!} + \frac{x^8}{8!} + \cdots$.
The key idea is to differentiate: $\frac{du}{dx} = \frac{3x^2}{3!} + \frac{6x^5}{6!} + \frac{9x^8}{9!} + \cdots = \frac{x^2}{2!} + \frac{x^5}{5!} + \frac{x^8}{8!} + \cdots = w$.
Similarly, $\frac{dv}{dx} = 1 + \frac{x^3}{3!} + \frac{x^6}{6!} + \frac{x^9}{9!} + \cdots = u$, and $\frac{dw}{dx} = x + \frac{x^4}{4!} + \frac{x^7}{7!} + \frac{x^{10}}{10!} + \cdots = v$.
So $u' = w$, $v' = u$, and $w' = v$. Now differentiate the left hand side of the desired equation:
$\frac{d}{dx}(u^3 + v^3 + w^3 - 3uvw) = 3u^2u' + 3v^2v' + 3w^2w' - 3(u'vw + uv'w + uvw')$
$= 3u^2w + 3v^2u + 3w^2v - 3(vw^2 + u^2w + uv^2) = 0 \Rightarrow u^3 + v^3 + w^3 - 3uvw = C$.
To find the value of the constant C, we put $x = 1$ in the equation and get
$1^3 + 0 + 0 - 3(1 \cdot 0 \cdot 0) = C \Rightarrow C = 1$, so $u^3 + v^3 + w^3 - 3uvw = 1$.

11. We start with the geometric series $\sum_{n=0}^{\infty} x^n = \frac{1}{1-x}$, $|x| < 1$, and differentiate:
$\sum_{n=1}^{\infty} nx^{n-1} = \frac{d}{dx}\left[\sum_{n=0}^{\infty} x^n\right] = \frac{d}{dx}\left[\frac{1}{1-x}\right] = \frac{1}{(1-x)^2}$ for $|x| < 1 \Rightarrow$

Problems Plus

$$\sum_{n=1}^{\infty} nx^n = x \sum_{n=1}^{\infty} nx^{n-1} = \frac{x}{(1-x)^2} \text{ for } |x| < 1. \text{ Differentiate again:}$$

$$\sum_{n=1}^{\infty} n^2 x^{n-1} = \frac{d}{dx} \frac{x}{(1-x)^2} = \frac{(1-x)^2 - x \cdot 2(1-x)(-1)}{(1-x)^4} = \frac{x+1}{(1-x)^3} \Rightarrow \sum_{n=1}^{\infty} n^2 x^n = \frac{x^2 + x}{(1-x)^3}$$

$$\Rightarrow \sum_{n=1}^{\infty} n^3 x^{n-1} = \frac{d}{dx} \frac{x^2 + x}{(1-x)^3} = \frac{(1-x)^3(2x+1) - (x^2+x)3(1-x)^2(-1)}{(1-x)^6} = \frac{x^2 + 4x + 1}{(1-x)^4}$$

$$\Rightarrow \sum_{n=1}^{\infty} n^3 x^n = \frac{x^3 + 4x^2 + x}{(1-x)^4}, \quad |x| < 1.$$

The radius of convergence is 1 because that is the radius of convergence for the geometric series we started with. If $x = \pm 1$, the series is $\sum_{n=1}^{\infty} n^3 (\pm 1)^n$, which diverges by the Test For Divergence, so the interval of convergence is $(-1, 1)$.

13. (a) Let $a = \arctan x$ and $b = \arctan y$. Then, by Equation 14(b) in Appendix B,

    $$\tan(a - b) = \frac{\tan a - \tan b}{1 + \tan a \tan b} = \frac{\tan(\arctan x) - \tan(\arctan y)}{1 + \tan(\arctan x) \tan(\arctan y)} \Rightarrow \tan(a - b) = \frac{x - y}{1 + xy}$$

    $$\Rightarrow \arctan x - \arctan y = a - b = \arctan \frac{x - y}{1 + xy} \text{ since } -\frac{\pi}{2} < \arctan x - \arctan y < \frac{\pi}{2}.$$

    (b) From part (a) we have

    $$\arctan \frac{120}{119} - \arctan \frac{1}{239} = \arctan \frac{\frac{120}{119} - \frac{1}{239}}{1 + \frac{120}{119} \cdot \frac{1}{239}} = \arctan \frac{\frac{28,561}{28,441}}{\frac{28,561}{28,441}} = \arctan 1 = \frac{\pi}{4}.$$

    (c) Replacing y by $-y$ in the formula of part (a), we get $\arctan x + \arctan y = \arctan \frac{x + y}{1 - xy}$.

    So $4 \arctan \frac{1}{5} = 2(\arctan \frac{1}{5} + \arctan \frac{1}{5}) = 2 \arctan \frac{\frac{1}{5} + \frac{1}{5}}{1 - \frac{1}{5} \cdot \frac{1}{5}} = 2 \arctan \frac{5}{12}$

    $$= \arctan \frac{5}{12} + \arctan \frac{5}{12} = \arctan \frac{\frac{5}{12} + \frac{5}{12}}{1 - \frac{5}{12} \cdot \frac{5}{12}} = \arctan \frac{120}{119}.$$

    Thus, from part (b) we have $4 \arctan \frac{1}{5} - \arctan \frac{1}{239} = \arctan \frac{120}{119} - \arctan \frac{1}{239} = \frac{\pi}{4}$.

    (d) From Example 9 in Section 10.9 we have $\arctan x = x - \frac{x^3}{3} + \frac{x^5}{5} - \frac{x^7}{7} + \frac{x^9}{9} - \frac{x^{11}}{11} + \cdots$, so

    $$\arctan \frac{1}{5} = \frac{1}{5} - \frac{1}{3 \cdot 5^3} + \frac{1}{5 \cdot 5^5} - \frac{1}{7 \cdot 5^7} + \frac{1}{9 \cdot 5^9} - \frac{1}{11 \cdot 5^{11}} + \cdots.$$ This is an alternating series and the size of the terms decreases to 0, so by Theorem 10.28, the sum lies between s_5 and s_6, i.e., $0.197395560 < \arctan \frac{1}{5} < 0.197395562$.

    (e) From the series in part (d) we get $\arctan \frac{1}{239} = \frac{1}{239} - \frac{1}{3 \cdot 239^3} + \frac{1}{5 \cdot 239^5} - \cdots$. The third term is less than 2.6×10^{-13}, so by Theorem 10.28 we have, to 9 decimal places, $\arctan \frac{1}{239} \approx s_2 \approx 0.004184076$. Thus $0.004184075 < \arctan \frac{1}{239} < 0.004184077$.

    (f) From part (c) we have $\pi = 16 \arctan \frac{1}{5} - 4 \arctan \frac{1}{239}$, so from parts (d) and (e) we have
    $16(0.197395560) - 4(0.004184077) < \pi < 16(0.197395562) - 4(0.004184075) \Rightarrow$
    $3.141592652 < \pi < 3.141592692$. So, to 7 decimal places, $\pi \approx 3.1415927$.

Problems Plus

15. $(a^n + b^n + c^n)^{1/n} = \{c^n[(a/c)^n + (b/c)^n + 1]\}^{1/n} = c[(a/c)^n + (b/c)^n + 1]^{1/n}$. Since $0 \le a \le c$, we have $0 \le a/c \le 1$, so $(a/c)^n \to 0$ or 1 as $n \to \infty$. Similarly, $(b/c)^n \to 0$ or 1 as $n \to \infty$. Thus $(a/c)^n + (b/c)^n + 1 \to d$, where $d = 1, 2,$ or 3 and so $[(a/c)^n + (b/c)^n + 1]^{1/n} \to 1$. Therefore $\lim_{n \to \infty} (a^n + b^n + c^n)^{1/n} = c$.

17. As in Section 10.9 we have to integrate the function x^x by integrating a series. Writing $x^x = (e^{\ln x})^x = e^{x \ln x}$ and using the Maclaurin series for e^x, we have
$$x^x = e^{x \ln x} = \sum_{n=0}^{\infty} \frac{(x \ln x)^n}{n!} = \sum_{n=0}^{\infty} \frac{x^n (\ln x)^n}{n!}$$

As with power series, it turns out that we can integrate this series term-by-term:
$$\int_0^1 x^x \, dx = \sum_{n=0}^{\infty} \int_0^1 \frac{x^n (\ln x)^n}{n!} \, dx = \sum_{n=0}^{\infty} \frac{1}{n!} \int_0^1 x^n (\ln x)^n \, dx$$

We integrate by parts with $u = (\ln x)^n$, $dv = x^n \, dx$, so $du = \frac{n(\ln x)^{n-1}}{x} dx$ and $v = \frac{x^{n+1}}{n+1}$:

$$\int_0^1 x^n (\ln x)^n \, dx = \lim_{t \to 0^+} \int_t^1 x^n (\ln x)^n \, dx$$
$$= \lim_{t \to 0^+} \left[\frac{x^{n+1}}{n+1}(\ln x)^n\right]_t^1 - \lim_{t \to 0^+} \int_t^1 \frac{n}{n+1} x^n (\ln x)^{n-1} \, dx = 0 - \frac{n}{n+1} \int_0^1 x^n (\ln x)^{n-1} \, dx$$

(where l'Hospital's Rule was used to help evaluate the first limit). Further integration by parts gives $\int_0^1 x^n (\ln x)^k \, dx = -\frac{k}{n+1} \int_0^1 x^n (\ln x)^{k-1} \, dx$ and, combining these steps, we get

$$\int_0^1 x^n (\ln x)^n \, dx = \frac{(-1)^n n!}{(n+1)^n} \int_0^1 x^n \, dx = \frac{(-1)^n n!}{(n+1)^{n+1}} \Rightarrow$$
$$\int_0^1 x^x \, dx = \sum_{n=0}^{\infty} \frac{1}{n!} \int_0^1 x^n (\ln x)^n \, dx = \sum_{n=0}^{\infty} \frac{1}{n!} \frac{(-1)^n n!}{(n+1)^{n+1}}$$
$$= \sum_{n=0}^{\infty} \frac{(-1)^n}{(n+1)^{n+1}} = \sum_{n=1}^{\infty} \frac{(-1)^{n-1}}{n^n}.$$

19. If L is the length of a side of the equilateral triangle, then the area is $A = \frac{1}{2} \cdot L \cdot \frac{\sqrt{3}}{2} L = \frac{\sqrt{3}}{4} L^2$ and so $L^2 = (4/\sqrt{3})A$. Let r be the radius of one of the circles when there are n rows of circles. The figure shows that $L = \sqrt{3}r + r + (n-2)(2r) + r + \sqrt{3}r = r(2n - 2 + 2\sqrt{3})$, so $r = \frac{L}{2(n + \sqrt{3} - 1)}$.

The number of circles is $1 + 2 + \cdots + n = \frac{n(n+1)}{2}$ and so the total area of the circles is
$$A_n = \frac{n(n+1)}{2} \pi r^2 = \frac{n(n+1)}{2} \pi \frac{L^2}{4(n + \sqrt{3} - 1)^2}$$

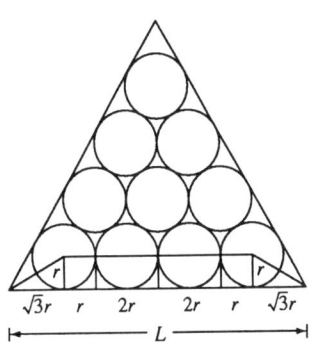

45

$$= \frac{n(n+1)}{2}\pi \cdot \frac{(4/\sqrt{3})A}{4(n+\sqrt{3}-1)^2} = \frac{n(n+1)}{(n+\sqrt{3}-1)^2} \cdot \frac{\pi A}{2\sqrt{3}}$$

$$\Rightarrow \frac{A_n}{A} = \frac{n(n+1)}{(n+\sqrt{3}-1)^2} \cdot \frac{\pi}{2\sqrt{3}} = \frac{1+\frac{1}{n}}{\left(1+\frac{\sqrt{3}-1}{n}\right)^2} \cdot \frac{\pi}{2\sqrt{3}} \to \frac{\pi}{2\sqrt{3}} \text{ as } n \to \infty.$$

21. (a) f is clearly continuous when $x \neq 0$ since x, $\sin x$, and π/x are continuous when $x \neq 0$. Also $|\sin(\pi/x)| \leq 1 \Rightarrow |x\sin(\pi/x)| \leq |x|$ and $\lim_{x \to 0} |x| = 0$, so by the Squeeze Theorem we have $\lim_{x \to 0} |x\sin(\pi/x)| = 0$ and so $\lim_{x \to 0} f(x) = \lim_{x \to 0} x\sin(\pi/x) = 0 = f(0)$. This shows that f is continuous at 0, so it is continuous on $(-1, 1)$.

(b) Note that $f(x) = 0$ when $x = 0$ and when
$\frac{\pi}{x} = n\pi \Rightarrow x = \frac{1}{n}$, n an integer. Since
$-1 \leq \sin(\pi/x) \leq 1$, the graph of f lies between
the lines $y = x$ and $y = -x$ and touches these
lines when $\frac{\pi}{x} = \frac{\pi}{2} + n\pi \Rightarrow x = \frac{1}{n+\frac{1}{2}}$.

(c) The enlargement of the portion of the graph
between $x = \frac{1}{n}$ and $x = \frac{1}{n-1}$ (the case where n
is odd is illustrated) shows that the arc length
from $x = \frac{1}{n}$ to $x = \frac{1}{n-1}$ is greater than
$|PQ| = \frac{1}{n-\frac{1}{2}} = \frac{2}{2n-1}$. Thus the total length
of the graph is greater than $2 \sum_{n=1}^{\infty} \frac{2}{2n-1}$.
This is a divergent series (by comparison with
the harmonic series), so the graph has infinite length.

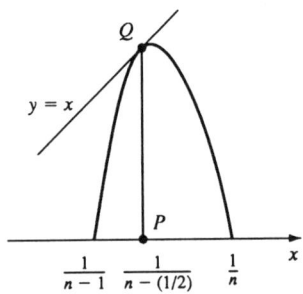

CHAPTER 11

EXERCISES 11.1

1. (a)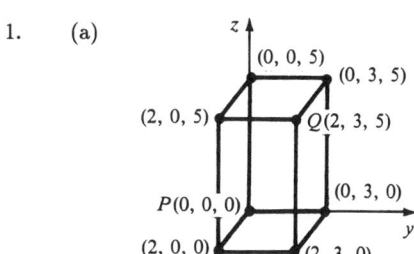

    (b) $|PQ| = \sqrt{(2-0)^2 + (3-0)^2 + (5-0)^2}$
    $= \sqrt{38}$

3. (a)

    (b) $|PQ| = \sqrt{(3-1)^2 + (4-1)^2 + (5-2)^2}$
    $= \sqrt{22}$

5. (a)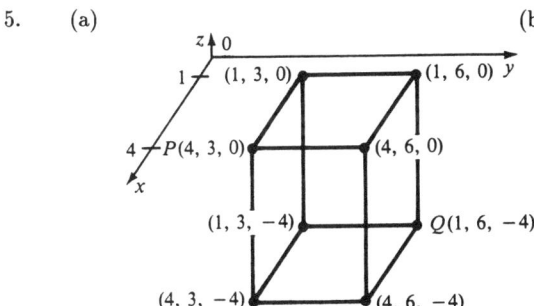

    (b) $|PQ| = \sqrt{(1-4)^2 + (6-3)^2 + (-4-0)^2}$
    $= \sqrt{34}$

7. $|AB| = \sqrt{(3-2)^2 + (3-1)^2 + (4-0)^2} = \sqrt{21}$

    $|BC| = \sqrt{(5-3)^2 + (4-3)^2 + (3-4)^2} = \sqrt{6}$

    $|CA| = \sqrt{(5-2)^2 + (4-1)^2 + (3-0)^2} = \sqrt{27} = 3\sqrt{3}$

    Since no two of the sides are equal in length the triangle isn't isosceles. But $|AB|^2 + |BC|^2 = 27 = |CA|^2$ so that the triangle is a right triangle.

Section 11.1

9. $|AB| = \sqrt{[5-(-2)]^2 + (4-6)^2 + (-3-1)^2} = \sqrt{49+4+16} = \sqrt{69}$

    $|BC| = \sqrt{(2-5)^2 + (-6+4)^2 + [4-(-3)]^2} = \sqrt{9+100+49} = \sqrt{158}$

    $|CA| = \sqrt{(-2-2)^2 + [6-(-6)]^2 + (1-4)^2} = \sqrt{16+144+9} = \sqrt{169} = 13$

    Since no two sides are of equal length and since $|AB|^2 + |BC|^2 \neq |CA|^2$ the triangle is neither isosceles nor a right triangle.

11. $|PQ| = \sqrt{(0-1)^2 + (3-2)^2 + (7-3)^2} = \sqrt{18} = 3\sqrt{2}$

    $|PR| = \sqrt{(3-1)^2 + (5-2)^2 + (11-3)^2} = \sqrt{4+9+64} = \sqrt{77}$

    $|QR| = \sqrt{(3-0)^2 + (5-3)^2 + (11-7)^2} = \sqrt{9+4+16} = \sqrt{29}$

    Since the sum of the two shortest distances isn't equal to the longest distance, the points aren't collinear.

    [To show that $\sqrt{18} + \sqrt{29} \neq \sqrt{77}$, assume they are equal. Then squaring both sides gives $18 + 29 + 2\sqrt{(18)(29)} = 77$; solving for the radical and squaring again gives $(18)(29) = (15)^2$ or $522 = 225$ which of course is not true, so the values can't be equal.]

13. $(x-0)^2 + (y-1)^2 + [z-(-1)]^2 = 4^2$ or $x^2 + (y-1)^2 + (z+1)^2 = 16$.

15. $(x+6)^2 + (y+1)^2 + (z-2)^2 = 12$.

17. Completing the squares in the equation gives
    $(x^2 + 2x + 1) + (y^2 + 8y + 16) + (z^2 - 4z + 4) = 28 + 1 + 16 + 4$
    $\Rightarrow (x+1)^2 + (y+4)^2 + (z-2)^2 = 49$
    $\Rightarrow C(-1, -4, 2)$, and $r = 7$.

19. $\left(x^2 + x + \frac{1}{4}\right) + (y^2 - 2y + 1) + (z^2 + 6z + 9) = 2 + \frac{1}{4} + 1 + 9$
    $\Rightarrow \left(x + \frac{1}{2}\right)^2 + (y-1)^2 + (z+3)^2 = \frac{49}{4} \Rightarrow C\left(-\frac{1}{2}, 1, -3\right)$, and $r = \frac{7}{2}$.

21. $\left(x^2 - x + \frac{1}{4}\right) + y^2 + z^2 = 0 + \frac{1}{4} \Rightarrow \left(x - \frac{1}{2}\right)^2 + (y-0)^2 + (z-0)^2 = \frac{1}{4}$
    $\Rightarrow C\left(\frac{1}{2}, 0, 0\right)$, $r = \frac{1}{2}$.

23. We need to find the set $\{P(x,y,z) \mid |AP| = |BP|\}$.

    $\sqrt{(x+1)^2 + (y-5)^2 + (z-3)^2} = \sqrt{(x-6)^2 + (y-2)^2 + (z+2)^2}$
    $\Rightarrow (x+1)^2 + (y-5)^2 + (z-3)^2 = (x-6)^2 + (y-2)^2 + (z+2)^2$
    $\Rightarrow x^2 + 2x + 1 + y^2 - 10y + 25 + z^2 - 6z + 9 = x^2 - 12x + 36 + y^2 - 4y + 4 + z^2 + 4z + 4$
    $\Rightarrow 14x - 6y - 10z = 9$.

    Thus the locus of points is a plane perpendicular to the line segment joining A and B (since this

Section 11.1

plane must contain the perpendicular bisector of the line segment AB).

25. Call the given point Q and show that $|P_1Q| = |QP_2| = \frac{1}{2}|P_1P_2|$.

$|P_1P_2| = \sqrt{(x_2-x_1)^2 + (y_2-y_1)^2 + (z_2-z_1)^2}$

$|P_1Q| = \sqrt{[(x_1+x_2)/2 - x_1]^2 + [(y_1+y_2)/2 - y_1]^2 + [(z_1+z_2)/2 - z_1]^2}$

$= \frac{1}{2}\sqrt{(x_2-x_1)^2 + (y_2-y_1)^2 + (z_2-z_1)^2} = \frac{1}{2}|P_1P_2|$

Similarly $|QP_2| = \frac{1}{2}|P_1P_2|$.

27. From Exercise 25, the midpoints of sides AB, BC and CA are respectively

$P_1\left(-\frac{1}{2}, 1, 4\right)$, $P_2\left(1, \frac{1}{2}, 5\right)$ and $P_3\left(\frac{5}{2}, \frac{3}{2}, 4\right)$. Then the lengths of the medians are:

$|AP_2| = \sqrt{0^2 + \left(\frac{1}{2} - 2\right)^2 + (5-3)^2} = \sqrt{\frac{9}{4} + 4} = \sqrt{\frac{25}{4}} = \frac{5}{2}$

$|BP_3| = \sqrt{\left(\frac{5}{2} + 2\right)^2 + \left(\frac{3}{2}\right)^2 + (4-5)^2} = \sqrt{\frac{81}{4} + \frac{9}{4} + 1} = \sqrt{\frac{94}{4}} = \frac{1}{2}\sqrt{94}$

$|CP_1| = \sqrt{\left(-\frac{1}{2} - 4\right)^2 + (1-1)^2 + (4-5)^2} = \sqrt{\frac{81}{4} + 1} = \frac{1}{2}\sqrt{85}$.

29. (a) Since the sphere touches the xy–plane, its radius is the distance from its center, $(2, -3, 6)$, to the xy–plane, namely 6. Therefore $r = 6$ and the equation of the sphere is
$(x-2)^2 + (y+3)^2 + (z-6)^2 = 36$.

(b) Here r = distance from center to yz–plane = 2. Therefore, the equation is
$(x-2)^2 + (y+3)^2 + (z-6)^2 = 4$.

(c) Here r = distance from center to xz–plane = 3 Therefore, the equation is
$(x-2)^2 + (y+3)^2 + (z-6)^2 = 9$.

31. A plane parallel to the yz–plane and 9 units in front of it.

33. A half–space containing all points to the right of the plane $y = 2$.

35. A plane perpendicular to the xz–plane and intersecting the xz–plane in the line $x = z$, $y = 0$.

37. A right circular cylinder with radius one and axis the z–axis.

39. All points outside of the sphere with radius one and center $(0, 0, 0)$.

41. Completing the square in the z–variable gives $x^2 + y^2 + (z^2 - 2z + 1) < 3 + 1$ or $x^2 + y^2 + (z-1)^2 < 4$, all points inside of the sphere with radius two and center $(0, 0, 1)$.

43. In the xy–plane the equation $xy = 1$ represents a hyperbola with center at the origin. Since z can assume any value, the region in R^3 is a hyperbolic cylinder.

49

Section 11.2

45. All points on and between the two horizontal planes $z = 2$ and $z = -2$.

EXERCISES 11.2

1. $\vec{a} = <4-1, 4-3> = <3, 1>$

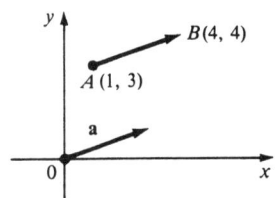

3. $\vec{a} = <3-3, -3+1> = <0, -2>$

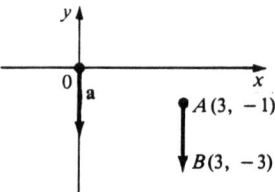

5. $\vec{a} = <2, 0, -2>$

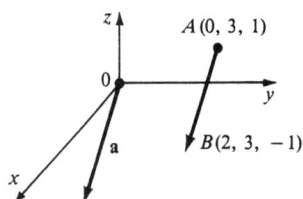

(using position vectors and the parallelogram law)

7. $<2, 3> + <3, -4> = <5, -1>$

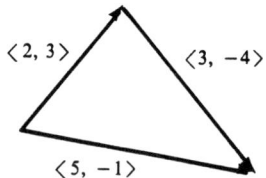

(using the triangle law)

9. $<1, 0, 1> + <0, 0, 1> = <1, 0, 2>$

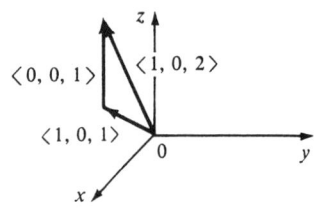

11. $|\vec{a}| = \sqrt{5^2 + (-12)^2} = \sqrt{169} = 13$
$\vec{a} + \vec{b} = <5-2, -12+8> = <3, -4>$
$\vec{a} - \vec{b} = <5-(-2), -12-8> = <7, -20>$
$2\vec{a} = <2(5), 2(-12)> = <10, -24>$

Section 11.2

$3\vec{a} + 4\vec{b} = <15, -36> + <-8, 32> = <7, -4>$.

13. $|\vec{a}| = \sqrt{2^2 + (-3)^2 + 6^2} = \sqrt{49} = 7$
$\vec{a} + \vec{b} = <3, -2, 10>$
$\vec{a} - \vec{b} = <1, -4, 2>$
$2\vec{a} = <4, -6, 12>$
$3\vec{a} + 4\vec{b} = <6, -9, 18> + <4, 4, 16> = <10, -5, 34>$.

15. $|\vec{a}| = \sqrt{1^2 + (-1)^2} = \sqrt{2}$
$\vec{a} + \vec{b} = (\vec{i} - \vec{j}) + (\vec{i} + \vec{j}) = 2\vec{i}$
$\vec{a} - \vec{b} = (\vec{i} - \vec{j}) - (\vec{i} + \vec{j}) = -2\vec{j}$
$2\vec{a} = 2(\vec{i} - \vec{j}) = 2\vec{i} - 2\vec{j}$
$3\vec{a} + 4\vec{b} = 3(\vec{i} - \vec{j}) + 4(\vec{i} + \vec{j}) = 3\vec{i} - 3\vec{j} + 4\vec{i} + 4\vec{j} = 7\vec{i} + \vec{j}$.

17. $|\vec{a}| = \sqrt{1^2 + 1^2 + 1^2} = \sqrt{3}$
$\vec{a} + \vec{b} = 3\vec{i} + 4\vec{k}$
$\vec{a} - \vec{b} = \vec{i} + \vec{j} + \vec{k} - 2\vec{i} + \vec{j} - 3\vec{k} = -\vec{i} + 2\vec{j} - 2\vec{k}$
$2\vec{a} = 2\vec{i} + 2\vec{j} + 2\vec{k}$
$3\vec{a} + 4\vec{b} = (3\vec{i} + 3\vec{j} + 3\vec{k}) + (8\vec{i} - 4\vec{j} + 12\vec{k}) = 11\vec{i} - \vec{j} + 15\vec{k}$.

19. $|<1, -2>| = \sqrt{1^2 + 2^2} = \sqrt{5}$. Thus $\vec{u} = \frac{1}{\sqrt{5}}<1, 2> = <\frac{1}{\sqrt{5}}, \frac{2}{\sqrt{5}}>$.

21. $|<-2, 4, 3>| = \sqrt{(-2)^2 + 4^2 + 3^2} = \sqrt{29}$.

    Thus $\vec{u} = \frac{1}{\sqrt{29}}<-2, 4, 3> = <-\frac{2}{\sqrt{29}}, \frac{4}{\sqrt{29}}, \frac{3}{\sqrt{29}}>$.

23. $|\vec{i} + \vec{j}| = \sqrt{1^2 + 1^2} = \sqrt{2}$. Thus $\vec{u} = \frac{1}{\sqrt{2}}(\vec{i} + \vec{j}) = \frac{1}{\sqrt{2}}\vec{i} + \frac{1}{\sqrt{2}}\vec{j}$.

25. $\vec{a} = 2\vec{i} + 3\vec{j}, \vec{b} = \vec{i} - \vec{j} \Rightarrow \vec{a} + 3\vec{b} = 2\vec{i} + 3\vec{j} + 3\vec{i} - 3\vec{j} = 5\vec{i} \Rightarrow \vec{i} = \frac{1}{5}\vec{a} + \frac{3}{5}\vec{b}$.
Substituting this expression for $\vec{i}$ into $\vec{a} = 2\vec{i} + 3\vec{j}$ gives
$\vec{a} = 2\left(\frac{1}{5}\vec{a} + \frac{3}{5}\vec{b}\right) + 3\vec{j} \Rightarrow \frac{3}{5}\vec{a} - \frac{6}{5}\vec{b} = 3\vec{j} \Rightarrow \vec{j} = \frac{1}{5}\vec{a} - \frac{2}{5}\vec{b}$.

27. $|\vec{F}_1| = 10$ lb and $|\vec{F}_2| = 12$ lb.
$\vec{F}_1 = -|\vec{F}_1|\cos 45°\vec{i} + |\vec{F}_1|\sin 45°\vec{j} = -10\cos 45°\vec{i} + 10\sin 45°\vec{j} = -5\sqrt{2}\vec{i} + 5\sqrt{2}\vec{j}$
$\vec{F}_2 = |\vec{F}_2|\cos 30°\vec{i} + |\vec{F}_2|\sin 30°\vec{j} = 12\cos 30°\vec{i} + 12\sin 30°\vec{j} = 6\sqrt{3}\vec{i} + 6\vec{j}$
$\vec{F} = \vec{F}_1 + \vec{F}_2 = (6\sqrt{3} - 5\sqrt{2})\vec{i} + (6 + 5\sqrt{2})\vec{j} \approx 3.32\vec{i} + 13.07\vec{j}$
$|\vec{F}| \approx \sqrt{(3.32)^2 + (13.07)^2} \approx 13.5$ lb
$\tan\theta = \frac{6 + 5\sqrt{2}}{6\sqrt{3} - 5\sqrt{2}} \Rightarrow \theta = \tan^{-1}\left(\frac{6 + 5\sqrt{2}}{6\sqrt{3} - 5\sqrt{2}}\right) \approx 76°$.

Section 11.3

29. With respect to the water's surface, the woman's velocity is the vector sum of the velocity of the ship with respect to the water, and her velocity with respect to the ship. If we let North be the positive y direction, then $\vec{v} = <0, 22> + <-3, 0> = <-3, 22>$. The woman's speed is $|\vec{v}| = \sqrt{9 + 484} \approx 22.2$ mi/h. The vector $\vec{v}$ makes an angle θ with the East, where $\theta = \tan^{-1}\left(\frac{22}{-3}\right) \approx 98°$. Therefore, the woman's direction is about $N(98 - 90)°W = N\,8°\,W$.

31. $\vec{a} + (\vec{b} + \vec{c}) = <a_1, a_2> + (<b_1, b_2> + <c_1, c_2>)$

   $= <a_1, a_2> + <b_1 + c_1, b_2 + c_2>$

   $= <a_1 + b_1 + c_1, a_2 + b_2 + c_2> = <(a_1 + b_1) + c_1, (a_2 + b_2) + c_2>$

   $= <a_1 + b_1, a_2 + b_2> + <c_1, c_2> = (<a_1, a_2> + <b_1, b_2>) + <c_1, c_2>$

   $= (\vec{a} + \vec{b}) + \vec{c}$.

33. $(c + d)\vec{a} = (c + d)<a_1, a_2, a_3> = <(c+d)a_1, (c+d)a_2, (c+d)a_3>$

   $= <ca_1 + da_1, ca_2 + da_2, ca_3 + da_3>$

   $= <ca_1, ca_2, ca_3> + <da_1, da_2, da_3> = c\vec{a} + d\vec{a}$.

35. Consider quadrilateral ABCD with sides AB and CD parallel and of equal length; that is, $\vec{AB} = \vec{DC}$. Thus $\vec{AD} = \vec{AB} + \vec{BD} = \vec{DC} + \vec{BD}$ [since $\vec{AB} = \vec{DC}$] $= \vec{BD} + \vec{DC} = \vec{BC}$. This shows that sides AD and BC are parallel and have equal lengths.

EXERCISES 11.3

1. $\vec{a} \cdot \vec{b} = (2)(-3) + (5)(1) = -1$.

3. $\vec{a} \cdot \vec{b} = (4)(-2) + (7)(1) + (-1)(4) = -5$.

5. $\vec{a} \cdot \vec{b} = (1)(\pi) + (-1)(2\pi) + (1)(3\pi) = 2\pi$.

7. $\vec{a} \cdot \vec{b} = (2)(1) + (3)(-3) + (-4)(1) = -11$.

9. $\vec{a} \cdot \vec{i} = <a_1, a_2, a_3> \cdot <1, 0, 0> = (a_1)(1) + (a_2)(0) + (a_3)(0) = a_1$
   Similarly $\vec{a} \cdot \vec{j} = (a_1)(0) + (a_2)(1) + (a_3)(0) = a_2$ and
   $\vec{a} \cdot \vec{k} = (a_1)(0) + (a_2)(0) + (a_3)(1) = a_3$.

11. $|\vec{a}| = \sqrt{1^2 + 2^2 + 2^2} = 3$. $|\vec{b}| = \sqrt{3^2 + 4^2 + 0^2} = 5$
    $\vec{a} \cdot \vec{b} = 3 + 8 + 0 = 11$. $\cos\theta = \frac{11}{(3)(5)}$ so $\theta = \cos^{-1}\left(\frac{11}{15}\right) \approx 43°$.

Section 11.3

13. $|\vec{a}| = \sqrt{1^2+2^2} = \sqrt{5}, |\vec{b}| = \sqrt{12^2+(-5)^2} = \sqrt{13},$
    $\vec{a}\cdot\vec{b} = 12-10 = 2, \quad \cos\theta = \dfrac{2}{13\sqrt{5}}$ so $\theta = \cos^{-1}\left(\dfrac{2}{13\sqrt{5}}\right) \approx 86°.$

15. $|\vec{a}| = \sqrt{36+4+9} = 7, |\vec{b}| = \sqrt{3}, \vec{a}\cdot\vec{b} = 6-2-3 = 1$
    $\cos\theta = \dfrac{1}{7\sqrt{3}}$ so $\theta = \cos^{-1}\left(\dfrac{1}{7\sqrt{3}}\right) \approx 85°.$

17. Let a, b and c be the angles at vertex A, B and C respectively. Then a is the angle between vectors $\vec{AB}$ and $\vec{AC}$, b is the angle between vectors $\vec{BA}$ and $\vec{BC}$ while c is the angle between vectors $\vec{CA}$ and $\vec{CB}$. Thus
$\cos a = \dfrac{\vec{AB}\cdot\vec{AC}}{|\vec{AB}||\vec{AC}|}$
$= <5,-1,2>\cdot<-2,-4,-3>\dfrac{1}{\sqrt{30\cdot 29}}$
$= \dfrac{1}{\sqrt{870}}(-10+4-6) = -\dfrac{12}{\sqrt{870}}$ and
$a = \cos^{-1}\left(-\dfrac{12}{\sqrt{870}}\right) \approx 114°.$

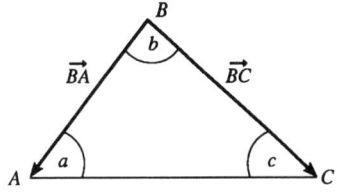

Similarly $\cos b = \dfrac{\vec{BA}\cdot\vec{BC}}{|\vec{BA}||\vec{BC}|}$
$= \dfrac{1}{\sqrt{30\cdot 83}}<-5,1,-2>\cdot<-7,-3,-5>$
$= \dfrac{1}{\sqrt{2490}}(35-3+10) = \dfrac{42}{\sqrt{2490}}$ so $b = \cos^{-1}\left(\dfrac{42}{\sqrt{2490}}\right) \approx 33°.$
And $\cos c = \dfrac{\vec{CA}\cdot\vec{CB}}{|\vec{CA}||\vec{CB}|} = \dfrac{1}{\sqrt{29\cdot 83}}<2,4,3>\cdot<7,3,5>$
$= \dfrac{1}{\sqrt{2407}}(14+12+15) = \dfrac{41}{\sqrt{2407}}$ so $c = \cos^{-1}\left(\dfrac{41}{\sqrt{2407}}\right) \approx 33°.$

Alternate solution: Apply the law of cosines three times as follows:

$\cos a = \dfrac{|\vec{BC}|^2-|\vec{AB}|^2-|\vec{AC}|^2}{2|\vec{AB}||\vec{AC}|}, \quad \cos b = \dfrac{|\vec{AC}|^2-|\vec{AB}|^2-|\vec{BC}|^2}{2|\vec{AB}||\vec{BC}|}$ and

$\cos c = \dfrac{|\vec{AB}|^2-|\vec{AC}|^2-|\vec{BC}|^2}{2|\vec{AC}||\vec{BC}|}.$

19. Since $\vec{a} = -2\vec{b}$, $\vec{a}$ and $\vec{b}$ are parallel vectors (and thus not orthogonal).

21. $\vec{a}\cdot\vec{b} = -2+16+(-15) \neq 0$ so $\vec{a}$ and $\vec{b}$ aren't orthogonal. Also since $\vec{a}$ isn't a scalar multiple of $\vec{b}$, $\vec{a}$ and $\vec{b}$ aren't parallel.

23. $\vec{a}\cdot\vec{b} = 3+(-1)+(-2) = 0$ so $\vec{a}$ and $\vec{b}$ are orthogonal.

Section 11.3

25. For the two vectors to be orthogonal we need $(x\vec{i} - 2\vec{j}) \cdot (x\vec{i} + 8\vec{j}) = 0 \Rightarrow x^2 - 16 = 0$. So $x = \pm 4$.

27. For the two vectors to be orthogonal, we need
$$0 = <x, 1, 2> \cdot <3, 4, x> = 3x + 4 + 2x = 5x + 4 \text{ implies } x = -\tfrac{4}{5}.$$

29. Let $\vec{a} = a_1\vec{i} + a_2\vec{j} + a_3\vec{k}$ be a vector orthogonal to both $\vec{i} + \vec{j}$ and $\vec{i} + \vec{k}$. Then by definition $a_1 + a_2 = 0$ and $a_1 + a_3 = 0$, so $a_1 = -a_2 = -a_3$. Furthermore $\vec{a}$ is to be a unit vector, so $1 = a_1^2 + a_2^2 + a_3^2 = 3a_1^2$ implies $a_1 = \pm\tfrac{1}{\sqrt{3}}$. Thus
$\vec{a} = \tfrac{1}{\sqrt{3}}\vec{i} - \tfrac{1}{\sqrt{3}}\vec{j} - \tfrac{1}{\sqrt{3}}\vec{k}$ and $\vec{a} = \tfrac{-1}{\sqrt{3}}\vec{i} + \tfrac{1}{\sqrt{3}}\vec{j} + \tfrac{1}{\sqrt{3}}\vec{k}$ are the two such unit vectors.

31. $|<1, 2, 2>| = \sqrt{1 + 4 + 4} = 3$ so $\cos \alpha = \tfrac{1}{3}$, $\cos \beta = \tfrac{2}{3}$ and
$\cos \gamma = \tfrac{2}{3}$ while $\alpha = \cos^{-1}(\tfrac{1}{3}) \approx 71°$ and $\beta = \gamma = \cos^{-1}(\tfrac{2}{3}) \approx 48°$.

33. $|-8\vec{i} + 3\vec{j} + 2\vec{k}| = \sqrt{64 + 9 + 4} = \sqrt{77}$ so $\cos \alpha = -\tfrac{8}{\sqrt{77}}$,
$\cos \beta = \tfrac{3}{\sqrt{77}}$ and $\cos \gamma = \tfrac{2}{\sqrt{77}}$ while $\alpha = \cos^{-1}(-\tfrac{8}{\sqrt{77}}) \approx 156°$,
$\beta = \cos^{-1}(\tfrac{3}{\sqrt{77}}) \approx 70°$ and $\gamma = \cos^{-1}(\tfrac{2}{\sqrt{77}}) \approx 77°$.

35. $|<2, 1.2, 0.8>| = \sqrt{4 + 1.44 + 0.64} = \tfrac{5}{5}\sqrt{6.08} = \tfrac{\sqrt{152}}{5}$ so $\cos \alpha = \tfrac{10}{\sqrt{152}} = \tfrac{5}{\sqrt{38}}$,
$\cos \beta = \tfrac{6}{\sqrt{152}} = \tfrac{3}{\sqrt{38}}$ and $\cos \gamma = \tfrac{4}{\sqrt{152}} = \tfrac{2}{\sqrt{38}}$ while $\alpha = \cos^{-1}(\tfrac{5}{\sqrt{38}}) \approx 36°$,
$\beta = \cos^{-1}(\tfrac{3}{\sqrt{38}}) \approx 61°$ and $\gamma = \cos^{-1}(\tfrac{2}{\sqrt{38}}) \approx 71°$.

37. $|\vec{a}| = \sqrt{4 + 9} = \sqrt{13}$. The scalar projection of $\vec{b}$ onto $\vec{a}$ is $\tfrac{\vec{a} \cdot \vec{b}}{|\vec{a}|} = \tfrac{2 \cdot 4 + 3 \cdot 1}{\sqrt{13}} = \tfrac{11}{\sqrt{13}}$.
The vector projection of $\vec{b}$ onto $\vec{a}$ is $\tfrac{\vec{a} \cdot \vec{b}}{|\vec{a}|^2}\vec{a} = \tfrac{11}{\sqrt{13}} \cdot \tfrac{1}{\sqrt{13}}<2, 3> = \tfrac{11}{13}<2, 3> = <\tfrac{22}{13}, \tfrac{33}{13}>$.

39. $|\vec{a}| = \sqrt{16 + 4 + 0} = 2\sqrt{5}$ so the scalar projection of $\vec{b}$ onto $\vec{a}$ is $\tfrac{\vec{a} \cdot \vec{b}}{|\vec{a}|} = \tfrac{1}{2\sqrt{5}}(4 + 2 + 0) = \tfrac{3}{\sqrt{5}}$.
The vector projection of $\vec{b}$ onto $\vec{a}$ is $\tfrac{\vec{a} \cdot \vec{b}}{|\vec{a}|^2}\vec{a} = \tfrac{3}{\sqrt{5}} \cdot \tfrac{1}{2\sqrt{5}}<4, 2, 0> = \tfrac{3}{10}<6, 3, 0> = <\tfrac{6}{5}, \tfrac{3}{5}, 0>$.

41. $|\vec{a}| = \sqrt{1 + 0 + 1} = \sqrt{2}$ so the scalar projection of $\vec{b}$ onto $\vec{a}$ is $\tfrac{\vec{a} \cdot \vec{b}}{|\vec{a}|} = \tfrac{1}{\sqrt{2}}(1 + 0 + 0) = \tfrac{1}{\sqrt{2}}$ while
the vector projection is $\tfrac{\vec{a} \cdot \vec{b}}{|\vec{a}|^2}\vec{a} = \tfrac{1}{\sqrt{2}} \cdot \tfrac{1}{\sqrt{2}}(\vec{i} + \vec{k}) = \tfrac{1}{2}(\vec{i} + \vec{k})$.

Section 11.3

43. Here $\vec{D} = (4-2)\vec{i} + (9-3)\vec{j} + (15-0)\vec{k} = 2\vec{i} + 6\vec{j} + 15\vec{k}$ so
$W = \vec{F} \cdot \vec{D} = 20 + 108 - 90 = 38$ joules.

45. $W = |\vec{F}||\vec{D}|\cos\theta = (25)(10)\cos 20° \approx 235$ ft-lb.

47. $\vec{v} \cdot \vec{a} = \left(\vec{b} - \frac{\vec{a}\cdot\vec{b}}{|\vec{a}|^2}\vec{a}\right) \cdot \vec{a} = \vec{b} \cdot \vec{a} - (\vec{a}\cdot\vec{b})\frac{|\vec{a}|^2}{|\vec{a}|^2} = \vec{b}\cdot\vec{a} - \vec{b}\cdot\vec{a} = 0.$

Therefore, $\vec{v}$ is perpendicular to $\vec{a}$.

49. For convenience, consider the unit cube positioned so that its back left corner is at the origin, and its edges lie along the coordinate axes. The diagonal of the cube that begins at the origin and ends at (1, 1, 1) has vector representation $<1, 1, 1>$. The angle θ, between this vector and the vector of the edge which which also begins at the origin and runs along the x-axis [i.e. $<1,0,0>$] is given by $\cos\theta = \frac{<1, 1, 1>\cdot<1, 0, 0>}{|<1, 1, 1>||<1, 0, 0>|} = \frac{1}{\sqrt{3}} \Rightarrow \theta = \cos^{-1}\left(\frac{1}{\sqrt{3}}\right) \approx 55°.$

51. Consider the H−C−H combination consisting of the sole carbon atom and the two hydrogen atoms that are at (1, 0, 0) and (0, 1, 0) (or any H−C−H combination for that matter). Vector representations of the line segments emanating from the carbon atom and extending to these two hydrogen atoms are
$<1-\frac{1}{2}, 0-\frac{1}{2}, 0-\frac{1}{2}> = <\frac{1}{2}, -\frac{1}{2}, -\frac{1}{2}>$ and $<0-\frac{1}{2}, 1-\frac{1}{2}, 0-\frac{1}{2}>$
$= <-\frac{1}{2}, \frac{1}{2}, -\frac{1}{2}>$. The bond angle, θ, is therefore given by

$\cos\theta = \frac{<\frac{1}{2}, -\frac{1}{2}, -\frac{1}{2}>\cdot<-\frac{1}{2}, \frac{1}{2}, -\frac{1}{2}>}{|<\frac{1}{2}, -\frac{1}{2}, -\frac{1}{2}>||<-\frac{1}{2}, \frac{1}{2}, -\frac{1}{2}>|}$

$= \frac{-\frac{1}{4}-\frac{1}{4}+\frac{1}{4}}{\sqrt{\frac{3}{4}}\sqrt{\frac{3}{4}}} = -\frac{1}{3} \Rightarrow \theta = \cos^{-1}\left(-\frac{1}{3}\right) \approx 109.5°.$

53. Let $\vec{a} = <a_1, a_2, a_3>$ and $\vec{b} = <b_1, b_2, b_3>$.
   (i) Property 2: $\vec{a} \cdot \vec{b} = <a_1, a_2, a_3>\cdot<b_1, b_2, b_3> = a_1b_1 + a_2b_2 + a_3b_3$
   $= b_1a_1 + b_2a_2 + b_3a_3 = <b_1, b_2, b_3>\cdot<a_1, a_2, a_3> = \vec{b}\cdot\vec{a}.$
   (ii) Property 5: $\vec{0}\cdot\vec{a} = <0, 0, 0>\cdot<a_1, a_2, a_3> = (0)(a_1) + (0)(a_2) + (0)(a_3) = 0.$

55. $|\vec{a} \cdot \vec{b}| = ||\vec{a}||\vec{b}|\cos\theta| = |\vec{a}||\vec{b}||\cos\theta|$. Since $|\cos\theta| \leq 1$,
$|\vec{a} \cdot \vec{b}| = |\vec{a}||\vec{b}||\cos\theta| \leq |\vec{a}||\vec{b}|$. Note: we have equality in the case of $\cos\theta = \pm 1$, so $\theta = 0$ or $\theta = \pi$, thus equality when $\vec{a}$ and $\vec{b}$ are parallel.

Section 11.4

57. (a) 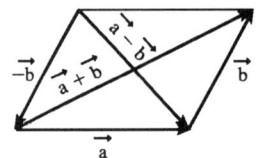 The Parallelogram Law states that the sum of the squares of the lengths of the diagonals of a parallelogram equals the sum of the squares of its (four) sides.

(b) $|\vec{a}+\vec{b}|^2 = (\vec{a}+\vec{b})\cdot(\vec{a}+\vec{b}) = |\vec{a}|^2 + 2(\vec{a}\cdot\vec{b}) + |\vec{b}|^2$ and
$|\vec{a}-\vec{b}|^2 = (\vec{a}-\vec{b})\cdot(\vec{a}-\vec{b}) = |\vec{a}|^2 - 2(\vec{a}\cdot\vec{b}) + |\vec{b}|^2$. Adding these two equations gives: $|\vec{a}+\vec{b}|^2 + |\vec{a}-\vec{b}|^2 = 2|\vec{a}|^2 + 2|\vec{b}|^2$.

EXERCISES 11.4

1. $\vec{a}\times\vec{b} = \begin{vmatrix} \vec{i} & \vec{j} & \vec{k} \\ 1 & 0 & 1 \\ 0 & 1 & 0 \end{vmatrix} = \begin{vmatrix} 0 & 1 \\ 1 & 0 \end{vmatrix}\vec{i} - \begin{vmatrix} 1 & 1 \\ 0 & 0 \end{vmatrix}\vec{j} + \begin{vmatrix} 1 & 0 \\ 0 & 1 \end{vmatrix}\vec{k} = -\vec{i} + \vec{k}.$

3. $\vec{a}\times\vec{b} = \begin{vmatrix} \vec{i} & \vec{j} & \vec{k} \\ -2 & 3 & 4 \\ 3 & 0 & 1 \end{vmatrix} = \begin{vmatrix} 3 & 4 \\ 0 & 1 \end{vmatrix}\vec{i} - \begin{vmatrix} -2 & 4 \\ 3 & 1 \end{vmatrix}\vec{j} + \begin{vmatrix} -2 & 3 \\ 3 & 0 \end{vmatrix}\vec{k} = 3\vec{i} + 14\vec{j} - 9\vec{k}.$

5. $\vec{a}\times\vec{b} = \begin{vmatrix} \vec{i} & \vec{j} & \vec{k} \\ 1 & 1 & 1 \\ 1 & 1 & -1 \end{vmatrix} = \begin{vmatrix} 1 & 1 \\ 1 & -1 \end{vmatrix}\vec{i} - \begin{vmatrix} 1 & 1 \\ 1 & -1 \end{vmatrix}\vec{j} + \begin{vmatrix} 1 & 1 \\ 1 & 1 \end{vmatrix}\vec{k} = -2\vec{i} + 2\vec{j}.$

7. $\vec{a}\times\vec{b} = \begin{vmatrix} \vec{i} & \vec{j} & \vec{k} \\ 2 & 0 & -1 \\ 1 & 2 & 0 \end{vmatrix} = \begin{vmatrix} 0 & -1 \\ 2 & 0 \end{vmatrix}\vec{i} - \begin{vmatrix} 2 & -1 \\ 1 & 0 \end{vmatrix}\vec{j} + \begin{vmatrix} 2 & 0 \\ 1 & 2 \end{vmatrix}\vec{k} = 2\vec{i} - \vec{j} + 4\vec{k}.$

9. $\vec{a}\times\vec{b} = \begin{vmatrix} \vec{i} & \vec{j} & \vec{k} \\ 0 & 1 & 2 \\ 3 & 1 & 0 \end{vmatrix} = \begin{vmatrix} 1 & 2 \\ 1 & 0 \end{vmatrix}\vec{i} - \begin{vmatrix} 0 & 2 \\ 3 & 0 \end{vmatrix}\vec{j} + \begin{vmatrix} 0 & 1 \\ 3 & 1 \end{vmatrix}\vec{k} = -2\vec{i} + 6\vec{j} - 3\vec{k}.$

Section 11.4

$$\vec{b} \times \vec{a} = \begin{vmatrix} \vec{i} & \vec{j} & \vec{k} \\ 3 & 1 & 0 \\ 0 & 1 & 2 \end{vmatrix} = \begin{vmatrix} 1 & 0 \\ 1 & 2 \end{vmatrix} \vec{i} - \begin{vmatrix} 3 & 0 \\ 0 & 2 \end{vmatrix} \vec{j} + \begin{vmatrix} 3 & 1 \\ 0 & 1 \end{vmatrix} \vec{k} = 2\vec{i} - 6\vec{j} + 3\vec{k}$$

(and notice $\vec{a} \times \vec{b} = -\vec{b} \times \vec{a}$ here, as we know is always true by Thm. 11.31).

11. $\begin{vmatrix} \vec{i} & \vec{j} & \vec{k} \\ 1 & -1 & 1 \\ 0 & 4 & 4 \end{vmatrix} = \begin{vmatrix} -1 & 1 \\ 4 & 4 \end{vmatrix} \vec{i} - \begin{vmatrix} 1 & 1 \\ 0 & 4 \end{vmatrix} \vec{j} + \begin{vmatrix} 1 & -1 \\ 0 & 4 \end{vmatrix} \vec{k} = -8\vec{i} - 4\vec{j} + 4\vec{k}$

and by Thm. 11.28 this cross product is orthogonal to both the original vectors. Thus two unit vectors orthogonal to both are

$\pm \dfrac{<-8, -4, 4>}{\sqrt{64 + 16 + 16}} = \pm \dfrac{<-8, -4, 4>}{4\sqrt{6}}$ or $<-\dfrac{2}{\sqrt{6}}, -\dfrac{1}{\sqrt{6}}, \dfrac{1}{\sqrt{6}}>$ and $<\dfrac{2}{\sqrt{6}}, \dfrac{1}{\sqrt{6}}, \dfrac{-1}{\sqrt{6}}>$.

13. Let $\vec{a} = <a_1, a_2, a_3>$, then
$\vec{0} \times \vec{a} =$
$\begin{vmatrix} \vec{i} & \vec{j} & \vec{k} \\ 0 & 0 & 0 \\ a_1 & a_2 & a_3 \end{vmatrix} = \begin{vmatrix} 0 & 0 \\ a_2 & a_3 \end{vmatrix} \vec{i} - \begin{vmatrix} 0 & 0 \\ a_1 & a_3 \end{vmatrix} \vec{j} + \begin{vmatrix} 0 & 0 \\ a_1 & a_2 \end{vmatrix} \vec{k} = \vec{0}$

$\vec{a} \times \vec{0} =$
$\begin{vmatrix} \vec{i} & \vec{j} & \vec{k} \\ a_1 & a_2 & a_3 \\ 0 & 0 & 0 \end{vmatrix} = \begin{vmatrix} a_2 & a_3 \\ 0 & 0 \end{vmatrix} \vec{i} - \begin{vmatrix} a_1 & a_3 \\ 0 & 0 \end{vmatrix} \vec{j} + \begin{vmatrix} a_1 & a_2 \\ 0 & 0 \end{vmatrix} \vec{k} = \vec{0}.$

15. $\vec{a} \times \vec{b} = <a_2 b_3 - a_3 b_2, a_3 b_1 - a_1 b_3, a_1 b_2 - a_2 b_1>$
$= <(-1)(b_2 a_3 - b_3 a_2), (-1)(b_3 a_1 - b_1 a_3), (-1)(b_1 a_2 - b_2 a_1)>$
$= -<b_2 a_3 - b_3 a_2, b_3 a_1 - b_1 a_3, b_1 a_2 - b_2 a_1> = -\vec{b} \times \vec{a}.$

17. $\vec{a} \times (\vec{b} + \vec{c}) = \vec{a} \times <b_1 + c_1, b_2 + c_2, b_3 + c_3>$
$= <a_2(b_3 + c_3) - a_3(b_2 + c_2), a_3(b_1 + c_1) - a_1(b_3 + c_3), a_1(b_2 + c_2) - a_2(b_1 + c_1)>$
$= <a_2 b_3 + a_2 c_3 - a_3 b_2 - a_3 c_2, a_3 b_1 + a_3 c_1 - a_1 b_3 - a_1 c_3,$
$\quad a_1 b_2 + a_1 c_2 - a_2 b_1 - a_2 c_1>$
$= <(a_2 b_3 - a_3 b_2) + (a_2 c_3 - a_3 c_2), (a_3 b_1 - a_1 b_3) + (a_3 c_1 - a_1 c_3),$
$\quad (a_1 b_2 - a_2 b_1) + (a_1 c_2 - a_2 c_1)>$
$= <a_2 b_3 - a_3 b_2, a_3 b_1 - a_1 b_3, a_1 b_2 - a_2 b_1>$
$\quad + <a_2 c_3 - a_3 c_2, a_3 c_1 - a_1 c_3, a_1 c_2 - a_2 c_1> = (\vec{a} \times \vec{b}) + (\vec{a} \times \vec{c}).$

Section 11.4

19. The vectors corresponding to $\vec{AB}$ and $\vec{AD}$ are $\vec{a} = <3, -1, 0>$ and $\vec{b} = <2, -2, 0>$. The area of the parallelogram with the given vertices is

$$|\vec{a} \times \vec{b}| = \begin{Vmatrix} \vec{i} & \vec{j} & \vec{k} \\ 3 & -1 & 0 \\ 2 & -2 & 0 \end{Vmatrix} = \left|(0)\vec{i} - (0)\vec{j} + (-6+2)\vec{k}\right| = \left|-4\vec{k}\right| = 4$$

21. (a) A vector orthogonal to the plane through the points P, Q and R is a vector orthogonal to both $\vec{PQ}$ and $\vec{PR}$ and thus by Thm. 11.28 is $\vec{PQ} \times \vec{PR}$. Here $\vec{PQ} = <-1, 2, 0>$ and $\vec{PR} = <-1, 0, 3>$
so $\vec{PQ} \times \vec{PR} = <(2)(3) - (0)(0), (0)(-1) - (-1)(3), (-1)(0) - (2)(-1)>$
$= <6, 3, 2>$ (or any nonzero scalar multiple of $<6, 3, 2>$) is the desired vector.

(b) From (a) $|\vec{PQ} \times \vec{PR}| = |<6, 3, 2>| = \sqrt{36 + 9 + 4} = 7$, so the area of the triangle equals $\frac{7}{2}$.

23. (a) $\vec{PQ} = <1, -1, 1>$ and $\vec{PR} = <4, 3, 7>$ and the desired vector is
$\vec{PQ} \times \vec{PR} = <(-1)(7) - (1)(3), (1)(4) - (1)(7), (1)(3) - (-1)(4)>$
$= <-10, -3, 7>$.

(b) $|\vec{PQ} \times \vec{PR}| = \sqrt{100 + 9 + 49} = \sqrt{158}$ and the area of the triangle equals $\frac{1}{2}\sqrt{158}$.

25. $\vec{a} \cdot (\vec{b} \times \vec{c}) = \begin{vmatrix} 1 & 0 & 6 \\ 2 & 3 & -8 \\ 8 & -5 & 6 \end{vmatrix} = 1\begin{vmatrix} 3 & -8 \\ -5 & 6 \end{vmatrix} - 0 + 6\begin{vmatrix} 2 & 3 \\ 8 & -5 \end{vmatrix}$

$= (18 - 40) + 6(-10 - 24) = -226$. Thus the volume is $|-226| = 226$ cubic units.

27. $\vec{a} = \vec{PQ} = <1, -1, 2>, \vec{b} = \vec{PR} = <3, 0, 6>$ and $\vec{c} = \vec{PS} = <2, -2, -3>$.

$\vec{a} \cdot (\vec{b} \times \vec{c}) = \begin{vmatrix} 1 & -1 & 2 \\ 3 & 0 & 6 \\ 2 & -2 & -3 \end{vmatrix} = 1\begin{vmatrix} 0 & 6 \\ -2 & -3 \end{vmatrix} - (-1)\begin{vmatrix} 3 & 6 \\ 2 & -3 \end{vmatrix} + 2\begin{vmatrix} 3 & 0 \\ 2 & -2 \end{vmatrix}$

$= 12 - 21 - 12 = -21$ and the volume is 21 cubic units.

29. $\vec{a} \cdot (\vec{b} \times \vec{c}) = \begin{vmatrix} 2 & 3 & 1 \\ 1 & -1 & 0 \\ 7 & 3 & 2 \end{vmatrix} = 2\begin{vmatrix} -1 & 0 \\ 3 & 2 \end{vmatrix} - 3\begin{vmatrix} 1 & 0 \\ 7 & 2 \end{vmatrix} + 1\begin{vmatrix} 1 & -1 \\ 7 & 3 \end{vmatrix}$

$= -4 - 6 + 10 = 0$ which says that the volume of the parallelepiped determined by $\vec{a}, \vec{b}$ and $\vec{c}$ is 0 and thus these three vectors are coplanar.

31. The magnitude of the torque is $|\vec{\tau}| = |\vec{r} \times \vec{F}| = |\vec{r}||\vec{F}|\sin\theta$
$= (0.18 \text{ m})(60 \text{ N})\sin(180 - (70 + 10))° = 10.8\sin 100° \approx 10.6$ J.

Section 11.5

33. (a) The distance between a point and a line is the length of the perpendicular from the point to the line, here $|\vec{PS}| = d$. But referring to triangle PQS, $d = |\vec{PS}| = |\vec{QP}|\sin\theta = |\vec{b}|\sin\theta$. But θ is the angle between $\vec{QP} = \vec{b}$ and $\vec{QR} = \vec{a}$. Thus by Thm. 11.29, $\sin\theta = \dfrac{|\vec{a}\times\vec{b}|}{|\vec{a}||\vec{b}|}$ and so $d = |\vec{b}|\sin\theta = \dfrac{|\vec{b}||\vec{a}\times\vec{b}|}{|\vec{a}||\vec{b}|} = \dfrac{|\vec{a}\times\vec{b}|}{|\vec{a}|}$.

(b) $\vec{a} = \vec{QR} = <-1,-2,-1>$ and $\vec{b} = \vec{QP} = <1,-5,-7>$. Then
$\vec{a}\times\vec{b} = <(-2)(-7)-(-1)(-5), (-1)(1)-(-1)(-7), (-1)(-5)-(-2)(1)>$
$= <9,-8,7>$. Thus the distance is $d = \dfrac{|\vec{a}\times\vec{b}|}{|\vec{a}|} = \dfrac{1}{\sqrt{6}}\sqrt{81+64+49}$
$= \sqrt{\dfrac{194}{6}} = \sqrt{\dfrac{97}{3}}$.

35. $(\vec{a}-\vec{b})\times(\vec{a}+\vec{b}) = (\vec{a}-\vec{b})\times\vec{a} + (\vec{a}-\vec{b})\times\vec{b}$ by Thm. 11.31.3
$= \vec{a}\times\vec{a} + (-\vec{b})\times\vec{a} + \vec{a}\times\vec{b} + (-\vec{b})\times\vec{b}$ by Thm. 11.31.4
$= (\vec{a}\times\vec{a}) - (\vec{b}\times\vec{a}) + (\vec{a}\times\vec{b}) - (\vec{b}\times\vec{b})$ by Thm. 11.31.2
$= \vec{0} - (\vec{b}\times\vec{a}) + (\vec{a}\times\vec{b}) - \vec{0}$ by example 2
$= (\vec{a}\times\vec{b}) + (\vec{a}\times\vec{b})$ by Thm. 11.31.1
$= 2(\vec{a}\times\vec{b})$.

37. $\vec{a}\times(\vec{b}\times\vec{c}) + \vec{b}\times(\vec{c}\times\vec{a}) + \vec{c}\times(\vec{a}\times\vec{b})$
$= [(\vec{a}\cdot\vec{c})\vec{b} - (\vec{a}\cdot\vec{b})\vec{c}] + [(\vec{b}\cdot\vec{a})\vec{c} - (\vec{b}\cdot\vec{c})\vec{a}] + [(\vec{c}\cdot\vec{b})\vec{a} - (\vec{c}\cdot\vec{a})\vec{b}]$ (by Exercise 36)
$= (\vec{a}\cdot\vec{c})\vec{b} - (\vec{a}\cdot\vec{b})\vec{c} + (\vec{a}\cdot\vec{b})\vec{c} - (\vec{b}\cdot\vec{c})\vec{a} + (\vec{b}\cdot\vec{c})\vec{a} - (\vec{a}\cdot\vec{c})\vec{b} = \vec{0}$.

39. (a) No. If $\vec{a}\cdot\vec{b} = \vec{a}\cdot\vec{c}$, then $\vec{a}\cdot(\vec{b}-\vec{c}) = 0$, so $\vec{a}$ is perpendicular to $\vec{b}-\vec{c}$, which can happen if $\vec{b} \neq \vec{c}$. For example, let $\vec{a} = <1,1,1>$, $\vec{b} = <1,0,0>$ and $\vec{c} = <0,1,0>$.

(b) No. If $\vec{a}\times\vec{b} = \vec{a}\times\vec{c}$ then $\vec{a}\times(\vec{b}-\vec{c}) = \vec{0}$, which implies that $\vec{a}$ is parallel to $\vec{b}-\vec{c}$ which of course can happen if $\vec{b} \neq \vec{c}$.

(c) Yes. Since $\vec{a}\cdot\vec{c} = \vec{a}\cdot\vec{b}$, $\vec{a}$ is perpendicular to $\vec{b}-\vec{c}$, by part (a). From part (b), $\vec{a}$ is also parallel to $\vec{b}-\vec{c}$. Thus since $\vec{a} \neq \vec{0}$ but is both parallel and perpendicular to $\vec{b}-\vec{c}$, we have $\vec{b}-\vec{c} = \vec{0}$ or $\vec{b} = \vec{c}$.

EXERCISES 11.5

1. $\vec{r}_0 = 3\vec{i} - \vec{j} + 8\vec{k}$ and $\vec{v} = \vec{a}$ so the vector equation is
$\vec{r} = (3\vec{i} - \vec{j} + 8\vec{k}) + t(2\vec{i} + 3\vec{j} + 5\vec{k}) = (3+2t)\vec{i} + (-1+3t)\vec{j} + (8+5t)\vec{k}$ and the parametric equations are: $x = 3+2t$, $y = -1+3t$, $z = 8+5t$.

Section 11.5

3. $\vec{r} = (\vec{j} + 2\vec{k}) + t(6\vec{i} + 3\vec{j} + 2\vec{k}) = (6t)\vec{i} + (1+3t)\vec{j} + (2+2t)\vec{k}$ is the vector equation while $x = 6t, y = 1 + 3t, z = 2 + 2t$ are the parametric equations.

5. The parallel vector is $\vec{v} = <6-2, 0-1, 3-8> = <4, -1, -5>$ so the direction numbers are $a = 4, b = -1, c = -5$. Letting $P_0 \equiv (2, 1, 8)$, the parametric equations are $x = 2 + 4t, y = 1 - t, z = 8 - 5t$ and symmetric equations are $\frac{x-2}{4} = \frac{y-1}{-1} = \frac{z-8}{-5}$.

7. $\vec{v} = <0, 1, -5>$ and letting $P_0 = (3, 1, -1)$, the parametric equations are $x = 3, y = 1 + t, z = -1 - 5t$ while the symmetric equations are $x = 3, y - 1 = \frac{z+1}{-5}$. Notice here that the direction number $a = 0$, so rather than writing $\frac{x-3}{0}$ in the symmetric equation we write the equation $x = 3$ separately.

9. $\vec{v} = \langle \frac{1}{3}, 4, -9 \rangle$ and letting $P_0 = (-\frac{1}{3}, 1, 1)$, the parametric equations are $x = -\frac{1}{3} + \frac{1}{3}t, y = 1 + 4t, z = 1 - 9t$ while the symmetric equations are $\frac{x + 1/3}{1/3} = \frac{y-1}{4} = \frac{z-1}{-9}$.

11. Direction vectors of the lines are respectively $\vec{v}_1 = <6, 9, 12>$ and $\vec{v}_2 = <4, 6, 8>$ and since $\vec{v}_1 = \frac{3}{2}\vec{v}_2$ the direction vectors and thus the lines are parallel.

13. (a) A direction vector of the line with given parametric equations is $\vec{v} = <2, 3, -7>$ and the desired parallel line must also have $\vec{v}$ as a direction vector. Here $P_0 = (0, 2, -1)$ so the symmetric equations for the line are $\frac{x}{2} = \frac{y-2}{3} = \frac{z+1}{-7}$.

    (b) The line intersects the xy-plane when $z = 0$, so we need $\frac{x}{2} = \frac{y-2}{3} = \frac{1}{-7}$ or $x = -\frac{2}{7}, y = \frac{11}{7}$. Thus the point of intersection with the xy-plane is $\left(-\frac{2}{7}, \frac{11}{7}, 0\right)$.
    Similarly for the yz- and xz-planes, we need respectively $x = 0$ and $y = 0$ or
    $0 = \frac{y-2}{3} = \frac{z+1}{-7}$ and $\frac{x}{2} = -\frac{2}{3} = \frac{z+1}{-7}$ or $y = 2, z = -1$ and $x = -\frac{4}{3}, z = \frac{11}{3}$.
    Thus the line intersects the yz-plane at $(0, 2, -1)$ and the xz-plane at $\left(-\frac{4}{3}, 0, \frac{11}{3}\right)$.

15. The lines aren't parallel since the direction vectors $<2, 4, -3>$ and $<1, 3, 2>$ aren't parallel so we check to see if the lines intersect. The parametric equations of the lines are $L_1: x = 4 + 2t, y = -5 + 4t, z = 1 - 3t$ and $L_2: x = 2 + s, y = -1 + 3s, z = 2s$. For the lines to intersect we must be able to find one value of t and one value of s satisfying the following three equations: $4 + 2t = 2 + s, -5 + 4t = -1 + 3s, 1 - 3t = 2s$. Solving the first two equations we get $t = -5, s = -8$ and checking we see that these values don't satisfy the third equation. Thus L_1 and L_2 aren't parallel and don't intersect, so they must be skew lines.

Section 11.5

17. Since the direction vectors are $\vec{v}_1 = <-6, 9, -3>$ and $\vec{v}_2 = <2, -3, 1>$, we have $\vec{v}_1 = -3\vec{v}_2$ so the lines are parallel.

19. Setting $a = 7$, $b = 1$, $c = 4$, $x_0 = 1$, $y_0 = 4$, $z_0 = 5$ in equation 11.40 gives $7(x-1) + 1(y-4) + 4(z-5) = 0$ or $7x + y + 4z = 31$ is the equation of the plane (using 11.40).

21. The equation of the plane is $15x + 9y - 12z = 15 + 18 - 36$ or $15x + 9y - 12z = -3$ or $5x + 3y - 4z = -1$ (using 11.41).

23. Since the two planes are parallel, they will have the same normal vectors. Thus $\vec{n} = <1, 1, -1>$ and the equation of the plane is $x + y - z = 6 + 5 + 2$ or $x + y - z = 13$.

25. The equation is $3x - 4y - 6z = -3 - 12 + 48$ or $3x - 4y - 6z = 33$.

27. Here the vectors $\vec{a} = <1, 1, 1>$ and $\vec{b} = <1, 2, 3>$ lie in the plane so $\vec{a} \times \vec{b}$ is a normal vector to the plane. Thus $\vec{n} = \vec{a} \times \vec{b} = <3-2, 1-3, 2-1>$ $= <1, -2, 1>$ and the equation of the plane is $x - 2y + z = 0$.

29. $\vec{a} = <-1, -2, -1>$ and $\vec{b} = <3, 1, 9>$ so a normal vector to the plane is $\vec{n} = \vec{a} \times \vec{b} = <-18+1, -3+9, -1+6> = <-17, 6, 5>$ and the equation of the plane is $-17x + 6y + 5z = -17 + 0 - 15$ or $-17x + 6y + 5z = -32$.

31. To find the equation of the plane we must first find two nonparallel vectors in the plane, then their cross product will be a normal vector to the plane. But since the given line lies in the plane, its direction vector $\vec{a} = <2, -3, -1>$ is one vector in the plane. To find another nonparallel vector $\vec{b}$ which lies in the plane pick ANY point on the line, say $(1, 2, 3)$ found by setting $t = 0$, and let $\vec{b}$ be the vector connecting this point to the given point in the plane. (BUT BEWARE we should first check that the given point is NOT on the given line. If it were on the line the plane wouldn't be uniquely determined. What would $\vec{n}$ then be when we set $\vec{n} = \vec{a} \times \vec{b}$?) Here $\vec{b} = <0, 4, -7>$ so $\vec{n} = \vec{a} \times \vec{b} = <21+4, 0+14, 8-0> = <25, 14, 8>$ and the equation of the plane is $25x + 14y + 8z = 25 + 84 - 32$ or $25x + 14y + 8z = 77$.

33. $(0, 0, 0)$ is a point on $x = y = z$. $<1, 1, 1>$ is the direction of this line, and thus also of the plane. $<0-0, 1-0, 2-0> = <0, 1, 2>$ is also a vector in the plane. Therefore, $\vec{n} = <1, 1, 1> \times <0, 1, 2> = <2-1, -2+0, 1-0> = <1, -2, 1>$. Choosing $(x_0, y_0, z_0) = (0, 0, 0)$, the equation of the plane is, by 11.41, $x - 2y + z = 1 \cdot 0 - 2 \cdot 0 + 1 \cdot 0 \Leftrightarrow x - 2y + z = 0$.

35. Substituting the parametric equations of the line into the equation of the plane gives $x + y + z = 1 + t + 2t + 3t = 1 \Rightarrow t = 0$. This value of t corresponds to the point of

Section 11.5

intersection $(1, 0, 0)$, obtained by substitution of $t = 0$ into the equations of the line.

37. Here, substitution gives $2x + y - z + 5 = 2(1 + 2t) + (-1) - t + 5 = 0$
$\Rightarrow 3t + 6 = 0 \Rightarrow t = -2$. Therefore, the point of intersection is $x = 1 + 2(-2) = -3$, $y = -1$ and $z = -2$ and the point of intersection is $(-3, -1, -2)$.

39. Setting $x = 0$, we see that $(0, 1, 0)$ satisfies the equations of both planes, so that they do in fact have a line of intersection. $\vec{v} = \vec{n}_1 \times \vec{n}_2 = <1, 1, 1> \times <1, 0, 1> = <1, 0, -1>$ is the direction of this line. Therefore, direction numbers of the intersecting line are $1, 0, -1$.

41. The normal vectors to the planes are respectively $\vec{n}_1 = <1, 0, 1>$ and $\vec{n}_2 = <0, 1, 1>$. Thus the normal vectors and subsequently the planes aren't parallel. Furthermore $\vec{n}_1 \cdot \vec{n}_2 = 1 \neq 0$ so the planes aren't perpendicular. Letting θ be the angle between the two planes, we have $\cos \theta = \frac{\vec{n}_1 \cdot \vec{n}_2}{|\vec{n}_1||\vec{n}_2|} = \frac{1}{\sqrt{2}\sqrt{2}} = \frac{1}{2}$ and $\theta = \cos^{-1}\left(\frac{1}{2}\right) = 60°$.

43. The respective normals are $\vec{n}_1 = <1, 4, -3>$ and $\vec{n}_2 = <-3, 6, 7>$ so the normals and thus the planes aren't parallel. But $\vec{n}_1 \cdot \vec{n}_2 = -3 + 24 - 21 = 0$ so the normals and thus the planes are perpendicular.

45. The normals are $\vec{n}_1 = <2, 4, -2>$ and $\vec{n}_2 = <-3, -6, 3>$ respectively. So $\vec{n}_1 = -\frac{3}{2}\vec{n}_2$. The normals, and hence the planes, are parallel. The planes are not the same because $-\frac{3}{2}(1) \neq 10$.

47. (a) To find a point on the line of intersection, set one of the variables equal to a constant, say $z = 0$. (This will only work if the line of intersection crosses the xy-plane, otherwise try setting x or y equal to 0.) Then the equations of the planes reduce to $x + y = 2$ and $3x - 4y = 6$. Solving these two equations gives $x = 2$, $y = 0$. So a point on the line of intersection is $(2, 0, 0)$. The direction of the line is $\vec{v} = \vec{n}_1 \times \vec{n}_2$ $= <5 - 4, -3 - 5, -4 - 3> = <1, -8, -7>$ and symmetric equations for the line are $x - 2 = \frac{y}{-8} = \frac{z}{-7}$.

(b) The angle between the planes satisfies $\cos \theta = \frac{|\vec{n}_1 \cdot \vec{n}_2|}{|\vec{n}_1||\vec{n}_2|} = \left|\frac{3 - 4 - 5}{\sqrt{3}\sqrt{50}}\right| = \left|-\frac{\sqrt{6}}{5}\right| = \frac{\sqrt{6}}{5}$

Therefore, $\theta = \cos^{-1}\left(\frac{\sqrt{6}}{5}\right) \approx 61°$.

49. Setting $x = 0$, the equations of the two planes become $z = y$ and $5y + z = -1$, which intersect at $y = -\frac{1}{6}$ and $z = -\frac{1}{6}$. Thus we can choose $(x_0, y_0, z_0) = \left(0, -\frac{1}{6}, -\frac{1}{6}\right)$. The direction of the line of intersection is $\vec{v} = \vec{n}_1 \times \vec{n}_2 = <2, -5, -1> \times <1, 1, -1> = <6, 1, 7>$. Parametric equations for this line are, by 11.36, $x = 6t$, $y = -\frac{1}{6} + t$, $z = -\frac{1}{6} + 7t$.

Section 11.5

51. The plane will contain all perpendicular bisectors of the line segment joining (1, 1, 0) and (0, 1, 1). All of these bisectors pass through the midpoint of this segment $\left(\frac{1}{2}, \frac{1+1}{2}, \frac{1}{2}\right) = \left(\frac{1}{2}, \frac{1}{2}, \frac{1}{2}\right)$. The direction of this line segment $<1-0, 1-1, 0-1>$ $= <1, 0, -1>$ is perpendicular to the plane so that we can choose this to be $\vec{n}$. Therefore the equation of the plane is $x + 0y - z = 1 \cdot \frac{1}{2} + 0 \cdot 1 - 1 \cdot \frac{1}{2} \Leftrightarrow x = z$.

53. A direction vector for the line of intersection is $\vec{a} = \vec{n}_1 \times \vec{n}_2 = <1, 1, -1> \times <2, -1, 3>$ $= <2, -5, -3>$ and $\vec{a}$ is parallel to the desired plane. Another vector parallel to the plane is the vector connecting any point on the line of intersection to the given point $(-1, 2, 1)$ in the plane. Setting $x = 0$, the equation of the planes reduce to $y - z = 2$ and $-y + 3z = 1$ with simultaneous solution $y = \frac{7}{2}$ and $z = \frac{3}{2}$. So a point on the line is $\left(0, \frac{7}{2}, \frac{3}{2}\right)$ and another vector parallel to the plane is $\left\langle -1, -\frac{3}{2}, -\frac{1}{2}\right\rangle$. Then a normal vector to the plane is
$\vec{n} = <2, -5, -3> \times \left\langle -1, -\frac{3}{2}, -\frac{1}{2}\right\rangle = <-2, 4, -8>$ and an equation of the plane is
$-2x + 4y - 8z = 2 + 8 - 8$ or $x - 2y + 4z = 1$.

55. The plane contains the points (a, 0, 0), (0, b, 0) and (0, 0, c). Thus the vectors
$\vec{a} = <-a, b, 0>$ and $\vec{b} = <-a, 0, c>$ lie in the plane and
$\vec{n} = \vec{a} \times \vec{b} = <bc - 0, 0 + ac, 0 + ab> = <bc, ac, ab>$ is a normal vector to the plane.
The equation of the plane is therefore $bcx + acy + abz = abc + 0 + 0$ or
$bcx + acy + abz = abc$. Notice that if $a \neq 0$, $b \neq 0$ and $c \neq 0$ then we can rewrite the equation as $\frac{x}{a} + \frac{y}{b} + \frac{z}{c} = 1$. This is a good equation to remember!

57. $D = \frac{1}{\sqrt{1+4+4}}|(1)(2) + (-2)(8) + (-2)(5) - 1| = \frac{25}{3}$.

59. $D = \frac{1}{\sqrt{4+9+36}}|(2)(-1) + (3)(4) + (6)(6) - 4| = \frac{42}{7} = 6$.

61. Substituting $t = 0$ and $t = 1$ into the parametric equations of the line gives the points $Q = (2, 2, 0)$ and $R = (3, -1, 5)$ respectively. Let $P = (1, 2, 3)$. In the notation of Exercise 33 in Section 11.4, $\vec{a} = \vec{QR} = <1, -3, 5>$ and $\vec{b} = \vec{QP} = <-1, 0, 3>$.
$\vec{a} \times \vec{b} = <-9, -5-3, -3> = <-9, -8, -3>$.

Therefore, $d = \frac{|\vec{a} \times \vec{b}|}{|\vec{a}|} = \frac{|<-9, -8, -3>|}{|<1, -3, 5>|} = \frac{\sqrt{81+64+9}}{\sqrt{1+9+25}} = \sqrt{\frac{22}{5}}$.

63. $z = x + 2y + 1 \Rightarrow 3x + 6y - 3z = -3$

Using the equation of the other plane, $3x + 6y - 3z = 4$, we now have $a = 3$, $b = 6$, $c = -3$, $d_1 = -3$ and $d_2 = 4$. Therefore,

$D = \frac{|d_1 - d_2|}{\sqrt{a^2 + b^2 + c^2}} = \frac{|-3-4|}{\sqrt{9+36+9}} = \frac{7}{\sqrt{54}} = \frac{7}{3\sqrt{6}} = \frac{7\sqrt{6}}{18}$.

65. L_1: $x = y = z \Rightarrow x = y$ (1)

L_2: $x + 1 = \frac{y}{2} = \frac{z}{3} \Rightarrow x + 1 = \frac{y}{2}$ (2)

The solution of (1) and (2) is $x = y = -2$. However, when $x = -2$, $x = z \Rightarrow z = -2$, but $x + 1 = \frac{z}{3} \Rightarrow z = -3$, which values do not agree. Hence the lines do not intersect. For L_1, $\vec{v} = <1, 1, 1>$, and for L_2, $\vec{v} = <1, 2, 3>$, so that the lines are not parallel. Thus the lines are skew lines. If two lines are skew, they can be viewed as lying in two parallel planes and so the distance between the skew lines would be the same as the distance between these parallel planes. The common normal vector to the planes must be perpendicular to both $<1, 1, 1>$ and $<1, 2, 3>$, the direction vectors of the two lines. So, set $\vec{n} = <1, 1, 1> \times <1, 2, 3>$
$= <3-2, -3+1, 2-1> = <1, -2, 1>$. From above, we know that $(-2, -2, -2)$ and $(-2, -2, -3)$ are points of L_1 and L_2 respectively.

So $d_1 = 1(-2) - 2(-2) + 1(-2) = 0$ and $d_2 = 1(-2) - 2(-2) + 1(-3) = -1$.

By Exercise 62, the distance between these two skew lines is

$D = \frac{|0 - (-1)|}{\sqrt{1 + 4 + 1}} = \frac{1}{\sqrt{6}}$.

Alternate solution (without reference to planes). A vector which is perpendicular to both of the lines is $\vec{n} = <1, 1, 1> \times <1, 2, 3> = <1, -2, 1>$. Pick any point on each of the lines, say $(-2, -2, -2)$ and $(-2, -2, -3)$, and form the vector $\vec{b} = <0, 0, 1>$ connecting the two points. The distance between the two skew lines is the absolute value of the scalar projection of $\vec{b}$ along $\vec{n}$, that is

$D = \frac{|\vec{n} \cdot \vec{b}|}{|\vec{n}|} = \frac{|1 \cdot 0 - 2 \cdot 0 + 1 \cdot 1|}{\sqrt{1 + 4 + 1}} = \frac{1}{\sqrt{6}}$.

EXERCISES 11.6

1. The trace in any plane $x = k$ is given by $z^2 - y^2 = 1 - k^2$, $x = k$ whose graph is a hyperbola. The trace in any plane $y = k$ is the circle given by $x^2 + z^2 = 1 + k^2$, $y = k$. And the trace in any plane $z = k$ is the hyperbola given by $x^2 - y^2 = 1 - k^2$, $z = k$. Thus the surface is a hyperboloid of one sheet with axis the y–axis.

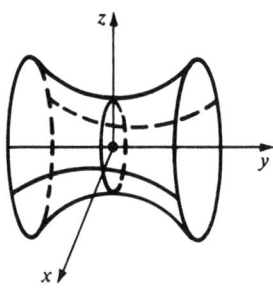

Section 11.6

3. Traces: $x = k$, $9y^2 + 36z^2 = 36 - 4k^2$, an ellipse for $|k| < 3$; $y = k$, $4x^2 + 36z^2 = 36 - 9k^2$, an ellipse for $|k| < 2$; $z = k$, $4x^2 + 9y^2 = 36(1 - k^2)$, an ellipse for $|k| < 1$. Thus the surface is an ellipsoid with center at the origin and axes along the x-, y- and z-axes.

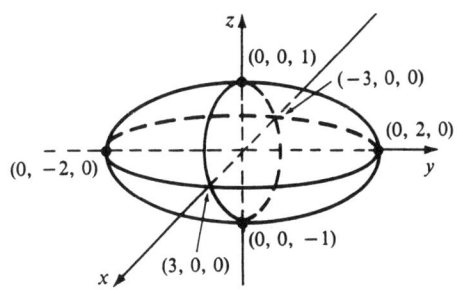

5. Traces: $x = k$, $4z^2 - y^2 = 1 + k^2$, a hyperbola; $y = k$, $4z^2 - x^2 = 1 + k^2$, a hyperbola; $z = k$, $-x^2 - y^2 = 1 - 4k^2$ or $x^2 + y^2 = 4k^2 - 1$, a circle for $k > \frac{1}{2}$ or $k < -\frac{1}{2}$. Thus the surface is a hyperboloid of two sheets with axis the z-axis.

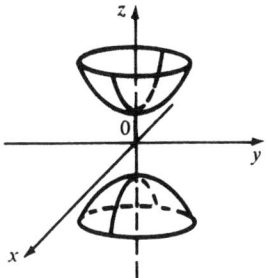

7. Traces: $x = k$, $z = y^2$, a parabola; $y = k$, $z = k^2$, a line; $z = k$, $y^2 = k$ or $y = \pm\sqrt{k}$, two parallel lines for $k > 0$. Thus the surface is a parabolic cylinder opening upward.

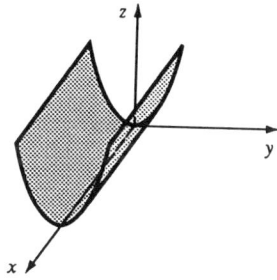

9. Traces: $x = k$, $y^2 = k^2 + z^2$ or $y^2 - z^2 = k^2$, a hyperbola for $k \neq 0$ and two intersecting lines for $k = 0$; $y = k$, $x^2 + z^2 = k^2$, a circle for $k \neq 0$; $z = k$, $y^2 = x^2 + k^2$ or $y^2 - x^2 = k^2$, a hyperbola for $k \neq 0$ and two intersecting lines for $k = 0$. Thus the surface is a cone (right circular) with axis the y-axis and vertex the origin.

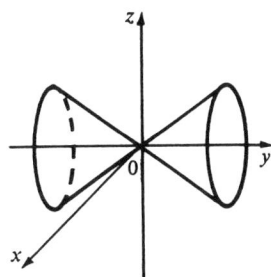

11. Traces: $x = k$, $k^2 + 4z^2 - y = 0$
or $y - k^2 = 4z^2$, a parabola; $y = k$,
$x^2 + 4z^2 = k$, an ellipse for $k > 0$;
$z = k$, $x^2 + 4k^2 - y = 0$ or
$y - 4k^2 = x^2$, a parabola. Thus the
surface is an elliptic paraboloid
with axis the y-axis and vertex the
origin.

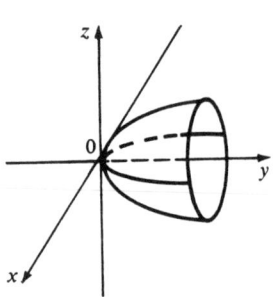

13. Traces: $x = k \Rightarrow \frac{y^2}{9} + \frac{z^2}{1} = 1$, ellipses;
$y = k$, $|k| \leq 3 \Rightarrow 9z^2 = 9 - k^2$
$\Rightarrow z = \pm\sqrt{1 - (k^2/9)}$, pairs of lines;
$z = k$, $|k| \leq 1 \Rightarrow y^2 = 9(1 - k^2) \Rightarrow$
$y = \pm 3\sqrt{1 - k^2}$, pairs of lines. This is
the equation of an elliptic cylinder, centered
at the origin, whose axis is the x-axis.

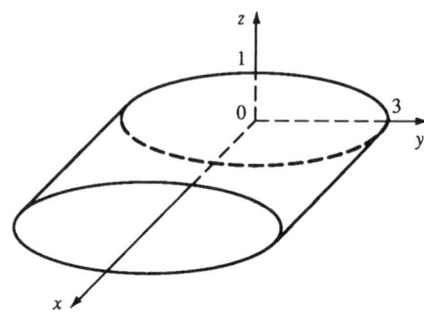

15. Traces: $x = k \Rightarrow y = z^2 - k^2$, parabolas;
$y = k \Rightarrow k = z^2 - x^2$, hyperbolas on the z-axis
for $k > 0$, and hyperbolas on the x-axis for $k < 0$;
$z = k \Rightarrow y = k^2 - x^2$, parabolas. Thus,
$\frac{y}{1} = \frac{z^2}{1^2} - \frac{x^2}{1^2}$ is a hyperbolic paraboloid.

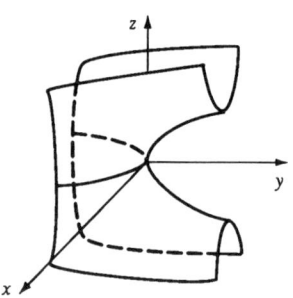

17. $z^2 = 3x^2 + 4y^2 - 12$ or $3x^2 + 4y^2 - z^2$
$= 12$ or $\frac{x^2}{4} + \frac{y^2}{3} - \frac{z^2}{12} = 1$
or $\frac{x^2}{2^2} + \frac{y^2}{(\sqrt{3})^2} - \frac{z^2}{(\sqrt{12})^2} = 1$
represents a hyperboloid of one
sheet with axis the z-axis.

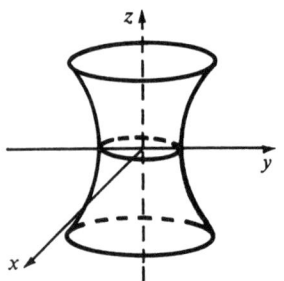

Section 11.6

19. $z = x^2 + y^2 + 1$ or $z - 1 = x^2 + y^2$,
a circular paraboloid with axis
the z-axis and vertex (0, 0, 1).

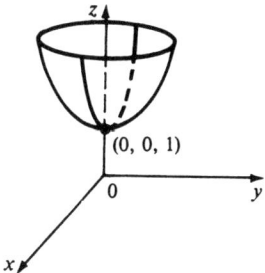

21. Completing the square in all
three variables gives
$(x + 2)^2 + (y - 3)^2 - 4(z + 1)^2 =$
$13 + 9$ or $\dfrac{(x+2)^2}{(\sqrt{22})^2} + \dfrac{(y-3)^2}{(\sqrt{22})^2} - \dfrac{(z+1)^2}{(\frac{1}{2}\sqrt{22})^2}$
$= 1$, a hyperboloid of one sheet
with center $(-2, 3, -1)$ and axis
the vertical line $y = 3$, $x = -2$.

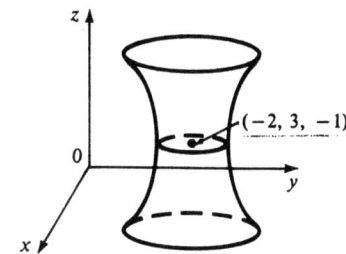

23. $x^2 + 4y^2 = 100$ or
$\dfrac{x^2}{10^2} + \dfrac{y^2}{5^2} = 1$,
an elliptic cylinder with
axis the z-axis.

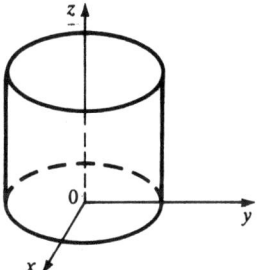

25. Completing the square in the
y-variable gives
$x^2 - (y - 2)^2 + z = 4 - 4 = 0$
or $z = (y - 2)^2 - x^2$, a
hyperbolic paraboloid with
center at (0, 2, 0).

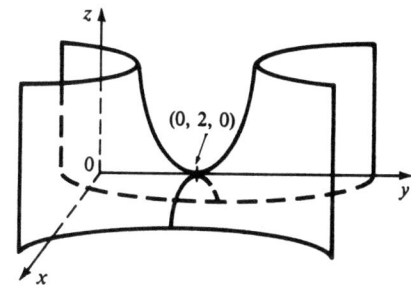

67

27. Completing the squares in the x and y variables gives
$(x-2)^2 - (y+2)^2 - z^2 = 4 - 4 = 0$
or $(x-2)^2 = (y+2)^2 + z^2$, a cone
with axis parallel to the x–axis
and vertex at $(2, -2, 0)$.

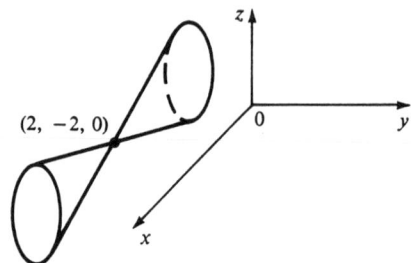

29.

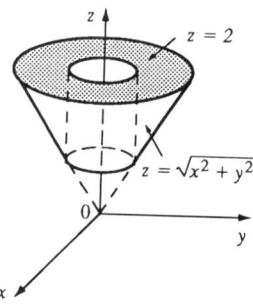

31. The surface is a paraboloid of
revolution (circular paraboloid)
with vertex at the origin, axis
the y–axis and opens to the right.
Thus the trace in the yz–plane is
also a parabola $y = z^2$, $x = 0$.
The equation is $y = x^2 + z^2$.

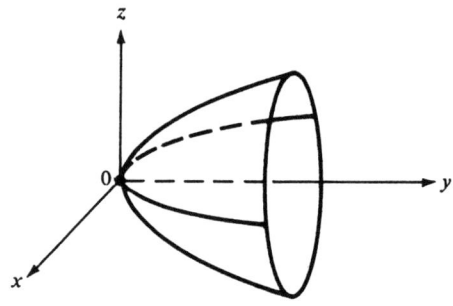

33. Let $P \equiv (x, y, z)$ be an arbitrary point equidistant from $(-1, 0, 0)$ and the plane $x = 1$.
Then the distance from P to $(-1, 0, 0)$ equals $\sqrt{(x+1)^2 + y^2 + z^2}$ and the distance from
P to the plane $x = 1$ is $\dfrac{|x-1|}{\sqrt{1^2}} = |x - 1|$ (by the formula in 11.5 Example 8). And

$|x - 1| = \sqrt{(x+1)^2 + y^2 + z^2}$ or $(x-1)^2 = (x+1)^2 + y^2 + z^2$ or
$x^2 - 2x + 1 = x^2 + 2x + 1 + y^2 + z^2$ or $-4x = y^2 + z^2$. Thus the collection of all such
points P is a circular paraboloid with vertex at the origin, axis the x–axis and which
opens back.

Section 11.7

35. If (a, b, c) satisfies $z = y^2 - x^2$, then $c = b^2 - a^2$.
    L_1: $x = a + t$, $y = b + t$, $z = c + 2(b - a)t$, L_2: $x = a + t$, $y = b - t$, $z = c - 2(b + a)t$
    Substitute the parametric equations of L_1 into the equation of the hyperbolic paraboloid in order to find the points of intersection:
    $z = y^2 - x^2 \Rightarrow c + 2(b - a)t = (b + t)^2 - (a + t)^2 = b^2 - a^2 + 2(b - a)t \Rightarrow c = b^2 - a^2$.
    As this is true for all values of t, L_1 lies on $z = y^2 - x^2$.
    Performing similar operations with L_2 gives:
    $z = y^2 - x^2 \Rightarrow c - 2(b + a)t = (b - t)^2 - (a + t)^2 = b^2 - a^2 - 2(b + a)t \Rightarrow c = b^2 - a^2$.
    This tells us that all of L_2 also lies on $z = y^2 - x^2$.

EXERCISES 11.7

1. The corresponding parametric equations are $x = t$, $y = -t$, $z = 2t$ which are the parametric equations of a line through the origin and with direction vector $<1, -1, 2>$.

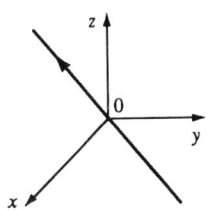

3. The parametric equations give $x^2 + z^2 = \sin^2 t + \cos^2 t = 1$, $y = 3$ which is a circle of radius 1, center $(0, 3, 0)$ in the plane $y = 3$.

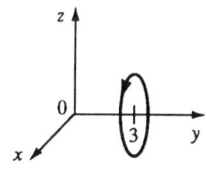

5. Eliminating the parameter t by substituting $z = t$ into $x = t^4 + 1$ gives $x = z^4 + 1$, which is a 4th degree curve in the xz–plane that opens along the positive x–axis with vertex $(1, 0, 0)$.

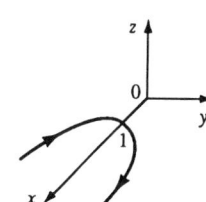

69

Section 11.7

7. Here $x = t$, $y = t^2$, $z = t^3$ are the parametric equations. For $0 \leq t < 1$, the x–coordinate increases faster than the y–coordinate which increases faster than the z–coordinate. For $t > 1$ the reverse is true. This curve is called a twisted cubic. The second component is always positive whereas the first and third have the same sign as t.

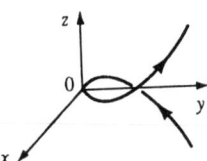

9. If $x = t\cos t$, $y = t\sin t$, and $z = t$, then $x^2 + y^2 = t^2 \cos^2 t + t^2 \sin^2 t = t^2 = z^2$, so the curve lies on the cone $z^2 = x^2 + y^2$. Thus the curve is a spiral on this cone.

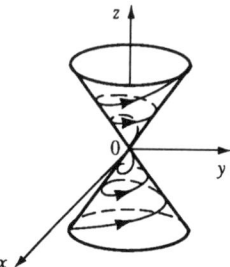

11. $\lim\limits_{t \to 0} <t, \cos t, 2> = <\lim\limits_{t \to 0} t, \lim\limits_{t \to 0} \cos t, \lim\limits_{t \to 0} 2> = <0, 1, 2>$.

13. $\lim\limits_{t \to 1} \sqrt{t+3} = 2$, $\lim\limits_{t \to 1} \frac{t-1}{t^2-1} = \lim\limits_{t \to 1} \frac{1}{t+1} = \frac{1}{2}$, $\lim\limits_{t \to 1} \left(\frac{\tan t}{t}\right) = \tan 1$.
Thus the given limit equals $\left<2, \frac{1}{2}, \tan 1\right>$.

15. The domain of $\vec{r}$ is R and $\vec{r}'(t) = <1, 2t, 3t^2>$.

17. Since $\tan t$ and $\sec t$ aren't defined for odd multiples of $\pi/2$, the domain of $\vec{r}$ is $\{t \mid t \neq (2n+1)\frac{\pi}{2}, n \in Z\}$. $\vec{r}'(t) = (\sec^2 t)\vec{j} + (\sec t \tan t)\vec{k}$.

19. We need $4 - t^2 > 0$ and $1 + t \geq 0$, so the domain of $\vec{r}$ is $\{t \mid -1 \leq t < 2\}$.
$\vec{r}'(t) = -\frac{2t}{4-t^2}\vec{i} + \frac{1}{2\sqrt{1+t}}\vec{j} - 12e^{3t}\vec{k}$.

21. The domain of $\vec{r}$ is R and $\vec{r}'(t) = 0 + \vec{b} + 2\vec{c}t = \vec{b} + 2t\vec{c}$ by Thm. 11.53.1.

Section 11.7

23. (a),(c) (b) $\vec{r}'(t) = <-\sin t, \cos t>$.

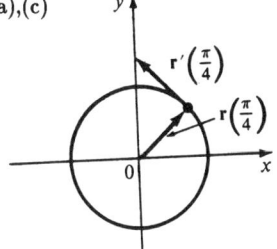

25. (a),(c) (b) $\vec{r}'(t) = \vec{i} + 2t\vec{j}$.
Since $(x-1)^2 = t^2 = y$,
the curve is a parabola.

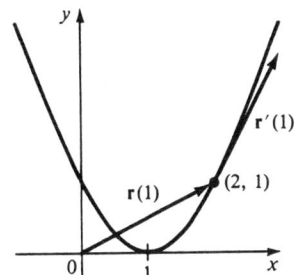

27. (a),(c) (b) $\vec{r}'(t) = e^t\vec{i} - 2e^{-2t}\vec{j}$
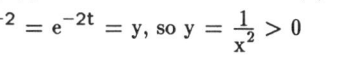
$x^{-2} = e^{-2t} = y$, so $y = \frac{1}{x^2} > 0$

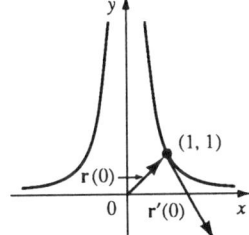

29. $\vec{r}'(t) = <2, 6t, 12t^2>$, $\vec{r}(1) = <2, 3, 4>$, $\vec{r}'(1) = <2, 6, 12>$. Thus
$\vec{T}(1) = \frac{\vec{r}'(1)}{|\vec{r}'(1)|} = \frac{1}{\sqrt{188}}<2, 6, 12> = \left\langle \frac{1}{\sqrt{46}}, \frac{3}{\sqrt{46}}, \frac{6}{\sqrt{46}} \right\rangle$.

31. $\vec{r}'(t) = \vec{i} + 2\cos t\vec{j} - 3\sin t\vec{k}$, $\vec{r}'(\pi/6) = \vec{i} + \sqrt{3}\vec{j} - \frac{3}{2}\vec{k}$. Thus
$\vec{T}(\pi/6) = \frac{1}{\sqrt{25/4}}(\vec{i} + \sqrt{3}\vec{j} - \frac{3}{2}\vec{k}) = \frac{2}{5}\vec{i} + \frac{2\sqrt{3}}{5}\vec{j} - \frac{3}{5}\vec{k}$.

33. The vector equation of the curve is $\vec{r}(t) = t\vec{i} + t^2\vec{j} + t^3\vec{k}$, so
$\vec{r}'(t) = \vec{i} + 2t\vec{j} + 3t^2\vec{k}$. At the point $(1, 1, 1)$, $t = 1$, so the tangent vector here is
$\vec{i} + 2\vec{j} + 3\vec{k}$. The tangent line goes through the point $(1, 1, 1)$ and has direction vector
$\vec{i} + 2\vec{j} + 3\vec{k}$. Thus parametric equations are $x = 1 + t$, $y = 1 + 2t$, $z = 1 + 3t$.

Section 11.7

35. $\vec{r}(t) = <t, \sqrt{2}\cos t, \sqrt{2}\sin t>$, $\vec{r}'(t) = <1, -\sqrt{2}\sin t, \sqrt{2}\cos t>$. At $(\pi/4, 1, 1)$, $t = \frac{\pi}{4}$ and $\vec{r}'(\pi/4) = <1, -1, 1>$. Thus the parametric equations of the tangent line are $x = \frac{\pi}{4} + t$, $y = 1 - t$, $z = 1 + t$.

37. $\vec{r}(t) = <t\cos 2\pi t, t\sin 2\pi t, 4t>$, $\vec{r}'(t) = <\cos 2\pi t - 2\pi t\sin 2\pi t, \sin 2\pi t + 2\pi t\cos 2\pi t, 4>$. At $(0, 1/4, 1)$, $t = \frac{1}{4}$ and $\vec{r}'(1/4) = \langle 0 - \frac{\pi}{2}, 1 + 0, 4 \rangle = \langle -\frac{\pi}{2}, 1, 4 \rangle$. Thus the parametric equations of the tangent line are $x = -\frac{\pi}{2}t$, $y = \frac{1}{4} + t$, $z = 1 + 4t$.

39. The angle of intersection of the two curves is the angle between the two tangent vectors to the curves at the point of intersection. Since $\vec{r}_1'(t) = <1, 2t, 3t^2>$ and at $(0, 0, 0)$ $t = 0$, $\vec{r}_1'(0) = <1, 0, 0>$ is a tangent vector to $\vec{r}_1$ at $(0, 0, 0)$. Also $\vec{r}_2' = <\cos t, 2\cos t, 1>$ so $\vec{r}_2'(0) = <1, 2, 1>$ is a tangent vector to $\vec{r}_2$ at $(0, 0, 0)$. If θ is the angle between these two tangent vectors, then
$\cos\theta = \frac{1}{\sqrt{1}\sqrt{6}}<1, 0, 0> \cdot <1, 2, 1> = \frac{1}{\sqrt{6}}$ and $\theta = \cos^{-1}\left(\frac{1}{\sqrt{6}}\right) \approx 66°$.

41. $\int_0^1 (t\vec{i} + t^2\vec{j} + t^3\vec{k})\,dt = \left(\int_0^1 t\,dt\right)\vec{i} + \left(\int_0^1 t^2\,dt\right)\vec{j} + \left(\int_0^1 t^3\,dt\right)\vec{k}$
$= \left[\frac{t^2}{2}\right]_0^1 \vec{i} + \left[\frac{t^3}{3}\right]_0^1 \vec{j} + \left[\frac{t^4}{4}\right]_0^1 \vec{k} = \frac{1}{2}\vec{i} + \frac{1}{3}\vec{j} + \frac{1}{4}\vec{k}$.

43. $\int_0^{\pi/4} (\cos 2t\,\vec{i} + \sin 2t\,\vec{j} + t\sin t\,\vec{k})\,dt$
$= \left[\frac{\sin 2t}{2}\vec{i} - \frac{\cos 2t}{2}\vec{j}\right]_0^{\pi/4} + \left[[-t\cos t]_0^{\pi/4} + \int_0^{\pi/4}\cos t\,dt\right]\vec{k}$
$= \frac{1}{2}\vec{i} + \frac{1}{2}\vec{j} + \left[-\frac{\pi}{4}\cos\frac{\pi}{4} + \sin\frac{\pi}{4}\right]\vec{k} = \frac{1}{2}\vec{i} + \frac{1}{2}\vec{j} + \frac{1}{\sqrt{2}}\left(1 - \frac{\pi}{4}\right)\vec{k}$
$= \frac{1}{2}\vec{i} + \frac{1}{2}\vec{j} + \frac{4 - \pi}{4\sqrt{2}}\vec{k}$

45. $\vec{r}(t) = \frac{t^3}{3}\vec{i} + t^4\vec{j} - \frac{t^3}{3}\vec{k} + \vec{c}$ where $\vec{c}$ is a constant vector. But $\vec{j} = \vec{r}(0) = (0)\vec{i} + (0)\vec{j} - (0)\vec{k} + \vec{c}$. Thus $\vec{c} = \vec{j}$ and $\vec{r}(t) = \frac{t^3}{3}\vec{i} + (t^4 + 1)\vec{j} - \frac{t^3}{3}\vec{k}$.

47. Let $\vec{u}(t) = <u_1(t), u_2(t), u_3(t)>$ and $\vec{v}(t) = <v_1(t), v_2(t), v_3(t)>$. In each part of this problem the basic procedure is to use (11.49) and then the individual component functions as the corresponding limit properties have already been developed for real–valued functions.

(a) By (11.49) $\lim_{t \to a} \vec{u}(t) + \lim_{t \to a} \vec{v}(t) = <\lim_{t \to a} u_1(t), \lim_{t \to a} u_2(t), \lim_{t \to a} u_3(t)>$
$+ <\lim_{t \to a} v_1(t), \lim_{t \to a} v_2(t), \lim_{t \to a} v_3(t)>$ and the limits of these component functions must each exist since the vector functions both possess limits as $t \to a$. Then adding the two vectors and using the addition property of limits for real–valued functions we have that $\lim_{t \to a} \vec{u}(t) + \lim_{t \to a} \vec{v}(t)$
$= <\lim_{t \to a} u_1(t) + \lim_{t \to a} v_1(t), \lim_{t \to a} u_2(t) + \lim_{t \to a} v_2(t), \lim_{t \to a} u_3(t) + \lim_{t \to a} v_3(t)>$

Section 11.7

$$= \,<\lim_{t\to a}(u_1(t)+v_1(t)),\,\lim_{t\to a}(u_2(t)+v_2(t)),\,\lim_{t\to a}(u_3(t)+v_3(t))>$$

$$= \lim_{t\to a}<u_1(t)+v_1(t),\,u_2(t)+v_2(t),\,u_3(t)+v_3(t)> \quad \text{(using 11.49 backwards)}$$

$$= \lim_{t\to a}[\vec{u}(t)+\vec{v}(t)].$$

(b) $\lim_{t\to a}c\vec{u}(t) = \lim_{t\to a}<cu_1(t),\,cu_2(t),\,cu_3(t)>$

$$= \,<\lim_{t\to a}cu_1(t),\,\lim_{t\to a}cu_2(t),\,\lim_{t\to a}cu_3(t)>$$

$$= \,<c\lim_{t\to a}u_1(t),\,c\lim_{t\to a}u_2(t),\,c\lim_{t\to a}u_3(t)>$$

$$= c<\lim_{t\to a}u_1(t),\,\lim_{t\to a}u_2(t),\,\lim_{t\to a}u_3(t)>$$

$$= c\lim_{t\to a}<u_1(t),\,u_2(t),\,u_3(t)> \,=\, c\lim_{t\to a}\vec{u}(t).$$

(c) $\lim_{t\to a}\vec{u}(t) \cdot \lim_{t\to a}\vec{v}(t)$

$$= \,<\lim_{t\to a}u_1(t),\,\lim_{t\to a}u_2(t),\,\lim_{t\to a}u_3(t)>\,\cdot\,<\lim_{t\to a}v_1(t),\,\lim_{t\to a}v_2(t),\,\lim_{t\to a}v_3(t)>$$

$$= [\lim_{t\to a}u_1(t)][\lim_{t\to a}v_1(t)] + [\lim_{t\to a}u_2(t)][\lim_{t\to a}v_2(t)] + [\lim_{t\to a}u_3(t)][\lim_{t\to a}v_3(t)]$$

$$= \lim_{t\to a}u_1(t)v_1(t) + \lim_{t\to a}u_2(t)v_2(t) + \lim_{t\to a}u_3(t)v_3(t)$$

$$= \lim_{t\to a}[u_1(t)v_1(t) + u_2(t)v_2(t) + u_3(t)v_3(t)]$$

$$= \lim_{t\to a}[\vec{u}(t)\cdot\vec{v}(t)].$$

(d) $\lim_{t\to a}\vec{u}(t) \times \lim_{t\to a}\vec{v}(t)$

$$= \,<\lim_{t\to a}u_1(t),\,\lim_{t\to a}u_2(t),\,\lim_{t\to a}u_3(t)>\,\times\,<\lim_{t\to a}v_1(t),\,\lim_{t\to a}v_2(t),\,\lim_{t\to a}v_3(t)>$$

$$= \,<[\lim_{t\to a}u_2(t)][\lim_{t\to a}v_3(t)] - [\lim_{t\to a}u_3(t)][\lim_{t\to a}v_2(t)],$$

$$[\lim_{t\to a}u_3(t)][\lim_{t\to a}v_1(t)] - [\lim_{t\to a}u_1(t)][\lim_{t\to a}v_3(t)],$$

$$[\lim_{t\to a}u_1(t)][\lim_{t\to a}v_2(t)] - [\lim_{t\to a}u_2(t)][\lim_{t\to a}v_1(t)]>$$

$$= \,<\lim_{t\to a}[u_2(t)v_3(t) - u_3(t)v_2(t)],\,\lim_{t\to a}[u_3(t)v_1(t) - u_1(t)v_3(t)],$$

$$\lim_{t\to a}[u_1(t)v_2(t) - u_2(t)v_1(t)]>$$

$$= \lim_{t\to a}<u_2(t)v_3(t) - u_3(t)v_2(t),\,u_3(t)v_1(t) - u_1(t)v_3(t),\,u_1(t)v_2(t) - u_2(t)v_1(t)]>$$

$$= \lim_{t\to a}[\vec{u}(t)\times\vec{v}(t)].$$

For Exercises 49–51, let $\vec{u}(t) = <u_1(t),u_2(t),u_3(t)>$ and $\vec{v}(t) = <v_1(t),v_2(t),v_3(t)>$. In each of these exercises the basic procedure is to apply Theorem 11.52 so the corresponding properties of derivatives of real-valued functions can be used.

Section 11.7

49. $\frac{d}{dt}[\vec{u}(t) + \vec{v}(t)] = \frac{d}{dt}<u_1(t) + v_1(t), u_2(t) + v_2(t), u_3(t) + v_3(t)>$
    $= <\frac{d}{dt}(u_1(t) + v_1(t)), \frac{d}{dt}(u_2(t) + v_2(t)), \frac{d}{dt}(u_3(t) + v_3(t))>$
    $= <u_1'(t) + v_1'(t), u_2'(t) + v_2'(t), u_3'(t) + v_3'(t)>$
    $= <u_1'(t), u_2'(t), u_3'(t)> + <v_1'(t), v_2'(t), v_3'(t)> = \vec{u}'(t) + \vec{v}'(t)$.

51. $\frac{d}{dt}[\vec{u}(t) \times \vec{v}(t)]$
    $= \frac{d}{dt}<u_2(t)v_3(t) - u_3(t)v_2(t), u_3(t)v_1(t) - u_1(t)v_3(t), u_1(t)v_2(t) - u_2(t)v_1(t)>$
    $= <u_2'v_3(t) + u_2(t)v_3'(t) - u_3'(t)v_2(t) - u_3(t)v_2'(t),$
    $\qquad u_3'(t)v_1(t) + u_3(t)v_1'(t) - u_1'(t)v_3(t) - u_1(t)v_3'(t),$
    $\qquad u_1'(t)v_2(t) + u_1(t)v_2'(t) - u_2'(t)v_1(t) - u_2(t)v_1'(t)>$
    $= <u_2'(t)v_3(t) - u_3'(t)v_2(t), u_3'(t)v_1(t) - u_1'(t)v_3(t), u_1'(t)v_2(t) - u_2'(t)v_1(t)>$
    $\quad + <u_2(t)v_3'(t) - u_3(t)v_2'(t), u_3(t)v_1'(t) - u_1(t)v_3'(t), u_1(t)v_2'(t) - u_2(t)v_1'(t)>$
    $= \vec{u}'(t) \times \vec{v}(t) + \vec{u}(t) \times \vec{v}'(t)$.

    <u>Alternate</u> <u>Solution</u>: Let $\vec{r}(t) = \vec{u}(t) \times \vec{v}(t)$, then
    $\vec{r}(t+h) - \vec{r}(t) = [\vec{u}(t+h) \times \vec{v}(t+h)] - [\vec{u}(t) \times \vec{v}(t)]$
    $= [\vec{u}(t+h) \times \vec{v}(t+h)] - [\vec{u}(t) \times \vec{v}(t)] + [\vec{u}(t+h) \times \vec{v}(t)] - [\vec{u}(t+h) \times \vec{v}(t)]$
    $= \vec{u}(t+h) \times [\vec{v}(t+h) - \vec{v}(t)] + [\vec{u}(t+h) - \vec{u}(t)] \times \vec{v}(t)$.
    (Remember to be careful of the order of the cross product.) Dividing through by h and taking the limit as h → 0 we have
    $\vec{r}'(t) = \lim_{h \to 0} \frac{\vec{u}(t+h) \times [\vec{v}(t+h) - \vec{v}(t)]}{h} + \lim_{h \to 0} \frac{[\vec{u}(t+h) - u(t)] \times v(t)}{h}$
    $= \vec{u}(t) \times \vec{v}'(t) + \vec{u}'(t) \times \vec{v}(t)$ by Exercise 47a and Definition 11.51.

53. $D_t[\vec{u}(t) \cdot \vec{v}(t)] = \vec{u}'(t) \cdot \vec{v}(t) + \vec{u}(t) \cdot \vec{v}'(t)$ by (11.53.4)
    $= (-4t\vec{j} + 9t^2\vec{k}) \cdot (t\vec{i} + \cos t\, \vec{j} + \sin t\, \vec{k}) + (\vec{i} - 2t^2\vec{j} + 3t^3\vec{k}) \cdot (\vec{i} - \sin t\, \vec{j} + \cos t\, \vec{k})$
    $= -4t\cos t + 9t^2 \sin t + 1 + 2t^2 \sin t + 3t^3 \cos t$
    $= 1 - 4t\cos t + 11t^2 \sin t + 3t^3 \cos t$.

55. $\frac{d}{dt}[\vec{r}(t) \times \vec{r}'(t)] = \vec{r}'(t) \times \vec{r}'(t) + \vec{r}(t) \times \vec{r}''(t)$ by (11.53.5). But $\vec{r}'(t) \times \vec{r}'(t) = \vec{0}$ by Example 2 of Section 11.4. Thus $\frac{d}{dt}[\vec{r}(t) \times \vec{r}'(t)] = \vec{r}(t) \times \vec{r}''(t)$.

57. $\frac{d}{dt}|\vec{r}(t)| = \frac{d}{dt}(\vec{r}(t) \cdot \vec{r}(t))^{1/2} = \frac{1}{2}(\vec{r}(t) \cdot \vec{r}(t))^{-1/2}(2\vec{r}(t) \cdot \vec{r}'(t))$
    $= \frac{\vec{r}(t) \cdot \vec{r}'(t)}{|\vec{r}(t)|}$.

59. Since $\vec{u}(t) = \vec{r}(t) \cdot [\vec{r}'(t) \times \vec{r}''(t)]$,
    $\vec{u}'(t) = \vec{r}'(t) \cdot [\vec{r}'(t) \times \vec{r}''(t)] + \vec{r}(t) \cdot \frac{d}{dt}[\vec{r}'(t) \times \vec{r}''(t)]$

Section 11.8

$$= 0 + \vec{r}(t) \cdot [\vec{r}''(t) \times \vec{r}''(t) + \vec{r}'(t) \times \vec{r}'''(t)] \qquad [\vec{r}'(t) \perp \vec{r}'(t) \times \vec{r}''(t)]$$
$$= \vec{r}(t) \cdot [\vec{r}'(t) \times \vec{r}'''(t)] \qquad [\vec{r}''(t) \times \vec{r}''(t) = \vec{0}]$$

EXERCISES 11.8

1. $\vec{r}'(t) = <2, 3\cos t, -3\sin t>$, $|\vec{r}'(t)| = \sqrt{4 + 9\cos^2 t + 9\sin^2 t} = \sqrt{13}$
   $L = \int_a^b \sqrt{13}\,dt = \sqrt{13}(b-a)$.

3. $\vec{r}'(t) = <6, 6\sqrt{2}t, 6t^2>$, $|\vec{r}'(t)| = 6\sqrt{1 + 2t^2 + t^4} = 6(1+t^2)$
   $L = \int_0^1 6(1+t^2)\,dt = \left[\frac{6(t+t^3)}{3}\right]_0^1 = \frac{24}{3} = 8$.

5. $\vec{r}'(t) = <2, 2t, 2t>$, $|\vec{r}'(t)| = 2\sqrt{1+2t^2}$
   $L = \int_0^1 2\sqrt{1+2t^2}\,dt = \int_a^b \sqrt{2}\sec^3\theta\,d\theta$
   $= \frac{\sqrt{2}}{2}\left[\ln|\sec\theta + \tan\theta| + \tan\theta\sec\theta\right]_a^b$
   $= \frac{\sqrt{2}}{2}\left[\ln\left|\sqrt{1+2t^2} + \sqrt{2}t\right| + \sqrt{2}t\sqrt{1+2t^2}\right]_0^1$ (or use Formula 21)
   $= \frac{1}{\sqrt{2}}\left[\ln|\sqrt{3} + \sqrt{2}| + \sqrt{2}\sqrt{3}\right] = \sqrt{3} + \frac{1}{\sqrt{2}}\ln(\sqrt{2} + \sqrt{3})$.

7. $\vec{r}'(t) = e^t(\cos t + \sin t)\vec{i} + e^t(\cos t - \sin t)\vec{j}$
   $\frac{ds}{dt} = |\vec{r}'(t)| = e^t\sqrt{(\cos t + \sin t)^2 + (\cos t - \sin t)^2} = e^t\sqrt{2\cos^2 t + 2\sin^2 t} = \sqrt{2}e^t$
   $s(t) = \int_0^t |\vec{r}'(u)|\,du = \int_0^t \sqrt{2}e^u\,du = \sqrt{2}(e^t - 1) \Rightarrow \frac{s}{\sqrt{2}} + 1 = e^t \Rightarrow t(s) = \ln\left(\frac{s}{\sqrt{2}} + 1\right)$
   Therefore, $\vec{r}(t(s)) = \left(\frac{s}{\sqrt{2}} + 1\right)\left[\sin\left[\ln\left(\frac{s}{\sqrt{2}} + 1\right)\right]\vec{i} + \cos\left[\ln\left(\frac{s}{\sqrt{2}} + 1\right)\right]\vec{j}\right]$.

9. $|\vec{r}'(t)| = \sqrt{(3\cos t)^2 + 16 + (-3\sin t)^2} = \sqrt{9 + 16} = 5$ and
   $s(t) = \int_0^t |\vec{r}'(u)|\,du = \int_0^t 5\,du = 5t \Rightarrow t(s) = \frac{s}{5}$
   Therefore, $\vec{r}(t(s)) = 3\sin\left(\frac{s}{5}\right)\vec{i} + \frac{4s}{5}\vec{j} + 3\cos\left(\frac{s}{5}\right)\vec{k}$.

11. (a) $\vec{T}(t) = \frac{1}{\sqrt{16+9}}<4\cos 4t, 3, -4\sin 4t> = \frac{1}{5}<4\cos 4t, 3, -4\sin 4t>$
    $\vec{N}(t) = \frac{5}{16\cdot 5}<-16\sin 4t, 0, -16\cos 4t> = <-\sin 4t, 0, -\cos 4t>$
    (b) $\kappa(t) = \frac{16}{5\cdot 5} = \frac{16}{25}$.

75

Section 11.8

13. (a) $\vec{T}(t) = \dfrac{1}{\sqrt{2 + e^{2t} + e^{-2t}}} <\sqrt{2}, e^t, -e^{-t}> = \dfrac{1}{e^t + e^t}<\sqrt{2}, e^t, e^{-t}>$

    $= \dfrac{1}{e^{2t}+1}<\sqrt{2}e^t, e^{2t}, -1>$

    $\vec{T}'(t) = \dfrac{-2e^{2t}}{(e^{2t}+1)^2}<\sqrt{2}e^t, e^{2t}, -1> + \dfrac{1}{e^{2t}+1}<\sqrt{2}e^t, 2e^{2t}, 0>$

    $= \dfrac{1}{(e^{2t}+1)^2}<-2\sqrt{2}e^{3t} + \sqrt{2}e^{3t} + \sqrt{2}e^t, -2e^{4t} + 2e^{4t} + 2e^{2t}, 2e^{2t}>$

    $= \dfrac{1}{(e^{2t}+1)^2}<\sqrt{2}(e^t - e^{3t}), 2e^{2t}, 2e^{2t}>$

    $|\vec{T}'(t)| = \dfrac{1}{(e^{2t}+1)^2}\sqrt{2(e^{2t} + e^{6t} - 2e^{4t}) + 8e^{4t}} = \dfrac{\sqrt{2}e^t}{(e^{2t}+1)^2}(e^{2t}+1) = \dfrac{\sqrt{2}e^t}{e^{2t}+1}$

    $\vec{N}(t) = \dfrac{1}{(e^{2t}+1)^2}\dfrac{e^{2t}+1}{\sqrt{2}e^t}<\sqrt{2}(e^t - e^{3t}), 2e^{2t}, 2e^{2t}>$

    $= \dfrac{1}{\sqrt{2}(e^{2t}+1)e^t}<\sqrt{2}(e^t - e^{3t}), 2e^{2t}, 2e^{2t}> = \dfrac{1}{e^{2t}+1}<1 - e^{2t}, \sqrt{2}e^t, \sqrt{2}e^t>$.

    (b) $\kappa(t) = \dfrac{\sqrt{2}e^t}{(e^{2t}+1)(e^t + e^{-t})} = \dfrac{\sqrt{2}e^{2t}}{(e^{2t}+1)^2}$.

15. (a) $\vec{T}(t) = \dfrac{1}{\sqrt{4t^2 + 4t^4 + 1}}<2t, 2t^2, 1> = \dfrac{1}{2t^2+1}<2t, 2t^2, 1>$

    $\vec{T}'(t) = -(2t^2+1)^{-2}(4t)<2t, 2t^2, 1> + (2t^2+1)^{-1}<2, 4t, 0>$

    $= \dfrac{1}{(2t^2+1)^2}<-8t^2 + 4t^2 + 2, -8t^3 + 8t^3 + 4t, -4t>$

    $= \dfrac{2}{(2t^2+1)^2}<1 - 2t^2, 2t, -2t>$

    $|\vec{T}'(t)| = \dfrac{2}{(2t^2+1)^2}\sqrt{1 - 4t^2 + 4t^4 + 8t^2} = \dfrac{2}{2t^2+1}$

    $\vec{N}(t) = \dfrac{2}{(2t^2+1)^2}\dfrac{(2t^2+1)}{2}<1 - 2t^2, 2t, -2t> = \dfrac{1}{(2t^2+1)}<1 - 2t^2, 2t, -2t>$

    (b) $\kappa(t) = \dfrac{2}{(2t^2+1)(2t^2+1)} = \dfrac{2}{(2t^2+1)^2}$.

17. $\vec{r}'(t) = \vec{j} - 2t\vec{k}, \ \vec{r}''(t) = -2\vec{k}, \ |\vec{r}'(t)|^3 = (4t^2+1)^{3/2}$,

    $|\vec{r}'(t) \times \vec{r}''(t)| = |-2\vec{i}| = 2, \ \kappa(t) = \dfrac{2}{(4t^2+1)^{3/2}}$.

19. $\vec{r}'(t) = <6t^2, -6t, 6>, \ \vec{r}''(t) = <12t, -6, 0>, \ |\vec{r}'(t)|^3 = 6^3(t^4 + t^2 + 1)^{3/2}$,

    $|\vec{r}'(t) \times \vec{r}''(t)| = |36<1, 2t, t^2>| = 36\sqrt{1 + 4t^2 + t^4}$

    $\kappa(t) = \dfrac{36\sqrt{1 + 4t^2 + t^4}}{6^3(t^4 + t^2 + 1)^{3/2}} = \dfrac{\sqrt{1 + 4t^2 + t^4}}{6(t^4 + t^2 + 1)^{3/2}}$.

Section 11.8

21. $\vec{r}'(t) = \langle \cos t, -\sin t, \cos t \rangle$, $\vec{r}''(t) = \langle -\sin t, -\cos t, -\sin t \rangle$,

$|\vec{r}'(t)|^3 = (\sqrt{\cos^2 t + 1})^3$, $|\vec{r}'(t) \times \vec{r}''(t)| = |\langle 1, 0, -1 \rangle| = \sqrt{2}$, $\kappa(t) = \dfrac{\sqrt{2}}{(1+\cos^2 t)^{3/2}}$.

23. $f'(x) = 3x^2$, $f''(x) = 6x$, $\kappa(x) = \dfrac{6|x|}{[1+9x^4]^{3/2}}$.

25. $y' = \cos x$, $y'' = -\sin x$, $\kappa(x) = \dfrac{|\sin x|}{[1+\cos^2 x]^{3/2}}$.

27. Since $y' = y'' = e^x$, the curvature is $\kappa(x) = e^x/[1+e^{2x}]^{3/2} = e^x[1+e^{2x}]^{-3/2}$.

$\kappa'(x) = e^x[1+e^{2x}]^{-3/2} + e^x\left(-\dfrac{3}{2}\right)[1+e^{2x}]^{-5/2}(2e^{2x})$

$= e^x \dfrac{(1+e^{2x}-3e^{2x})}{(1+e^{2x})^{5/2}} = e^x \dfrac{(1-2e^{2x})}{(1+e^{2x})^{5/2}}$. Then when

$\kappa'(x) = 0$, it must be that $1-2e^{2x} = 0$ or $e^{2x} = \tfrac{1}{2}$ or $x = -\tfrac{1}{2}\ln 2$. And since $1-2e^{2x} > 0$ for $x < -\tfrac{1}{2}\ln 2$ and $1-2e^{2x} < 0$ for $x > -\tfrac{1}{2}\ln 2$, the maximum curvature is attained at the point $\left(-\tfrac{1}{2}\ln 2, e^{(-1/2)\ln 2}\right) = \left(-\tfrac{1}{2}\ln 2, \tfrac{1}{\sqrt{2}}\right)$.

29. $\kappa(t) = \dfrac{|(3t^2)(2) - (6t)(2t)|}{[9t^4 + 4t^2]^{3/2}} = \dfrac{6t^2}{(t^2)^{3/2}[9t^2+4]^{3/2}}$

$= \dfrac{6t^2}{|t|^3[9t^2+4]^{3/2}} = \dfrac{6}{|t|[9t^2+4]^{3/2}}$.

31. $\kappa = \left|\dfrac{d\vec{T}}{ds}\right| = \left|\dfrac{d\vec{T}/dt}{ds/dt}\right| = \dfrac{|d\vec{T}/dt|}{ds/dt}$ and $\vec{N} = \dfrac{d\vec{T}/dt}{|d\vec{T}/dt|}$.

Thus $\kappa \vec{N} = \dfrac{|d\vec{T}/dt|(d\vec{T}/dt)}{|d\vec{T}/dt|(ds/dt)} = \dfrac{d\vec{T}/dt}{ds/dt} = \dfrac{d\vec{T}}{ds}$ by the chain rule.

33. For a plane curve, $\vec{T} = |\vec{T}|\cos\phi\,\vec{i} + |\vec{T}|\sin\phi\,\vec{j} = \cos\phi\,\vec{i} + \sin\phi\,\vec{j}$.

Then $\dfrac{d\vec{T}}{ds} = \left(\dfrac{d\vec{T}}{d\phi}\right)\left(\dfrac{d\phi}{ds}\right) = (-\sin\phi\,\vec{i} + \cos\phi\,\vec{j})\left(\dfrac{d\phi}{ds}\right)$ and

$\left|\dfrac{d\vec{T}}{ds}\right| = |-\sin\phi\,\vec{i} + \cos\phi\,\vec{j}|\left|\dfrac{d\phi}{ds}\right| = \left|\dfrac{d\phi}{ds}\right|$.

Hence for a plane curve, the curvature is $\kappa = \left|\dfrac{d\phi}{ds}\right|$.

35. (a) $|\vec{B}| = 1 \Rightarrow \vec{B}\cdot\vec{B} = 1 \Rightarrow \dfrac{d}{ds}(\vec{B}\cdot\vec{B}) = 0 \Rightarrow 2\dfrac{d\vec{B}}{ds}\cdot\vec{B} = 0 \Rightarrow \dfrac{d\vec{B}}{ds} \perp \vec{B}$

(b) $\vec{B} = \vec{T} \times \vec{N} \Rightarrow \dfrac{d\vec{B}}{ds} = \dfrac{d}{ds}(\vec{T}\times\vec{N}) = \dfrac{d}{dt}(\vec{T}\times\vec{N})\dfrac{1}{ds/dt} = \dfrac{d}{dt}(\vec{T}\times\vec{N})\dfrac{1}{|\vec{r}'(t)|}$

$= [(\vec{T}'\times\vec{N}) + (\vec{T}\times\vec{N}')]\dfrac{1}{|\vec{r}'(t)|} = \left[\left(\vec{T}'\times\dfrac{\vec{T}'}{|\vec{T}'|}\right) + (\vec{T}\times\vec{N}')\right]\dfrac{1}{|\vec{r}'(t)|} = \dfrac{\vec{T}\times\vec{N}'}{|\vec{r}'(t)|} \Rightarrow \dfrac{d\vec{B}}{ds} \perp \vec{T}$.

(c) $\vec{B} = \vec{T}\times\vec{N} \Rightarrow \vec{T}\perp\vec{N}, \vec{B}\perp\vec{T}$ and $\vec{B}\perp\vec{N}$. So $\vec{B}, \vec{T}$ and $\vec{N}$ form an orthogonal set of vectors in the 3-dimensional space $\mathbf{R}^3$, which makes them a basis for this space. From parts (a) and (b) $\frac{d\vec{B}}{ds}$ is perpendicular to both $\vec{B}$ and $\vec{T}$, so $\frac{d\vec{B}}{ds}$ is parallel to $\vec{N}$. Therefore, $\frac{d\vec{B}}{ds} = -\tau(s)\vec{N}$, where $\tau(s)$ is a number.

37. (a) $\vec{T} = \frac{\vec{r}'}{|\vec{r}'|} = \frac{\vec{r}'}{\sqrt{\vec{r}' \cdot \vec{r}'}} \Rightarrow \vec{T}' = \frac{\vec{r}''}{\sqrt{\vec{r}' \cdot \vec{r}'}} - \frac{2\vec{r}'' \cdot \vec{r}'}{2[\vec{r}' \cdot \vec{r}']^{3/2}}\vec{r}'$

$= \frac{\vec{r}''}{|\vec{r}'|} - \frac{\vec{r}'' \cdot \vec{r}'}{|\vec{r}'|^3}\vec{r}'$. Also $|\vec{T}'| = \frac{|\vec{r}' \times \vec{r}''|}{|\vec{r}'|^2}$ [see proof of Thm. 11.63].

Thus $\vec{N} = \frac{\vec{T}'}{|\vec{T}'|} = \left[\vec{r}'' - \frac{\vec{r}'' \cdot \vec{r}'}{|\vec{r}'|^2}\vec{r}'\right]\frac{|\vec{r}'|}{|\vec{r}' \times \vec{r}''|}$ and

$\vec{B} = \vec{T} \times \vec{N} = \frac{1}{|\vec{r}' \times \vec{r}''|}\vec{r}' \times \left[\vec{r}'' - \frac{\vec{r}'' \cdot \vec{r}'}{|\vec{r}'|^2}\vec{r}'\right] = \frac{\vec{r}' \times \vec{r}''}{|\vec{r}' \times \vec{r}''|}$

$= \frac{\vec{r}' \times \vec{r}''}{\sqrt{(\vec{r}' \times \vec{r}'') \cdot (\vec{r}' \times \vec{r}'')}}$ [since $\vec{r}' \times \vec{r}' = \vec{0}$]

$\Rightarrow \frac{d\vec{B}}{dt} = \frac{\vec{r}' \times \vec{r}'''}{|\vec{r}' \times \vec{r}''|} + \left[-\frac{1}{2}\frac{2(\vec{r}' \times \vec{r}'') \cdot (\vec{r}' \times \vec{r}''')}{[(\vec{r}' \times \vec{r}'') \cdot (\vec{r}' \times \vec{r}'')]^{3/2}}\right](\vec{r}' \times \vec{r}'')$

$= \frac{\vec{r}' \times \vec{r}'''}{|\vec{r}' \times \vec{r}''|} - \frac{(\vec{r}' \times \vec{r}'') \cdot (\vec{r}' \times \vec{r}''')}{|\vec{r}' \times \vec{r}''|^3}(\vec{r}' \times \vec{r}'')$

Also, $\frac{d\vec{B}}{ds} = \frac{d\vec{B}}{dt}\frac{1}{(ds/dt)} = \frac{d\vec{B}}{dt}\frac{1}{|\vec{r}'|}$. Since $\frac{d\vec{B}}{ds} = -\tau\vec{N}$, and since $|\vec{N}| = 1$, we have

$\tau = \tau\vec{N} \cdot \vec{N} = -\frac{d\vec{B}}{ds} \cdot \vec{N} = -\frac{1}{|\vec{r}'|}\frac{d\vec{B}}{dt} \cdot \vec{N}$

$= -\frac{1}{|\vec{r}' \times \vec{r}''|^2}(\vec{r}' \times \vec{r}''') \cdot \vec{r}'' = \frac{1}{|\vec{r}' \times \vec{r}''|^2}(\vec{r}''' \times \vec{r}') \cdot \vec{r}'' = \frac{(\vec{r}' \times \vec{r}'') \cdot \vec{r}'''}{|\vec{r}' \times \vec{r}''|^2}$

In taking the above inner product the following facts have been used;
$(\vec{r}' \times \vec{r}''') \cdot \vec{r}' = \vec{r}' \cdot (\vec{r}' \times \vec{r}''') = (\vec{r}' \times \vec{r}') \cdot \vec{r}''' = 0$, and, similarly,
$(\vec{r}' \times \vec{r}'') \cdot \vec{r}' = 0$ and $(\vec{r}' \times \vec{r}'') \cdot \vec{r}'' = \vec{r}' \cdot (\vec{r}'' \times \vec{r}'') = 0$

(b) $\vec{r}(t) = <\cos t, \sin t, t> \Rightarrow \vec{r}'(t) = <-\sin t, \cos t, 1>$
$\Rightarrow \vec{r}''(t) = <-\cos t, -\sin t, 0> \Rightarrow \vec{r}'''(t) = <\sin t, -\cos t, 0>$
$\vec{r}'(t) \times \vec{r}''(t) = <\sin t, -\cos t, 1> \Rightarrow |\vec{r}'(t) \times \vec{r}''(t)| = \sqrt{2}$

Thus $\tau = \frac{(\vec{r}' \times \vec{r}'') \cdot \vec{r}'''}{|\vec{r}' \times \vec{r}''|^2} = \frac{<\sin t, -\cos t, 1> \cdot <\sin t, -\cos t, 0>}{2} = \frac{1}{2}$.

Section 11.9

EXERCISES 11.9

1. $\vec{v}(t) = \vec{r}'(t) = \langle 2t, 1\rangle$ At $t=1$:
   $\vec{a}(t) = \vec{r}''(t) = \langle 2, 0\rangle$ $\vec{v}(1) = \langle 2, 1\rangle$
   $|\vec{v}(t)| = \sqrt{4t^2+1}$ $\vec{a}(1) = \langle 2, 0\rangle$

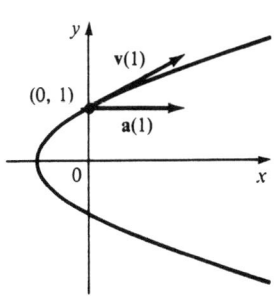

3. $\vec{v}(t) = \langle e^t, -e^{-t}\rangle$, $\vec{v}(0) = \langle 1, -1\rangle$
   $\vec{a}(t) = \langle e^t, e^{-t}\rangle$, $\vec{a}(0) = \langle 1, 1\rangle$
   $|\vec{v}(t)| = \sqrt{e^{2t}+e^{-2t}} = e^{-t}\sqrt{e^{4t}+1}$
   Since $x = e^t$, $t = \ln x$ and
   $y = e^{-t} = e^{-\ln x} = \frac{1}{x}$, and
   $x > 0$, $y > 0$.

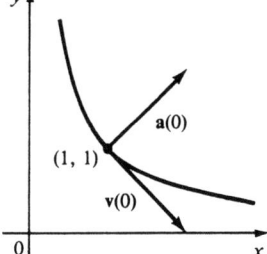

5. $\vec{v}(t) = \langle \cos t, 1, -\sin t\rangle$
   $\vec{v}(0) = \langle 1, 1, 0\rangle$
   $\vec{a}(t) = \langle -\sin t, 0, -\cos t\rangle$
   $\vec{a}(0) = \langle 0, 0, -1\rangle$
   $|\vec{v}(t)| = \sqrt{\cos^2 t + 1 + \sin^2 t} = \sqrt{2}$
   Since $x^2 + z^2 = 1$, $y = t$, the path
   of the particle is a helix about
   the y-axis.

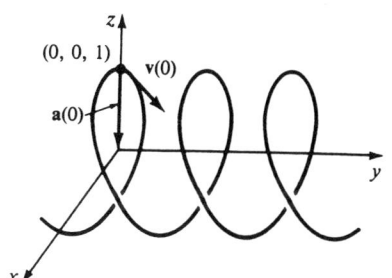

7. $\vec{v}(t) = \langle 3t^2, 2t, 3t^2\rangle$, $\vec{a}(t) = \langle 6t, 2, 6t\rangle$, $|\vec{v}(t)| = \sqrt{9t^4+4t^2+9t^4} = \sqrt{18t^4+4t^2}$
   $= |t|\sqrt{18t^2+4}$.

9. $\vec{v}(t) = \langle -t^{-2}, 0, 2t\rangle$, $\vec{a}(t) = \langle 2t^{-3}, 0, 2\rangle$, $|\vec{v}(t)| = \sqrt{t^{-4}+4t^2}$
   $= \frac{1}{t^2}\sqrt{4t^6+1}$.

11. $\vec{v}(t) = e^t\langle \cos t, \sin t, t\rangle + e^t\langle -\sin t, \cos t, 1\rangle$
    $= e^t\langle \cos t - \sin t, \sin t + \cos t, t+1\rangle$
    $\vec{a}(t) = e^t\langle \cos t - \sin t - \sin t - \cos t, \sin t + \cos t + \cos t - \sin t, t+1+1\rangle$
    $= e^t\langle -2\sin t, 2\cos t, t+2\rangle$
    $|\vec{v}(t)| = e^t\sqrt{\cos^2 t + \sin^2 t - 2\cos t \sin t + \sin^2 t + \cos^2 t + 2\sin t \cos t + t^2 + 2t + 1}$
    $= e^t\sqrt{t^2+2t+3}$.

79

Section 11.9

13. $\vec{v}(t) = \int (\vec{k}) dt = t\vec{k} + \vec{c}_1$ and $\vec{i} - \vec{j} = \vec{v}(0) = 0\vec{k} + \vec{c}_1$, so $\vec{c}_1 = \vec{i} - \vec{j}$ and $\vec{v}(t) = \vec{i} - \vec{j} + t\vec{k}$.
    $\vec{r}(t) = \int (\vec{i} - \vec{j} + t\vec{k}) dt = t\vec{i} - t\vec{j} + \frac{t^2}{2}\vec{k} + \vec{c}_2$. But $\vec{0} = \vec{r}(0) = \vec{0} + \vec{c}_2$, so $\vec{c}_2 = \vec{0}$ and
    $\vec{r}(t) = t\vec{i} - t\vec{j} + \frac{t^2}{2}\vec{k}$.

15. $\vec{v}(t) = \int (\vec{i} + 2\vec{j} + 2t\vec{k}) dt = t\vec{i} + 2t\vec{j} + t^2\vec{k} + \vec{c}_1$, and $\vec{0} = \vec{v}(0) = \vec{0} + \vec{c}_1$, so $\vec{c}_1 = \vec{0}$ and
    $\vec{v}(t) = t\vec{i} + 2t\vec{j} + t^2\vec{k}$. $\vec{r}(t) = \int (t\vec{i} + 2t\vec{j} + t^2\vec{k}) dt = \frac{t^2}{2}\vec{i} + t^2\vec{j} + \frac{t^3}{3}\vec{k} + \vec{c}_2$.
    But $\vec{i} + \vec{k} = \vec{r}(0) = \vec{0} + \vec{c}_2$, so $\vec{c}_2 = \vec{i} + \vec{k}$ and $\vec{r}(t) = [1 + \frac{t^2}{2}]\vec{i} + t^2\vec{j} + [1 + \frac{t^3}{3}]\vec{k}$.

17. $\vec{v}(t) = <2t, 5, 2t - 16>$, $|\vec{v}(t)| = \sqrt{4t^2 + 25 + 4t^2 - 64t + 256} = \sqrt{8t^2 - 64t + 281}$ and
    $\frac{d}{dt}|\vec{v}(t)| = \frac{1}{2}(8t^2 - 64t + 281)^{-1/2}(16t - 64)$. And this is zero if and only if the numerator is zero, that is, $16t - 64 = 0$ or $t = 4$. Since $\frac{d}{dt}|\vec{v}(t)| < 0$ for $t < 4$ and $\frac{d}{dt}|\vec{v}(t)| > 0$ for $t > 4$, the minimum speed of $\sqrt{153}$ is attained at $t = 4$ units of time.

19. $|\vec{F}(t)| = 20$ N in the direction of the positive z-axis, so $\vec{F}(t) = 20\vec{k}$. Also m = 4 kg,
    $\vec{r}(0) = \vec{0}$ and $\vec{v}(0) = \vec{i} - \vec{j}$. Since $20\vec{k} = \vec{F}(t) = 4\vec{a}(t)$, $\vec{a}(t) = 5\vec{k}$. Then $\vec{v}(t) = 5t\vec{k} + \vec{c}_1$
    where $\vec{c}_1 = \vec{i} - \vec{j}$ so $\vec{v}(t) = \vec{i} - \vec{j} + 5t\vec{k}$ and the speed is
    $|\vec{v}(t)| = \sqrt{1 + 1 + 25t^2} = \sqrt{25t^2 + 2}$. Also $\vec{r}(t) = t\vec{i} - t\vec{j} + \frac{5t^2}{2}\vec{k} + \vec{c}_2$ and $\vec{0} = \vec{r}(0)$, so
    $\vec{c}_2 = \vec{0}$ and $\vec{r}(t) = t\vec{i} - t\vec{j} + \frac{5}{2}t^2\vec{k}$.

21. $|\vec{v}(0)| = 500$ m/s and since the angle of elevation is 30° the direction of the velocity is
    $\frac{1}{2}(\sqrt{3}\vec{i} + \vec{j})$. Thus $\vec{v}(0) = 250(\sqrt{3}\vec{i} + \vec{j})$ and if we set up the axes so the projectile starts at the origin then $\vec{r}(0) = \vec{0}$. Ignoring air resistance, the only force is that due to gravity so $\vec{F}(t) = -mg\vec{j}$ where g ≈ 9.8 m/s². Thus $\vec{a}(t) = -g\vec{j}$ and $\vec{v}(t) = -gt\vec{j} + \vec{c}_1$. But
    $250(\sqrt{3}\vec{i} + \vec{j}) = \vec{v}(0) = \vec{c}_1$, so $\vec{v}(t) = 250\sqrt{3}\vec{i} + (250 - gt)\vec{j}$ and
    $\vec{r}(t) = 250\sqrt{3}t\vec{i} + (250t - \frac{1}{2}gt^2)\vec{j} + \vec{c}_2$ where $\vec{0} = \vec{r}(0) = \vec{c}_2$. Thus
    $\vec{r}(t) = 250\sqrt{3}t\vec{i} + (250t - \frac{1}{2}gt^2)\vec{j}$.

    (a) Setting $250t - \frac{1}{2}gt^2 = 0$ gives $t = 0$ or $t = \frac{500}{g} \approx 51.0$ s. So
    the range is $(250\sqrt{3})\frac{500}{g} \approx 22$ km.

    (b) $0 = \frac{d}{dt}(250t - \frac{1}{2}gt^2) = 250 - gt$ implies the maximum height is
    attained when $t = \frac{250}{g} \approx 25.5$ s and thus the maximum height is
    $(250)\frac{250}{g} - g\left(\frac{250}{g}\right)^2\frac{1}{2} = \frac{(250)^2}{2g} \approx 3.2$ km.

    (c) From part (a), impact occurs at $t = \frac{500}{g} \approx 51.0$. Thus the
    velocity at impact is $\vec{v}\left(\frac{500}{g}\right) = 250\sqrt{3}\vec{i} + (250 - g\frac{500}{g})\vec{j}$
    $= 250\sqrt{3}\vec{i} - 250\vec{j}$ and the speed is $|\vec{v}\frac{500}{g}| = 250\sqrt{3 + 1} = 500$ m/s.

Section 11.9

23. As in Example 5, $\vec{r}(t) = (v_0 \cos 45°)t\vec{i} + [(v_0 \sin 45°)t - \frac{1}{2}gt^2]\vec{j}$
$= \frac{1}{2}[v_0\sqrt{2}t\vec{i} + (v_0\sqrt{2}t - gt^2)]\vec{j}$. Then the ball lands at $t = \frac{v_0\sqrt{2}}{g}$ s.
And since it lands 90 m away, $90 = \frac{1}{2}v_0\sqrt{2}\frac{v_0\sqrt{2}}{g}$ or $v_0^2 = 90g$ and the initial velocity is $v_0 = \sqrt{90g} \approx 30$ m/s.

25. From (11.68) $x = (v_0 \cos \alpha)t$ or $t = \frac{x}{v_0 \cos \alpha}$. Thus
$y = (v_0 \sin \alpha)\frac{x}{v_0 \cos \alpha} - \frac{1}{2}g\left(\frac{x}{v_0 \cos \alpha}\right)^2 = (\tan \alpha)x - \frac{g}{2v_0^2 \cos^2 \alpha}x^2$. Thus the trajectory is a parabola. Continuing by completing the square, we see that
$y - \frac{(\tan^2 \alpha)v_0^2 \cos^2 \alpha}{2g} = -\frac{g}{2v_0^2 \cos^2 \alpha}\left(x - \frac{(\tan \alpha)v_0^2(\cos^2 \alpha)}{g}\right)^2$
or $y - \frac{(\sin^2 \alpha)v_0^2}{2g} = -\frac{g}{2v_0^2 \cos^2 \alpha}\left(x - \frac{(\sin \alpha \cos \alpha)v_0^2}{g}\right)^2$. Thus a parabola with
vertex at $\left(\frac{(\sin \alpha \cos \alpha)v_0^2}{g}, \frac{(\sin^2 \alpha)v_0^2}{2g}\right)$ so the maximum height is $y = \frac{(\sin^2 \alpha)v_0^2}{2g}$.

27. $\vec{r}'(t) = (1 - \cos t)\vec{i} + (\sin t)\vec{j}$, $|\vec{r}'(t)| = \sqrt{1 - 2\cos t + 1} = \sqrt{2(1 - \cos t)}$
$\vec{r}''(t) = (\sin t)\vec{i} + (\cos t)\vec{j}$. Thus $a_T = \frac{\sin t}{\sqrt{2(1 - \cos t)}}$ and
$a_N = \frac{|(\cos t - \cos^2 t - \sin^2 t)\vec{k}|}{\sqrt{2(1 - \cos t)}} = \frac{\sqrt{[(\cos t) - 1]^2}}{\sqrt{2}\sqrt{1 - \cos t}} = \frac{1}{\sqrt{2}}\sqrt{\frac{(1 - \cos t)^2}{(1 - \cos t)}} = \frac{\sqrt{1 - \cos t}}{\sqrt{2}}$.

29. $\vec{r}'(t) = 3t^2\vec{i} + 2t\vec{j} + \vec{k}$, $|\vec{r}'(t)| = \sqrt{9t^4 + 4t^2 + 1}$, $\vec{r}''(t) = 6t\vec{i} + 2\vec{j}$. Thus $a_T = \frac{(18t^3 + 4t)}{\sqrt{9t^4 + 4t^2 + 1}}$
and $a_N = \frac{|-2\vec{i} + 6t\vec{j} + (6t^2 - 12t^2)\vec{k}|}{\sqrt{9t^4 + 4t^2 + 1}}$
$= \frac{\sqrt{4 + 36t^2 + 36t^4}}{\sqrt{9t^4 + 4t^2 + 1}} = \frac{2\sqrt{9t^4 + 9t^2 + 1}}{\sqrt{9t^4 + 4t^2 + 1}}$.

31. $\vec{r}'(t) = e^t\vec{i} + \sqrt{2}\vec{j} - e^{-t}\vec{k}$, $|\vec{r}'(t)| = \sqrt{e^{2t} + 2 + e^{-2t}} = e^t + e^{-t}$,
$\vec{r}''(t) = e^t\vec{i} + e^{-t}\vec{k}$. Then $a_T = \frac{e^{2t} - e^{-2t}}{e^t + e^{-t}} = e^t - e^{-t} = 2\sinh t$ and
$a_N = \frac{|\sqrt{2}e^{-t}\vec{i} - 2\vec{j} - \sqrt{2}e^t\vec{k}|}{e^t + e^{-t}} = \frac{\sqrt{2(e^{-2t} + 2 + e^{2t})}}{(e^t + e^{-t})}$
$= \sqrt{2}\frac{e^t + e^{-t}}{e^t + e^{-t}} = \sqrt{2}$.

Section 11.10

33. With $\vec{r} = (r\cos\theta)\vec{i} + (r\sin\theta)\vec{j}$ and $\vec{h} = \alpha\vec{k}$ where $\alpha > 0$,

   (a) $\vec{h} = \vec{r}\times\vec{r}' = [(r\cos\theta)\vec{i} + (r\sin\theta)\vec{j}]\times[(r'\cos\theta - r\sin\theta\frac{d\theta}{dt})\vec{i} + (r'\sin\theta + r\cos\theta\frac{d\theta}{dt})\vec{j}]$

   $= [rr'\cos\theta\sin\theta + r^2\cos^2\theta\frac{d\theta}{dt} - rr'\cos\theta\sin\theta + r^2\sin^2\theta\frac{d\theta}{dt}]\vec{k} = r^2\frac{d\theta}{dt}\vec{k}$.

   (b) Since $\vec{h} = \alpha\vec{k}$, $\alpha > 0$, $\alpha = |\vec{h}|$. But by (a) $\alpha = |\vec{h}| = r^2\frac{d\theta}{dt}$.

   (c) $A(t) = \frac{1}{2}\int_{\theta_0}^{\theta}|\vec{r}|^2 d\theta = \frac{1}{2}\int_{t_0}^{t} r^2 \frac{d\theta}{dt} dt$ in polar coordinates. Thus

   by the Fundamental Theorem of Calculus $\frac{dA}{dt} = \frac{1}{2}r^2\frac{d\theta}{dt}$.

   (d) $\frac{dA}{dt} = \frac{1}{2}r^2\frac{d\theta}{dt} = \frac{h}{2} = $ constant since $\vec{h}$ is a constant vector and $h = |\vec{h}|$.

35. From Exercise 34, $T^2 = \frac{4\pi^2}{GM}a^3$.

   $T \approx 365.25$ days $\times 24 \frac{\text{hours}}{\text{day}} \times 60 \frac{\text{minutes}}{\text{hour}} \times 60 \frac{\text{seconds}}{\text{minute}} \approx 3.1558 \times 10^7$ seconds

   Therefore, $a^3 = \frac{GMT^2}{4\pi^2} \approx \frac{(6.67\times 10^{-11})(1.99\times 10^{30})(3.1558\times 10^7)^2}{4\pi^2} \approx 3.348\times 10^{33}$ m^3

   $\Rightarrow a \approx 1.496 \times 10^{11}$ m. Thus, the length of the major axis of the earth's orbit (i.e., 2a) is approximately 2.99×10^{11} m $= 2.99 \times 10^8$ km.

EXERCISES 11.10

1. $r = 3$, $\theta = \frac{\pi}{2}$, $z = 1$ so $x = 0$, $y = 3$ and the point is $(0, 3, 1)$.

3. $x = 2\cos(4\pi/3) = -1$, $z = 8$, $y = 2\sin(4\pi/3) = -\sqrt{3}$ so the point is $(-1, -\sqrt{3}, 8)$.

5. $x = 3\cos 0 = 3$, $y = 3\sin 0 = 0$ and $z = -6$ so the point is $(3, 0, -6)$.

7. $r^2 = (-1)^2 + (0)^2 = 1$ so $r = 1$; $z = 0$; $\tan\theta = 0$ so $\theta = 0$ or π. But $x = -1$ so $\theta = \pi$ and the point is $(1, \pi, 0)$.

9. $r^2 = 4$ so $r = 2$, $\tan\theta = \frac{1}{\sqrt{3}}$ so $\theta = \frac{\pi}{6}$ and $z = 4$. Thus the point in cylindrical coordinates is $(2, \pi/6, 4)$.

11. $r = \sqrt{4^2 + 4^2} = 4\sqrt{2}$; $z = 4$; $\tan\theta = 4/4$, so $\theta = \pi/4$ or $\theta = 5\pi/4$, but both x and y are positive, so $\theta = \pi/4$ and the point is $(4\sqrt{2}, \pi/4, 4)$.

13. $x = (1)\sin 0 \cos 0 = 0$, $y = (1)\sin 0 \sin 0 = 0$, $z = (1)\cos 0 = 1$ so the point in rectangular coordinates is $(0, 0, 1)$.

15. $x = \sin(\pi/6)\cos(\pi/6) = \sqrt{3}/4$, $y = \sin(\pi/6)\sin(\pi/6) = 1/4$ and $z = \cos(\pi/6) = \sqrt{3}/2$ so the point is $(\sqrt{3}/4, 1/4, \sqrt{3}/2)$.

Section 11.10

17. $x = 4\sin(\pi/6)\cos(\pi/4) = 4\left(\frac{1}{2}\right)\frac{1}{\sqrt{2}} = \sqrt{2}$, $y = 4\sin(\pi/6)\sin(\pi/4) = \sqrt{2}$

    and $z = 4\cos(\pi/6) = 4(\sqrt{3}/2) = 2\sqrt{3}$ so the point is $(\sqrt{2}, \sqrt{2}, 2\sqrt{3})$.

19. $\rho = \sqrt{9+0+0} = 3$, $\cos\phi = 0/3 = 0$ so $\phi = \pi/2$ and $\cos\theta = \dfrac{-3}{3\sin(\pi/2)} = -1$ so $\theta = \pi$

    and the spherical coordinates are $(3, \pi, \pi/2)$.

21. $\rho = \sqrt{3+1} = 2$, $\cos\phi = 1/2$ so $\phi = \pi/3$ and $\cos\theta = \dfrac{\sqrt{3}}{2\sin(\pi/3)} = \dfrac{\sqrt{3}\cdot 2}{2\cdot\sqrt{3}} = 1$ so $\theta = 0$ and

    the point is $(2, 0, \pi/3)$. (Note: it is also apparent that $\theta = 0$ since the point is in the xz-plane and $x > 0$.)

23. $\rho = \sqrt{1+1+2} = 2$; $z = -\sqrt{2} = 2\cos\phi$, so $\cos\phi = -1/\sqrt{2}$ which implies that $\phi = 3\pi/4$;

    $\cos\theta = \dfrac{1}{2\sin(3\pi/4)} = \dfrac{1}{\sqrt{2}}$, so $\theta = \pi/4$ or $\theta = 7\pi/4$, but $x > 0$ and $y < 0$ so $\theta = 7\pi/4$. Thus

    the point is $(2, 7\pi/4, 3\pi/4)$.

25. $\rho = \sqrt{r^2 + z^2} = \sqrt{2+0} = \sqrt{2}$; $\theta = \pi/4$; $z = 0 = \sqrt{2}\cos\phi$ so $\phi = \pi/2$ and

    the point is $(\sqrt{2}, \pi/4, \pi/2)$.

27. $\rho = \sqrt{r^2 + z^2} = \sqrt{4^2 + 4^2} = 4\sqrt{2}$; $\theta = \pi/3$; $z = 4 = 4\sqrt{2}\cos\phi$ so $\cos\phi = 1/\sqrt{2} \Rightarrow \phi = \pi/4$

    and the point is $(4\sqrt{2}, \pi/3, \pi/4)$.

29. $z = 2\cos 0 = 2$, $r = \sqrt{\rho^2 - z^2} = \sqrt{2^2 - 2^2} = 0$, (or $r = 2\sin 0 = 0$), $\theta = 0$ and the point is

    $(0, 0, 2)$.

31. $z = 8\cos(\pi/2) = 0$, $r = 8\sin(\pi/2) = 8$, $\theta = \pi/6$ and the point is $(8, \pi/6, 0)$.

33. Since $r = 3$, $x^2 + y^2 = 9$ and the surface is a cylinder of radius 3 and axis the z-axis.

35. Since $\phi = \pi/3$, the surface is one frustum of the right circular cone with vertex at the origin and axis the positive z-axis.

37. $z = r^2 = x^2 + y^2$ so the surface is a circular paraboloid with vertex at the origin and axis the positive z-axis.

39. $2 = \rho\cos\phi = z$ is a plane through the point $(0, 0, 2)$ and parallel to the xy-plane.

41. Since $\phi = 0$, $x = 0$ and $y = 0$ while $z = \rho \geq 0$. Thus the locus of points is the positive z-axis including the origin.

43. $r = 2\cos\theta \Rightarrow r^2 = x^2 + y^2 = 2r\cos\theta = 2x \Leftrightarrow (x-1)^2 + y^2 = 1$ which is the equation of a circular cylinder of radius 1, whose axis is the vertical line $x = 1$, $y = 0$, $z = z$.

45. Since $r^2 + z^2 = 25$ and $r^2 = x^2 + y^2$, we have $x^2 + y^2 + z^2 = 25$, a sphere of radius 5 and center at the origin.

Section 11.10

47. Since $x^2 = \rho^2 \sin^2\phi \cos^2\theta$ and $z^2 = \rho^2 \cos^2\phi$, the equation of the surface in rectangular coordinates is $x^2 + z^2 = 4$. Thus the surface is a right circular cylinder of radius 2 about the y–axis.

49. Since $r^2 - r = 0$, $r = 0$ or $r = 1$. But $x^2 + y^2 = r^2$. Thus the surface consists of the right circular cylinder of radius 1 and axis the z–axis along with the surface given by $x^2 + y^2 = 0$, that is, the z–axis.

51. (a) $r^2 = x^2 + y^2$, so $r^2 + z^2 = 16$.
    (b) $\rho^2 = x^2 + y^2 + z^2$, so $\rho^2 = 16$ or $\rho = 4$.

53. (a) $r\cos\theta + 2r\sin\theta + 3z = 6$.
    (b) $\rho \sin\phi\cos\theta + 2\rho\sin\phi\sin\theta + 3\rho\cos\phi = 6$ or $\rho(\sin\phi\cos\theta + 2\sin\phi\sin\theta + 3\cos\phi) = 6$.

55. (a) $r^2(\cos^2\theta - \sin^2\theta) - 2z^2 = 4$ or $2z^2 = r^2\cos 2\theta - 4$.
    (b) $\rho^2(\sin^2\phi\cos^2\theta - \sin^2\phi\sin^2\theta - 2\cos^2\phi) = 4$ or $\rho^2(\sin^2\phi \cos 2\theta - 2\cos^2\phi) = 4$.

57. (a) $r^2 = 2r\sin\theta$ or $r = 2\sin\theta$.
    (b) $\rho^2 \sin^2\phi(\cos^2\theta + \sin^2\theta) = 2\rho\sin\phi\sin\theta$ or $\rho\sin^2\phi = 2\sin\phi\sin\theta$ or $\rho\sin\phi = 2\sin\theta$.

59. $z = r^2 = x^2 + y^2$ is a circular paraboloid with vertex $(0, 0, 0)$, opening upward. $z = 2 - r^2$ $\Rightarrow z - 2 = -(x^2 + y^2)$ is a circular paraboloid with vertex $(0, 0, 2)$ opening downward. Thus $r^2 \leq z \leq 2 - r^2$ is the solid region enclosed by these two surfaces.

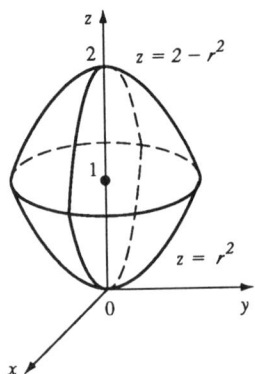

61. $-\frac{\pi}{2} \leq \theta \leq \frac{\pi}{2}$ restricts the solid to the 4 octants in which x is positive.
$\rho = \sec\phi \Rightarrow \rho\cos\phi = z = 1$, which is the equation of a horizontal plane.
$0 \leq \phi \leq \frac{\pi}{6}$ describes a cone, opening upward. So the solid lies above the cone $\phi = \frac{\pi}{6}$ and below the plane $z = 1$.

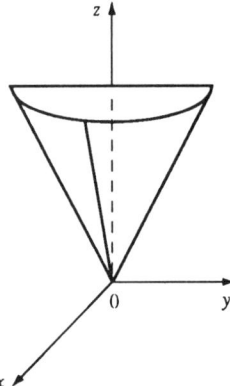

Chapter 11 Review

63. $z \geq \sqrt{x^2+y^2}$ because the solid lies above the cone. Squaring both sides of this inequality gives $z^2 \geq x^2+y^2 \Rightarrow 2z^2 \geq x^2+y^2+z^2 = \rho^2 \Rightarrow z^2 = \rho^2\cos^2\phi \geq \frac{1}{2}\rho^2 \Rightarrow \cos^2\phi \geq \frac{1}{2}$. The cone opens upward so that the inequality is $\cos\phi \geq \frac{1}{\sqrt{2}}$, or equivalently $0 \leq \phi \leq \frac{\pi}{4}$.

In spherical coordinates the sphere $z = x^2+y^2+z^2$ is $\rho\cos\phi = \rho^2 \Rightarrow \rho = \cos\phi$.

$0 \leq \rho \leq \cos\phi$ because the solid lies below the sphere. The solid can therefore be described as the region in spherical coordinates satisfying $0 \leq \rho \leq \cos\phi$, $0 \leq \phi \leq \frac{\pi}{4}$.

REVIEW EXERCISES FOR CHAPTER 11

1. By Thm. 11.13.2, this is true.

3. True. If θ is the angle between $\vec{u}$ and $\vec{v}$, then by Thm. 11.29 $|\vec{u}\times\vec{v}| = |\vec{u}||\vec{v}|\sin\theta = |\vec{v}||\vec{u}|\sin\theta = |\vec{v}\times\vec{u}|$. [Or, by Thm. 11.31, $|\vec{u}\times\vec{v}| = |-\vec{v}\times\vec{u}| = |-1||\vec{v}\times\vec{u}| = |\vec{v}\times\vec{u}|$.]

5. Thm. 11.31.2 tells us that this is true.

7. This is true by Thm. 11.31.5.

9. This is true by Theorem 11.28.

11. If $|\vec{u}| = 1$, $|\vec{v}| = 1$ and θ is the angle between these two vectors (so $0 \leq \theta \leq \pi$), then by Thm. 11.29 $|\vec{u}\times\vec{v}| = |\vec{u}||\vec{v}|\sin\theta = \sin\theta$, which is equal to 1 if and only if $\theta = \pi/2$ (i.e. the two vectors are orthogonal). Therefore, the assertion that the cross product of two unit vectors is a unit vector is false.

13. This is false because by 11.48, $\frac{x^2}{1} + \frac{y^2}{1} = 1$ is the equation of a circular cylinder.

15. $|AB| = \sqrt{9+16+144} = \sqrt{169} = 13$, $|BC| = \sqrt{1+1+36} = \sqrt{38}$, $|CA| = \sqrt{4+25+36} = \sqrt{65}$.

17. Completing the squares gives
$(x+2)^2 + (y+3)^2 + (z-5)^2 = -2+4+9+25 = 36$, thus $C(-2,-3,5)$ and radius 6.

19. $6\vec{a} - 5\vec{c} = (6-0)\vec{i} + (6-5)\vec{j} + (-12+25)\vec{k} = 6\vec{i} + \vec{j} + 13\vec{k}$.

21. $\vec{a}\cdot\vec{b} = (1)(3) + (1)(-2) + (-2)(1) = -1$.

23. $\vec{b}\times\vec{c} = \begin{vmatrix} \vec{i} & \vec{j} & \vec{k} \\ 3 & -2 & 1 \\ 0 & 1 & -5 \end{vmatrix} = 9\vec{i} + 15\vec{j} + 3\vec{k}$, $|\vec{b}\times\vec{c}| = 3\sqrt{9+25+1} = 3\sqrt{35}$.

25. $\vec{c}\times\vec{c} = \vec{0}$ for any $\vec{c}$.

27. $\cos\theta = \dfrac{\vec{a}\cdot\vec{b}}{|\vec{a}||\vec{b}|} = \dfrac{-1}{\sqrt{6}\sqrt{14}} = \dfrac{-1}{2\sqrt{21}}$ and $\theta = \cos^{-1}\dfrac{-1}{2\sqrt{21}} \approx 96°$.

29. $|\vec{b}|\cos\theta = \vec{a}\cdot\vec{b}/|\vec{a}| = -1/\sqrt{6}$.

31. We need $4x + 3x - 28 = 0$ or $x = 4$.

33. (a) $(\vec{u}\times\vec{v})\cdot\vec{w} = \vec{u}\cdot(\vec{v}\times\vec{w}) = 2$

    (b) $\vec{u}\cdot(\vec{w}\times\vec{v}) = \vec{u}\cdot[-(\vec{v}\times\vec{w})] = -\vec{u}\cdot(\vec{v}\times\vec{w}) = -2$

    (c) $\vec{v}\cdot(\vec{u}\times\vec{w}) = (\vec{v}\times\vec{u})\cdot\vec{w} = -(\vec{u}\times\vec{v})\cdot\vec{w} = -2$

    (d) $(\vec{u}\times\vec{v})\cdot\vec{v} = \vec{u}\cdot(\vec{v}\times\vec{v}) = \vec{u}\cdot\vec{0} = 0$

35. For simplicity, consider a unit cube positioned with its back left corner at the origin. Vector representations of the diagonals joining the points (0, 0, 0) to (1, 1, 1) and (1, 0, 0) to (0, 1, 1) are $<1, 1, 1>$ and $<-1, 1, 1>$ respectively. Let θ be the angle between these two vectors. $<1, 1, 1>\cdot<-1, 1, 1> = -1+1+1 = 1 = |<1, 1, 1>||<-1, 1, 1>|\cos\theta = 3\cos\theta$
$\Rightarrow \cos\theta = \tfrac{1}{3} \Rightarrow \theta = \cos^{-1}\left(\tfrac{1}{3}\right) \approx 71°$.

37. $\vec{AB} = <1, 0, -1>$, $\vec{AC} = <0, 4, 3>$, then

    (a) a vector perpendicular to the plane is
    $\vec{AB}\times\vec{AC} = <0+4, -(3+0), 4-0> = <4, -3, 4>$

    (b) $A = (1/2)|\vec{AB}\times\vec{AC}| = (1/2)\sqrt{16+9+16} = \sqrt{41}/2$.

39. Let F_1 be the magnitude of the force directed 20° away from the direction of shore. And let F_2 be the magnitude of the other force. Separating these forces into components parallel to the direction of the resultant force and perpendicular to it gives
$F_1\cos 20° + F_2\cos 30° = 255$ (1)
$F_1\sin 20° - F_2\sin 30° = 0 \Rightarrow F_1 = F_2\dfrac{\sin 30°}{\sin 20°}$ (2)

    Substituting (2) into (1) gives $F_2(\sin 30°\cot 20° + \cos 30°) = 255 \Rightarrow F_2 \approx 114$ N.
    Substituting this into (2) gives $F_1 \approx 166$ N.

41. $x = 1+2t, y = 2-t, z = 4+3t$.

43. $\vec{v} = <4, -3, 5>$ so $x = 1+4t, y = -3t, z = 1+5t$.

45. $x+2y+5z = -4+2+10$ or $x+2y+5z = 8$.

47. Substitution of the parametric equations into the equation of the plane gives
$2x - y + z = 2(2-t) - (1+3t) + 4t = 2 \Rightarrow -t + 3 = 2 \Rightarrow t = 1$.
When $t = 1$, the parametric equations yield $x = 2-1 = 1$, $y = 1+3 = 4$ and $z = 4$.
Therefore, the point of intersection is (1, 4, 4).

49. Since the direction vectors $<2, 3, 4>$ and $<6, -1, 2>$ aren't parallel, neither are the lines.

For the lines to intersect, the three equations $1 + 2t = -1 + 6s$, $2 + 3t = 3 - s$, $3 + 4t = -5 + 2s$ must be satisfied simultaneously. Solving the first two equations gives $t = 1/5$, $s = 2/5$ and checking we see these values don't satisfy the third equation. Thus the lines aren't parallel and they don't intersect, so they must be skew.

51. Use the formula in Example 11.5.8: $D = |(3)(6) + (1)(2) + (-4)(-1) - 2|/\sqrt{26} = 22/\sqrt{26}$.

53. A plane through the x–axis intersecting the yz–plane in the line $y = z$, $x = 0$.

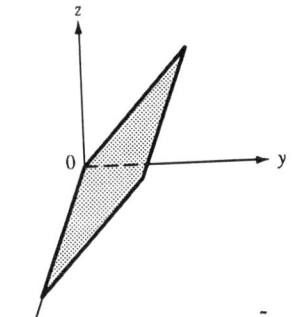

55. A circular paraboloid with vertex the origin and axis the y–axis.

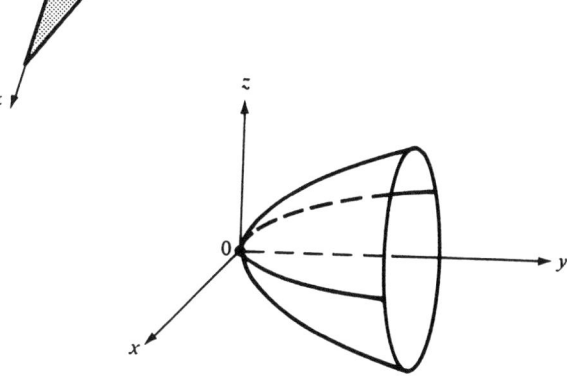

57. A (right elliptical) cone with vertex at the origin and axis the x–axis.

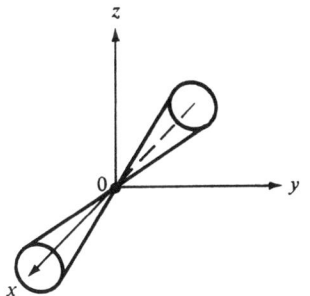

59. A hyperboloid of two sheets with axis the y–axis. Traces parallel to the xz–plane are circles for $|y| > 2$.

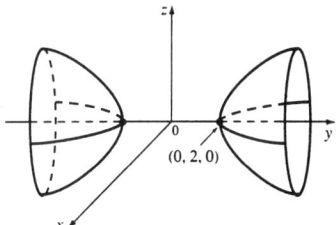

Chapter 11 Review

61. $4x^2 + y^2 = 16 \Leftrightarrow \frac{x^2}{4} + \frac{y^2}{16} = 1$. The equation of the ellipsoid is $\frac{x^2}{4} + \frac{y^2}{16} + \frac{z^2}{c^2} = 1$, since the horizontal trace in the plane $z = 0$ must be the original ellipse. The traces of the ellipsoid in the yz–plane must be circles since the surface is obtained by rotation about the x–axis. Therefore, $c^2 = 16$ and the equation of the ellipsoid is
$\frac{x^2}{4} + \frac{y^2}{16} + \frac{z^2}{16} = 1 \Leftrightarrow 4x^2 + y^2 + z^2 = 16$.

63. (a) Since $x = 2$ and $y^2 + z^2 = 1$, the curve is a circle in the plane $x = 2$ with center $(2, 0, 0)$ and radius 1.

    (b) $\vec{r}'(t) = \cos t\, \vec{j} - \sin t\, \vec{k}$
    $\vec{r}''(t) = -\sin t\, \vec{j} - \cos t\, \vec{k}$.

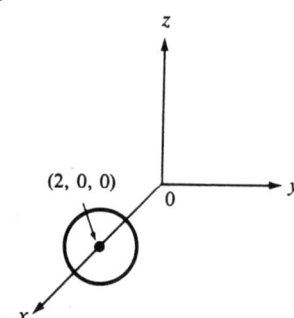

65. $\int_0^1 [(t+t^2)\vec{i} + (2+t^3)\vec{j} + t^4\vec{k}]\,dt$
    $= \{[(t^2/2) + (t^3/3)]\vec{i} + [2t + (t^4/4)]\vec{j} + (t^5/5)\vec{k}\}\Big|_0^1 = (5/6)\vec{i} + (9/4)\vec{j} + (1/5)\vec{k}$.

67. The angle of intersection of the two curves, θ, is the angle between their respective tangents at the point of intersection. For both curves the point $(1, 0, 0)$ occurs when $t = 0$.
    $\vec{r}_1'(t) = -\sin t\, \vec{i} + \cos t\, \vec{j} + \vec{k} \Rightarrow \vec{r}_1'(0) = \vec{j} + \vec{k}$ and $\vec{r}_2'(t) = \vec{i} + 2t\vec{j} + 3t^2\vec{k} \Rightarrow \vec{r}_2'(0) = \vec{i}$
    $\vec{r}_1'(0) \cdot \vec{r}_2'(0) = (\vec{j} + \vec{k}) \cdot \vec{i} = 0$. Therefore, the curves intersect at right angles to each other, i.e., $\theta = \pi/2$.

69. (a) $\vec{T}(t) = <t^2, t, 1>/\sqrt{t^4 + t^2 + 1}$.

    (b) $\vec{T}'(t) = -\frac{1}{2}(t^4 + t^2 + 1)^{-3/2}(4t^3 + 2t)<t^2, t, 1> + (t^4 + t^2 + 1)^{-1/2}<2t, 1, 0>$
    $= (-2t^3 - t)<t^2, t, 1>/(t^4 + t^2 + 1)^{3/2} + <2t, 1, 0>/(t^4 + t^2 + 1)^{1/2}$
    $= [<-2t^5 - t^3, -2t^4 - t^2, -2t^3 - t>$
    $\quad + <2t^5 + 2t^3 + 2t, t^4 + t^2 + 1, 0>]/(t^4 + t^2 + 1)^{3/2}$
    $= <2t, -t^4 + 1, -2t^3 - t>/(t^4 + t^2 + 1)^{3/2}$
    $|\vec{T}'(t)| = \sqrt{4t^2 + t^8 - 2t^4 + 1 + 4t^6 + 4t^4 + t^2}/(t^4 + t^2 + 1)^{3/2}$
    $= \sqrt{t^8 + 4t^6 + 2t^4 + 5t^2}/(t^4 + t^2 + 1)^{3/2}$ and
    $\vec{N}(t) = <2t, 1 - t^4, -2t^3 - t>/\sqrt{t^8 + 4t^6 + 2t^4 + 5t^2}$.

    (c) $\kappa(t) = |\vec{T}'(t)|/|\vec{r}'(t)| = \sqrt{t^8 + 4t^6 + 2t^4 + 5t^2}/(t^4 + t^2 + 1)^2$.

71. Using Exercise 28 of Section 11.8 we have
    $\vec{r}'(t) = <-3\sin t, 4\cos t>$, $\vec{r}''(t) = <-3\cos t, -4\sin t>$,
    $|\vec{r}'(t)|^3 = (\sqrt{9\sin^2 t + 4\cos^2 t})^3$ and then

88

Chapter 11 Review

$\kappa(t) = |(-3\sin t)(-4\sin t) - (-3\cos t)(4\cos t)|/[9\sin^2 t + 16\cos^2 t]^{3/2}$
$= 12/[9\sin^2 t + 16\cos^2 t]^{3/2}$.
At $(3, 0)$, $t = 0$ and $\kappa(0) = 12/16^{3/2} = 12/64 = 3/16$.
At $(0, 4)$, $t = \pi/2$ and $\kappa(\pi/2) = 12/9^{3/2} = 12/27 = 4/9$.

73. $\vec{v}(t) = \int(t\vec{i} + \vec{j} + t^2\vec{k})\,dt = (t^2/2)\vec{i} + t\vec{j} + (t^3/3)\vec{k} + \vec{c}_1$, but $\vec{i} + 2\vec{j} + \vec{k} = \vec{v}(0) = \vec{0} + \vec{c}_1$ so $\vec{c}_1 = \vec{i} + 2\vec{j} + \vec{k}$ and $\vec{v}(t) = (1 + (t^2/2))\vec{i} + (2 + t)\vec{j} + (1 + (t^3/3))\vec{k}$.
$\vec{r}(t) = \int\vec{v}(t)\,dt = (t + (t^3/6))\vec{i} + (2t + (t^2/2))\vec{j} + (t + (t^4/12))\vec{k} + \vec{c}_2$. But
$\vec{r}(0) = \vec{0}$ so $\vec{c}_2 = \vec{0}$ and $\vec{r}(t) = (t + (t^3/6))\vec{i} + (2t + (t^2/2))\vec{j} + (t + (t^4/12))\vec{k}$.

75. $x = 2\cos(\pi/6) = \sqrt{3}$, $y = 2\sin(\pi/6) = 1$, $z = 2$, so in rectangular coordinates the point is $(\sqrt{3}, 1, 2)$. $\rho = \sqrt{3 + 1 + 4} = 2\sqrt{2}$, $\theta = \pi/6$, and $\cos\phi = z/\rho = 1/\sqrt{2}$, so $\phi = \pi/4$ and the spherical coordinates are $(2\sqrt{2}, \pi/6, \pi/4)$.

77. $x = 4\sin(\pi/6)\cos(\pi/3) = 1$, $y = 4\sin(\pi/6)\sin(\pi/3) = \sqrt{3}$,
$z = 4\cos(\pi/6) = 2\sqrt{3}$ so in rectangular coordinates the point is
$(1, \sqrt{3}, 2\sqrt{3})$. $r^2 = x^2 + y^2 = 4$, $r = 2$, so the cylindrical coordinates are $(2, \pi/3, 2\sqrt{3})$.

79. A half–plane including the z–axis and intersecting the xy–plane in the half–line $x = y$, $x > 0$.

81. Since $\rho = 3\sec\phi$, $\rho\cos\phi = 3$ or $z = 3$. Thus the surface is a plane parallel to the xy–plane and through the point $(0, 0, 3)$.

83. (a) $r^2 + z^2 = 4$.

    (b) $\rho^2 = 4$ or $\rho = 2$.

85. The resulting surface is a paraboloid of revolution with equation $z = 4x^2 + 4y^2$. Changing to cylindrical coordinates we have $z = 4(x^2 + y^2) = 4r^2$.

87. (a) Instead of proceeding directly, we use part 3 of Theorem 11.53:
$$\vec{r}(t) = t\vec{R}(t) \Rightarrow \vec{v} = \vec{r}\,'(t) = \vec{R}(t) + t\vec{R}\,'(t) = \cos\omega t\,\vec{i} + \sin\omega t\,\vec{j} + t\vec{v}_d$$

(b) Using the same method as in part (a) and starting with $\vec{v} = \vec{R}(t) + t\vec{R}\,'(t)$, we have
$$\vec{a} = \vec{v}\,' = \vec{R}\,'(t) + \vec{R}\,'(t) + t\vec{R}\,''(t) = 2\vec{R}\,'(t) + t\vec{R}\,''(t) = 2\vec{v}_d + t\vec{a}_d$$

(c) Here we have $\vec{r}(t) = e^{-t}\cos\omega t\,\vec{i} + e^{-t}\sin\omega t\,\vec{j} = e^{-t}\vec{R}(t)$. So, as in parts (a) and (b),
$$\vec{v} = \vec{r}\,'(t) = e^{-t}\vec{R}\,'(t) - e^{-t}\vec{R}(t) = e^{-t}[\vec{R}\,'(t) - \vec{R}(t)] \Rightarrow$$
$$\vec{a} = \vec{v}\,' = e^{-t}[\vec{R}\,''(t) - \vec{R}\,'(t)] - e^{-t}[\vec{R}\,'(t) - \vec{R}(t)]$$
$$= e^{-t}[\vec{R}\,''(t) - 2\vec{R}\,'(t) + \vec{R}(t)] = e^{-t}\vec{a}_d - 2e^{-t}\vec{v}_d + e^{-t}\vec{R}.$$

Thus the Coriolis acceleration (the "extra" terms not involving $\vec{a}_d$) is $-2e^{-t}\vec{v}_d + e^{-t}\vec{R}$.

APPLICATIONS PLUS (after Chapter 11)

1. (a) The first step in the chain occurs when the local government spends D dollars. The people who receive it spend a fraction c of those D dollars, that is, Dc dollars. Those who receive the Dc dollars spend a fraction c of it, that is, Dc^2 dollars. Continuing in this way, we see that the total spending after n transactions is
$$S_n = D + Dc + Dc^2 + \cdots + Dc^{n-1} = \frac{D(1-c^n)}{1-c} \text{ by (10.14)}.$$

(b) $\lim_{n\to\infty} S_n = \lim_{n\to\infty} \frac{D(1-c^n)}{1-c} = \frac{D}{1-c} \lim_{n\to\infty}(1-c^n) = \frac{D}{1-c} = \frac{D}{s} = kD$, since
$0 < c < 1 \Rightarrow \lim_{n\to\infty} c^n = 0$.
If $c = .8$, then $s = 1 - c = .2$ and the multiplier $k = \frac{1}{s} = 5$.

3. (a) Initially, the ball falls a distance H, then rebounds a distance rH, falls rH, rebounds r^2H, falls r^2H, etc. The total distance it travels is $H + 2rH + 2r^2H + 2r^3H + \cdots$
$= H(1 + 2r + 2r^2 + 2r^3 + \cdots) = H[1 + 2r(1 + r + r^2 + \cdots)] = H[1 + 2r(\frac{1}{1-r})]$
$= H\left(\frac{1+r}{1-r}\right)$ meters.

(b) From Exercise 9.1.32 or Example 5 in section 11.9, we know that a ball falls $\frac{1}{2}gt^2$ meters in t seconds, where g is gravitational acceleration. Thus a ball falls h meters in $\sqrt{2h/g}$ seconds. For the bouncing ball, the total travel time in seconds is
$\sqrt{2H/g} + 2\sqrt{2rH/g} + 2\sqrt{2r^2H/g} + 2\sqrt{2r^3H/g} + \cdots$
$= \sqrt{2H/g}[1 + 2\sqrt{r} + 2(\sqrt{r})^2 + 2(\sqrt{r})^3 + \cdots] = \sqrt{2H/g}\{1 + 2\sqrt{r}[1 + \sqrt{r} + (\sqrt{r})^2 + \cdots]\}$
$= \sqrt{2H/g}[1 + 2\sqrt{r}(\frac{1}{1-\sqrt{r}})] = \sqrt{\frac{2H}{g}} \frac{1+\sqrt{r}}{1-\sqrt{r}}$.

(c) It will help to make a chart of the time for each descent and each rebound of the ball, together with the velocity just before and just after each bounce. Recall that the time in seconds needed to fall h meters is $\sqrt{2h/g}$. The ball hits the ground with velocity $-g\sqrt{2h/g}$ (*taking the upward direction to be positive*) and rebounds with velocity $kg\sqrt{2h/g}$, taking time $k\sqrt{2h/g}$ to reach the top of its bounce, where its velocity is 0. At that point, its height is k^2h. (All these results follow from the formulas for vertical motion with gravitational acceleration $-g$:
$\frac{d^2y}{dt^2} = -g \Rightarrow v = \frac{dy}{dt} = v_0 - gt \Rightarrow y = y_0 + v_0 t - \frac{1}{2}gt^2$)

Applications Plus

Number of descent	time of descent	speed before bounce	speed after bounce	time of ascent	peak height
1	$\sqrt{2H/g}$	$\sqrt{2Hg}$	$k\sqrt{2Hg}$	$k\sqrt{2H/g}$	k^2H
2	$\sqrt{2k^2H/g}$	$\sqrt{2k^2Hg}$	$k\sqrt{2k^2Hg}$	$k\sqrt{2k^2H/g}$	k^4H
3	$\sqrt{2k^4H/g}$	$\sqrt{2k^4Hg}$	$k\sqrt{2k^4Hg}$	$k\sqrt{2k^4H/g}$	k^6H
...	...	...	...	...	...

The total travel time in seconds is
$$\sqrt{2H/g} + k\sqrt{2H/g} + k\sqrt{2H/g} + k^2\sqrt{2H/g} + k^2\sqrt{2H/g} + \cdots$$
$$= \sqrt{2H/g}(1 + 2k + 2k^2 + 2k^3 + \cdots) = \sqrt{2H/g}[1 + 2k(1 + k + k^2 + \cdots)]$$
$$= \sqrt{2H/g}[1 + 2k(\tfrac{1}{1-k})] = \sqrt{\tfrac{2H}{g}}\tfrac{1+k}{1-k}.$$

<u>Another Method</u>: We could use part (b). At the top of the bounce, the height is $k^2h = rh$, so $\sqrt{r} = k$ and the result follows from part (b).

5. (a) The projectile reaches maximum height when $0 = \frac{dy}{dt} = \frac{d}{dt}[(v_0 \sin\alpha)t - \tfrac{1}{2}gt^2]$
$= v_0\sin\alpha - gt$; that is, when $t = (v_0\sin\alpha)/g$ and
$y = (v_0\sin\alpha)\left(\tfrac{v_0\sin\alpha}{g}\right) - \tfrac{1}{2}g\left(\tfrac{v_0\sin\alpha}{g}\right)^2 = \tfrac{v_0^2\sin^2\alpha}{2g}$. This is the maximum height attained when the projectile is fired with an angle of elevation α. This maximum height is largest when $\alpha = \tfrac{\pi}{2}$. In that case, $\sin\alpha = 1$ and the maximum height is $\tfrac{v_0^2}{2g}$.

(b) Let $R = v_0^2/g$. We are asked to consider the parabola $x^2 + 2Ry - R^2 = 0$ which can be rewritten as $y = \tfrac{-1}{2R}x^2 + \tfrac{R}{2}$. The points on or inside this parabola are those for which $-R \leq x \leq R$ and $0 \leq y \leq \tfrac{-1}{2R}x^2 + \tfrac{R}{2}$. When the projectile is fired at angle of elevation α, the points (x,y) along its path satisfy the relations
$x = (v_0\cos\alpha)t$ and $y = (v_0\sin\alpha)t - \tfrac{1}{2}gt^2$, where $0 \leq t \leq (2v_0\sin\alpha)/g$ (*as in Example* 5 *in Section* 11.9). Thus
$|x| \leq |(v_0\cos\alpha)(2v_0\sin\alpha)/g| = |(v_0^2/g)\sin 2\alpha| \leq \left|\tfrac{v_0^2}{g}\right| = |R|$. This shows that $-R \leq x \leq R$. For t in the specified range, we also have
$y = t(v_0\sin\alpha - \tfrac{1}{2}gt) = \tfrac{1}{2}gt\left(\tfrac{2v_0\sin\alpha}{g} - t\right) \geq 0$ and
$y = (v_0\sin\alpha)\tfrac{x}{v_0\cos\alpha} - \tfrac{1}{2}g\left(\tfrac{x}{v_0\cos\alpha}\right)^2 = (\tan\alpha)x - \tfrac{g}{2v_0^2\cos^2\alpha}x^2$
$= \tfrac{-1}{2R\cos^2\alpha}x^2 + (\tan\alpha)x$. Thus
$y - \left(\tfrac{-1}{2R}x^2 + \tfrac{R}{2}\right) = \tfrac{-1}{2R\cos^2\alpha}x^2 + \tfrac{1}{2R}x^2 + (\tan\alpha)x - \tfrac{R}{2}$
$= \tfrac{x^2}{2R}\left(1 - \tfrac{1}{\cos^2\alpha}\right) + (\tan\alpha)x - \tfrac{R}{2} = \tfrac{1}{2R}[x^2(1 - \sec^2\alpha) + 2R(\tan\alpha)x - R^2]$
$= \tfrac{1}{2R}[-(\tan^2\alpha)x^2 + 2R(\tan\alpha)x - R^2] = \tfrac{-1}{2R}[(\tan\alpha)x - R]^2 \leq 0.$

Applications Plus

We have shown that every target that can be hit by the projectile lies on or inside the parabola $y = \frac{-1}{2R}x^2 + \frac{R}{2}$.

Now let (a, b) be any point on or inside the parabola $y = \frac{-1}{2R}x^2 + \frac{R}{2}$. Then $-R \leq a \leq R$ and $0 \leq b \leq \frac{-1}{2R}a^2 + \frac{R}{2}$. We seek an angle α such that (a, b) lies in the path of the projectile; that is, we wish to find an angle α such that

$$b = \frac{-1}{2R\cos^2\alpha}a^2 + (\tan\alpha)a \text{ or equivalently } b = \frac{-1}{2R}(\tan^2\alpha + 1)a^2 + (\tan\alpha)a.$$

Rearranging this equation we get $\frac{a^2}{2R}\tan^2\alpha - a\tan\alpha + \left(\frac{a^2}{2R} + b\right) = 0$ or

(*) $a^2(\tan\alpha)^2 - 2aR(\tan\alpha) + (a^2 + 2bR) = 0$. This quadratic equation for $\tan\alpha$ has real solutions exactly when the discriminant is nonnegative. Now

$B^2 - 4AC \geq 0 \Leftrightarrow (-2aR)^2 - 4a^2(a^2 + 2bR) \geq 0 \Leftrightarrow 4a^2(R^2 - a^2 - 2bR) \geq 0$

$\Leftrightarrow -a^2 - 2bR + R^2 \geq 0 \Leftrightarrow b \leq \frac{1}{2R}(R^2 - a^2) \Leftrightarrow b \leq \frac{-1}{2R}a^2 + \frac{R}{2}$.

This condition is satisfied since (a, b) is on or inside the parabola $y = \frac{-1}{2R}x^2 + \frac{R}{2}$.

It follows that (a, b) lies in the path of the projectile when $\tan\alpha$ satisfies (*), that is, when $\tan\alpha = \dfrac{2aR \pm \sqrt{4a^2(R^2 - a^2 - 2bR)}}{2a^2} = \dfrac{R \pm \sqrt{R^2 - 2bR - a^2}}{a}$.

(c) If the gun is pointed at a target with height h at a distance D downrange, then $\tan\alpha = \frac{h}{D}$. When the projectile reaches a distance D downrange (*remember we are assuming that it doesn't hit the ground first*), we have $D = x = (v_0\cos\alpha)t$, so $t = \dfrac{D}{(v_0\cos\alpha)}$ and $y = (v_0\sin\alpha)t - \frac{1}{2}gt^2 = D\tan\alpha - \dfrac{gD^2}{2v_0^2\cos^2\alpha}$. Meanwhile, the target, whose x-coordinate is also D, has fallen from height h to height $h - \frac{1}{2}gt^2 = D\tan\alpha - \dfrac{gD^2}{2v_0^2\cos^2\alpha}$. Thus the projectile hits the target.

7. (a) $m\dfrac{d^2\vec{R}}{dt^2} = -mg\vec{j} - k\dfrac{d\vec{R}}{dt} \Rightarrow \dfrac{d}{dt}(m\dfrac{d\vec{R}}{dt} + k\vec{R} + mgt\vec{j}) = \vec{0} \Rightarrow m\dfrac{d\vec{R}}{dt} + k\vec{R} + mgt\vec{j} = \vec{c}$

($\vec{c}$ *is a constant vector in the xy-plane*). At $t = 0$, this says $m\vec{v}(0) + k\vec{R}(0) = \vec{c}$.

Since $\vec{v}(0) = \vec{v}_0$ and $\vec{R}(0) = \vec{0}$, we have $\vec{c} = m\vec{v}_0$. Therefore

$\dfrac{d\vec{R}}{dt} + \dfrac{k}{m}\vec{R} + gt\vec{j} = \vec{v}_0$, or $\dfrac{d\vec{R}}{dt} + \dfrac{k}{m}\vec{R} = \vec{v}_0 - gt\vec{j}$.

Applications Plus

(b) Multiplying by $e^{(k/m)t}$ gives $e^{(k/m)t}\frac{d\vec{R}}{dt} + \frac{k}{m}e^{(k/m)t}\vec{R} = e^{(k/m)t}\vec{v}_0 - gte^{(k/m)t}\vec{j}$

or $\frac{d}{dt}\left(e^{(k/m)t}\vec{R}\right) = e^{(k/m)t}\vec{v}_0 - gte^{(k/m)t}\vec{j}$. Integrating gives

$e^{(k/m)t}\vec{R} = \frac{m}{k}e^{(k/m)t}\vec{v}_0 - \left(\frac{mg}{k}te^{(k/m)t} - \frac{m^2g}{k^2}e^{(k/m)t}\right)\vec{j} + \vec{b}$ for some constant

vector $\vec{b}$. Setting $t = 0$ yields the relation $\vec{R}(0) = \frac{m}{k}\vec{v}_0 + \frac{m^2g}{k^2}\vec{j} + \vec{b}$, so

$\vec{b} = -\frac{m}{k}\vec{v}_0 - \frac{m^2g}{k^2}\vec{j}$. Thus

$e^{(k/m)t}\vec{R} = \frac{m}{k}\left[e^{(k/m)t} - 1\right]\vec{v}_0 - \left\{\frac{mg}{k}te^{(k/m)t} - \frac{m^2g}{k^2}\left[e^{(k/m)t} - 1\right]\right\}\vec{j}$

and $\vec{R}(t) = \frac{m}{k}\left[1 - e^{-kt/m}\right]\vec{v}_0 + \frac{mg}{k}\left[\frac{m}{k}(1 - e^{-kt/m}) - t\right]\vec{j}$

9. (a) For the function $F(x) = \begin{cases} 0 & \text{if } x \leq 0 \\ P(x) & \text{if } 0 < x < 1 \\ 1 & \text{if } x \geq 1 \end{cases}$ to be continuous,

we must have $P(0) = 0$ and $P(1) = 1$. For F' to be continuous, we must have $P'(0) = P'(1) = 0$. The curvature of the curve $y = F(x)$ at the point $(x, F(x))$ is

$\kappa(x) = \frac{|F''(x)|}{[1 + (F'(x))^2]^{3/2}}$. For $\kappa(x)$ to be continuous, we must have

$P''(0) = P''(1) = 0$. Write $P(x) = ax^5 + bx^4 + cx^3 + dx^2 + ex + f$. Then

$P'(x) = 5ax^4 + 4bx^3 + 3cx^2 + 2dx + e$ and $P''(x) = 20ax^3 + 12bx^2 + 6cx + 2d$.

Our six conditions are:

$P(0) = 0$: $f = 0$ (1)

$P(1) = 1$: $a + b + c + d + e + f = 1$ (2)

$P'(0) = 0$: $e = 0$ (3)

$P'(1) = 0$: $5a + 4b + 3c + 2d + e = 0$ (4)

$P''(0) = 0$: $d = 0$ (5)

$P''(1) = 0$: $20a + 12b + 6c + 2d = 0$ (6)

From (1), (3), and (5) we have $d = e = f = 0$. Thus (2), (4) and (6) become

(7) $a + b + c = 1$, (8) $5a + 4b + 3c = 0$, and (9) $10a + 6b + 3c = 0$.

Subtracting (8) from (9) gives (10) $5a + 2b = 0$. Multiplying (7) by 3 and

subtracting from (8) gives (11) $2a + b = -3$. Multiplying (11) by 2 and

subtracting from (10) gives $a = 6$. By (10), $b = -15$. By (7), $c = 10$. Thus

$P(x) = 6x^5 - 15x^4 + 10x^3$.

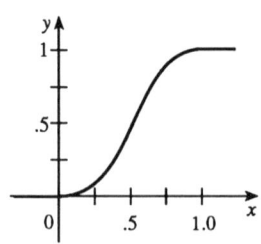

Applications Plus

(b) $$H(x) = \begin{cases} 1 & \text{if } x \leq 0 \\ \sqrt{1-x^2} & \text{if } 0 < x < 1/\sqrt{2} \\ -x + \sqrt{2} & \text{if } x \geq 1/\sqrt{2} \end{cases}$$

$\lim_{x \to 0^+} \sqrt{1-x^2} = 1 = H(0)$, and $\lim_{x \to (1/\sqrt{2})^-} \sqrt{1-x^2} = 1/\sqrt{2} = H(1/\sqrt{2})$, so H is continuous.

$$H'(x) = \begin{cases} 0 & \text{if } x < 0 \\ -x/\sqrt{1-x^2} & \text{if } 0 < x < 1/\sqrt{2} \\ -1 & \text{if } x > 1/\sqrt{2} \end{cases}$$

Since $\lim_{x \to 0^+} H'(x) = 0 = \lim_{x \to 0^-} H'(x)$ and $\lim_{x \to (1/\sqrt{2})^-} H'(x) = -1 = \lim_{x \to (1/\sqrt{2})^+} H'(x)$, the definitions $H'(0) = 0$ and $H'(1/\sqrt{2}) = -1$ make H' a continuous function.

$$H''(x) = \begin{cases} 0 & \text{if } x < 0 \\ -1/(1-x^2)^{3/2} & \text{if } 0 < x < 1/\sqrt{2} \\ 0 & \text{if } x > 1/\sqrt{2} \end{cases}$$

since $\frac{d}{dx}[-x(1-x^2)^{-1/2}] = -(1-x^2)^{-1/2} - x^2(1-x^2)^{-3/2} = -(1-x^2)^{-3/2}$.

Now $\lim_{x \to 0^+} H''(x) = -1 \neq 0 = \lim_{x \to 0^-} H''(x)$, so H'' cannot be made continuous at $x = 0$. (The same is true at $x = 1/\sqrt{2}$). So H does not have continuous curvature.

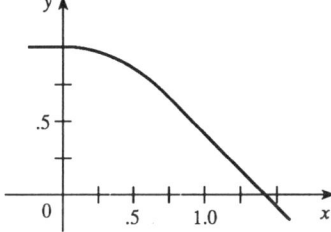

(c) As in part (a), set $P(x) = ax^5 + bx^4 + cx^3 + dx^2 + ex + f$. The continuity conditions on P are $P(0) = 0$, $P(1) = 1$, $P'(0) = 0$, $P'(1) = 1$, $P''(0) = 0$, and $P''(1) = 0$. The conditions $P(0) = P'(0) = P''(0) = 0 \Rightarrow d = e = f = 0$ as before. The other conditions $\Rightarrow a + b + c = 1$, $5a + 4b + 3c = 1$, and $10a + 6b + 3c = 0$. From these, we find that $a = 3$, $b = -8$, and $c = 6$. Therefore $P(x) = 3x^5 - 8x^4 + 6x^3$. Since there was no solution with $a = 0$, this could not have been done with a polynomial of degree 4.

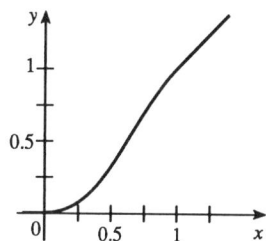

Applications Plus

11. (a) Dividing the equation $|\vec{F}|\sin\theta = mv_R^2/R$ by the equation $|\vec{F}|\cos\theta = mg$, we obtain $\tan\theta = v_R^2/(Rg)$, so $v_R^2 = Rg\tan\theta$.

    (b) $R = 400$ ft and $\theta = 12°$, so $v_R = \sqrt{Rg\tan\theta} \approx \sqrt{400 \cdot 32 \cdot \tan 12°}$ ≈ 52.16 ft/s ≈ 36 mi/h.

    (c) We want to choose a new radius R_1 for which the new rated speed is $\frac{3}{2}$ of the old one: $\sqrt{R_1 g\tan 12°} = \frac{3}{2}\sqrt{Rg\tan 12°}$. Squaring, we get $R_1 g\tan 12° = \frac{9}{4}Rg\tan 12°$, so $R_1 = \frac{9}{4}R = \frac{9}{4}(400) = 900$ ft.

CHAPTER 12

EXERCISES 12.1

1. (a) $f(2,1) = 4 - 1 + 4(2)(1) - 7(2) + 10 = 7$
   (b) $f(-3,5) = 9 - 25 + 4(-3)(5) + 21 + 10 = -45$
   (c) $f(x+h,y) = (x+h)^2 - y^2 + 4(x+h)y - 7(x+h) + 10$
       $= x^2 + 2xh + h^2 - y^2 + 4xy + 4hy - 7x - 7h + 10$
   (d) $f(x,y+k) = x^2 - (y+k)^2 + 4x(y+k) - 7x + 10$
       $= x^2 - y^2 - 2ky - k^2 + 4xy + 4xk - 7x + 10$
   (e) $f(x,x) = x^2 - x^2 + 4x^2 - 7x + 10 = 4x^2 - 7x + 10$.

3. (a) $F(1,1) = 3(1)(1)/(1+2) = 1$
   (b) $F(-1,2) = 3(-1)(2)/(1+8) = -2/3$
   (c) $F(t,1) = 3t/(t^2+2)$
   (d) $F(-1,y) = 3(-1)y/(1+2y^2) = -3y/(1+2y^2)$
   (e) $F(x,x^2) = 3xx^2/(x^2 + 2(x^2)^2) = 3x^3/(x^2 + 2x^4) = 3x/(1+2x^2)$.

5. $D = R^2$ and the range is R.

7. $x + y \neq 0$ so $D = \{(x,y) \mid x+y \neq 0\}$. Since $2/(x+y)$ can't be zero, the range is $\{z \mid z \neq 0\}$.

9. $D = R^2$ since the exponential function is defined everywhere and the range is $\{z \mid z > 0\}$.

11. For the logarithmic function to be defined, we need $x - y + z > 0$. Thus
    $D = \{(x,y,z) \mid x+z > y\}$ and the range is R.

13. $D = R^3$ while the range is R.

15. $y - 2x \geq 0$ so $D = \{(x,y) \mid y \geq 2x\}$.

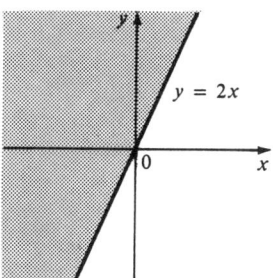

17. $x + 2y \neq 0$ and $9 - x^2 - y^2 \geq 0$, so
$D = \{(x,y) \mid y \neq -(1/2)x$ and
$x^2 + y^2 \leq 9\}$.

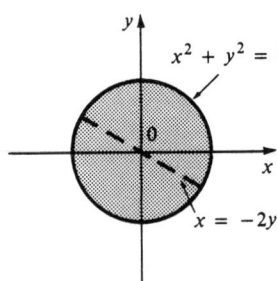

19. $D = \{(x,y) \mid x^2 + y \geq 0\}$
$= \{(x,y) \mid y \geq -x^2\}$.

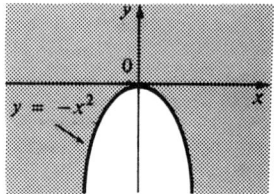

21. $D = \{(x,y) \mid xy > 1\}$.

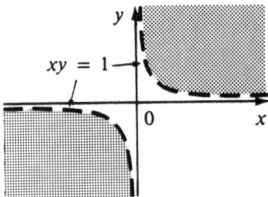

23. $D = \{(x,y) \mid y \neq (\pi/2) + n\pi, n \in \mathbb{Z}\}$.

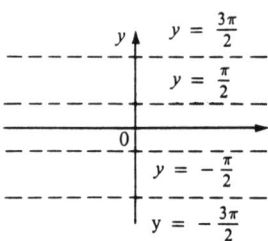

98

Section 12.1

25. $D = \{(x,y) | -1 \le x+y \le 1\}$
    $= \{(x,y) | -1-x \le y \le 1-x\}$.

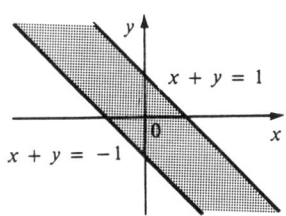

27. Since $\sin y > 0$ implies
    $2n\pi < y < (2n+1)\pi$,
    $n \in Z$, $D = \{(x,y) | x > 0$
    and $2n\pi < y < (2n+1)\pi$,
    $n \in Z\}$.

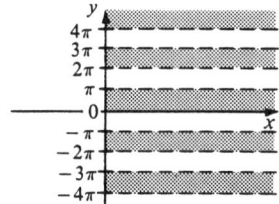

29. $D = \{(x,y,z) | x^2 + y^2 + z^2 \le 1\}$
    (the points inside or on the sphere of
    radius 1, center the origin).

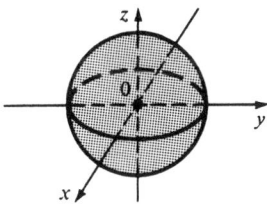

31. $z = 3$, a horizontal plane through
    the point $(0, 0, 3)$.

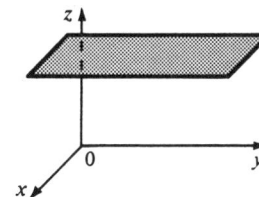

Section 12.1

33. $z = 1 - x - y$ or $x + y + z = 1$,
a plane with intercepts 1, 1, and 1.

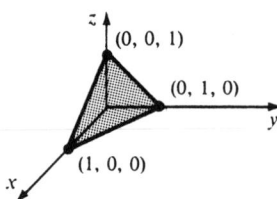

35. $z = x^2 + 9y^2$, an elliptic paraboloid with vertex the origin.

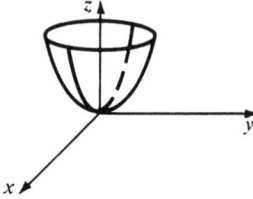

37. $z = \sqrt{x^2 + y^2}$, top half of right circular cone.

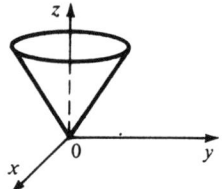

39. $z = y^2 - x^2$, a hyperbolic paraboloid.

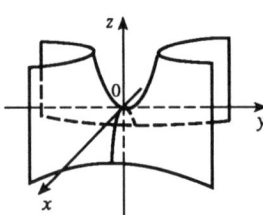

41. $z = 1 - x^2$, a parabolic cylinder.

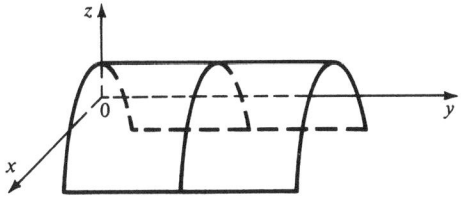

43. The level curves are $xy = k$. For $k = 0$ the curves are the coordinate axis; if $k > 0$, they are hyperbolas in the 1st and 3rd quadrants; if $k < 0$, they are hyperbolas in the 2nd and 4th quadrants.

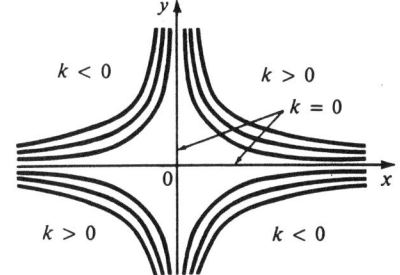

45. $k = x^2 + 9y^2$, a family of ellipses ($k = 0$, the origin) with major axis the x–axis.

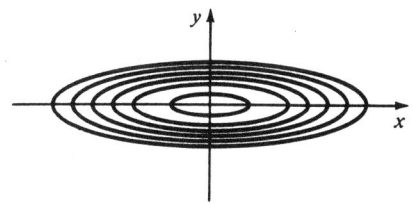

47. $k = \frac{x}{y}$ is a family of lines without the point $(0,0)$.

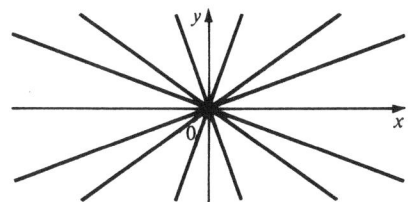

49. $k = \sqrt{x+y}$ or for $x+y \geq 0$,
$k^2 = x+y$, or $y = -x+k^2$
(note: $k \geq 0$ since $k = \sqrt{x+y}$.)

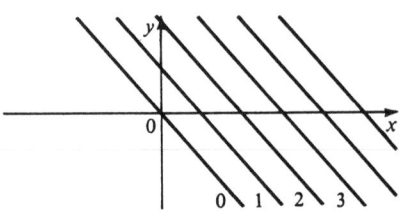

51. $k = x - y^2$, or $x - k = y^2$, a family of parabolas with vertex $(k, 0)$.

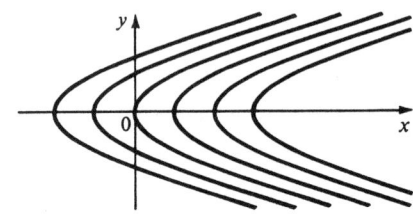

53. $k = x + 3y + 5z$ is a family of parallel planes with normal vector $<1, 3, 5>$.

55. $k = x^2 - y^2 + z^2$ are the equations of the level surfaces. For $k = 0$, the surface is a right circular cone with vertex the origin and axis the y-axis. For $k > 0$, we have a family of hyperboloids of one sheet with axis the y-axis. For $k < 0$, we have a family of hyperboloids of two sheets with axis the y-axis.

57. The isothermals are given by
$k = 100/(1 + x^2 + 2y^2)$ or
$x^2 + 2y^2 = (100 - k)/k$
$(0 < k \leq 100)$, a family of ellipses.

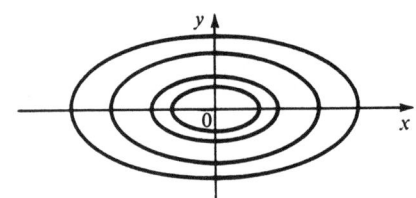

59. (a) B (b) III.
61. (a) F (b) V.
63. (a) D (b) IV.

EXERCISES 12.2

1. Since a polynomial, the limit equals $(2^2)(3^2) - 2(2)(3^5) + 3(3) = -927$.

3. Since a rational function defined at $(0, 0)$, the limit equals $(0 + 0 - 5)/(2 - 0) = -5/2$.

Section 12.2

5. The product of two functions continuous at (π, π), so the limit equals
$\pi \sin((\pi + \pi)/4) = \pi$.

7. This limit does not exist because the function is not defined on the line $y = -x$ and is therefore not defined on an entire disk with center $(0,0)$. (See Definition 12.5.)

9. Let $f(x,y) = (x-y)/(x^2+y^2)$. First approach $(0,0)$ along the x-axis. Then
$f(x,0) = x/x^2 = 1/x$ and $\lim_{x \to 0} f(x,0)$ doesn't exist. Thus $\lim_{(x,y) \to (0,0)} f(x,y)$ doesn't exist.

11. Let $f(x,y) = 8x^2y^2/(x^4+y^4)$. Approaching $(0,0)$ along the x-axis gives $f(x,y) \to 0$ as $(x,y) \to (0,0)$ along the x-axis. Approaching $(0,0)$ along the line $y = x$,
$f(x,x) = 8x^4/2x^4 = 4$ for $x \neq 0$, so along this line $f(x,y) \to 4$ as $(x,y) \to (0,0)$. Thus the limit doesn't exist.

13. Let $f(x,y) = 2xy/(x^2+2y^2)$. As $(x,y) \to (0,0)$ along the x-axis, $f(x,y) \to 0$.
But as $(x,y) \to (0,0)$ along the line $y = x$, $f(x,x) = 2x^2/3x^2$
so $f(x,y) \to 2/3$ as $(x,y) \to (0,0)$ along this line. Thus the limit doesn't exist.

15. We can show that the limit along any line through $(0,0)$ is 0 and that the limits along the paths $x = y^2$ and $y = x^2$ are also 0. So we suspect that the limit exists and equals 0. Let $\epsilon > 0$ be given. We need to find $\delta > 0$ such that $|xy/\sqrt{x^2+y^2} - 0| < \epsilon$ whenever $0 < \sqrt{x^2+y^2} < \delta$ or $|xy|/\sqrt{x^2+y^2} < \epsilon$ whenever $0 < \sqrt{x^2+y^2} < \delta$. But $|x| = \sqrt{x^2} \leq \sqrt{x^2+y^2}$ so $|xy|/\sqrt{x^2+y^2} \leq |y| = \sqrt{y^2} \leq \sqrt{x^2+y^2}$. Thus choose $\delta = \epsilon$ and let $0 < \sqrt{x^2+y^2} < \delta = \epsilon$, then $|xy/\sqrt{x^2+y^2} - 0| \leq \sqrt{x^2+y^2} < \delta = \epsilon$. Hence by definition $\lim_{(x,y) \to (0,0)} xy/\sqrt{x^2+y^2} = 0$.
[OR: Use the Squeeze Theorem:
$0 \leq \left|\dfrac{xy}{\sqrt{x^2+y^2}}\right| \leq |x|$ since $|y| \leq \sqrt{x^2+y^2}$, and $|x| \to 0$ as $(x,y) \to (0,0)$.]

17. Since $(xy+1)/(x^2+y^2+1)$ is a rational function defined at $(0,0)$ the limit is $\dfrac{0+1}{0+0+1} = 1$.

19. Let $f(x,y) = 2x^2y/(x^4+y^2)$. Then $f(x,0) = 0$ for $x \neq 0$,
so $f(x,y) \to 0$ as $(x,y) \to (0,0)$ along the x-axis. But $f(x,x^2) = 2x^4/2x^4 = 1$
for $x \neq 0$, so $f(x,y) \to 1$ as $(x,y) \to (0,0)$ along the parabola $y = x^2$. Thus the limit doesn't exist.

21. $\lim_{(x,y) \to (0,0)} (x^2+y^2)/[\sqrt{x^2+y^2+1} - 1]$
$= \lim_{(x,y) \to (0,0)} (x^2+y^2)(\sqrt{x^2+y^2+1} + 1)/(x^2+y^2)$
$= \lim_{(x,y) \to (0,0)} [\sqrt{x^2+y^2+1} + 1] = 2$.

23. Let $f(x,y) = (xy-x)/(x^2+y^2-2y+1)$. Then $f(0,y) = 0$ for $y \neq 1$,
so $f(x,y) \to 0$ as $(x,y) \to (0,1)$ along the y-axis.

103

Section 12.2

But $f(x, x+1) = x(x+1-1)/(x^2 + (x+1-1)^2) = 1/2$ for $x \neq 0$ so $f(x,y) \to 1/2$ as $(x,y) \to (0,1)$ along the line $y = x+1$. Thus the limit doesn't exist.

25. $\lim_{(x,y,z) \to (1,2,3)} (xz^2 - y^2z)/(xyz - 1) = [(1)(3^2) - (2^2)(3)]/[(1)(2)(3) - 1] = -3/5$
    since the function is continuous at $(1, 2, 3)$.

27. Let $f(x,y,z) = (x^2 - y^2 - z^2)/(x^2 + y^2 + z^2)$. Then $f(x, 0, 0) = 1$ for $x \neq 0$ and $f(0, y, 0) = -1$ for $y \neq 0$, so as $(x, y, z) \to (0, 0, 0)$ along the x-axis, $f(x, y, z) \to 1$ but as $(x, y, z) \to (0, 0, 0)$ along the y-axis, $f(x, y, z) \to -1$. Thus the limit doesn't exist.

29. Let $f(x,y,z) = (xy + yz^2 + xz^2)/(x^2 + y^2 + z^4)$. Then $f(x, 0, 0) = 0/x^2 = 0$ for $x \neq 0$, so as $(x, y, z) \to (0, 0, 0)$ along the x-axis, $f(x, y, z) \to 0$. But $f(x, x, 0) = x^2/(2x^2) = \frac{1}{2}$ for $x \neq 0$, so as $(x, y, z) \to (0, 0, 0)$ along the line $y = x$, $z = 0$, $f(x, y, z) \to \frac{1}{2}$. Thus the limit does not exist.

31. $h(x,y) = g(f(x,y)) = g(x^4 + x^2y^2 + y^4) = e^{-(x^4 + x^2y^2 + y^4)} \cos(x^4 + x^2y^2 + y^4)$. Since f is a polynomial it is continuous throughout R^2 and g is the product of two functions both of which are continuous on R, by Theorem 12.7, h is continuous on R^2.

33. $h(x,y) = g(f(x,y)) = (2x + 3y - 6)^2 + \sqrt{2x + 3y - 6}$. Since f is a polynomial, it is continuous on R^2 and g is continuous on its domain $\{t \mid t \geq 0\}$. Thus h is continuous on its domain $D = \{(x,y) \mid 2x + 3y - 6 \geq 0\} = \{(x,y) \mid y \geq (-2/3)x + 2\}$ which consists of all points on and above right of the line $y = (-2/3)x + 2$.

35. $F(x,y)$ is a rational function and thus is continuous on its domain $D = \{(x,y) \mid x^2 + y^2 - 1 \neq 0\}$, i.e., F is continuous except on the circle $x^2 + y^2 = 1$.

37. $F(x,y) = g(f(x,y))$ where $f(x,y) = x^4 - y^4$, a polynomial so continuous on R^2 and $g(t) = \tan t$, continuous on its domain $\{t \mid t \neq (2n+1)(\pi/2), n \in Z\}$. Thus F is continuous on its domain $D = \{(x,y) \mid x^4 - y^4 \neq (2n+1)(\pi/2), n \in Z\}$.

39. $G(x,y) = g(x,y)f(x,y)$ where $g(x,y) = e^{xy}$ and $f(x,y) = \sin(x+y)$ both of which are continuous on R^2. Thus G is continuous on R^2.

41. $G(x,y) = g_1(f_1(x,y)) - g_2(f_2(x,y))$ where $f_1(x,y) = x + y$ and $f_2(x,y) = x - y$ both of which are polynomials so continuous on R^2 and $g_1(t) = \sqrt{t}$, $g_2(s) = \sqrt{s}$ both of which are continuous on their respective domains $\{t \mid t \geq 0\}$ and $\{s \mid s \geq 0\}$. Thus $g_1 \circ f_1$ is continuous on its domain $D_1 = \{(x,y) \mid x + y \geq 0\} = \{(x,y) \mid y \geq -x\}$ and $g_2 \circ f_2$ is continuous on its domain $D_2 = \{(x,y) \mid x - y \geq 0\} = \{(x,y) \mid y \leq x\}$. Then G, being the difference of these two composite functions, is continuous on its domain $D = D_1 \cap D_2 = \{(x,y) \mid -x \leq y \leq x\} = \{(x,y) \mid |y| \leq x\}$.

43. $f(x,y,z) = xg(f(y,z))$ where $f(y,z) = yz$, continuous on R^2 and $g(t) = \ln t$, continuous on its domain $\{t \mid t > 0\}$. Since $h(x) = x$ is continuous on R,

Section 12.2

$f(x, y, z)$ is continuous on its domain $D = \{(x, y, z) \mid yz > 0\}$.

In Exercises 45–47 each f is a piecewise defined function where each first piece is a rational function defined everywhere except at the origin. Thus each f is continuous on R^2 except possibly at the origin. So for each we need only check $\lim_{(x,y) \to (0,0)} f(x,y)$.

45. Letting $z = \sqrt{2}x$, $\lim_{(x,y) \to (0,0)} (2x^2 - y^2)/(2x^2 + y^2) =$

    $\lim_{(z,y) \to (0,0)} (z^2 - y^2)/(z^2 + y^2)$ which doesn't exist by Example 12.1.

    Thus f is not continuous at $(0,0)$ and the largest set on which f is continuous is $\{(x,y) \mid (x,y) \neq (0,0)\}$.

47. Since $x^2 \leq 2x^2 + y^2$, we have $\left|x^2 y^3/(2x^2 + y^2)\right| \leq |y^3|$. We know that $|y^3| \to 0$ as $(x,y) \to (0,0)$. So, by the Squeeze Theorem, $\lim_{(x,y) \to (0,0)} f(x,y) = \lim_{(x,y) \to (0,0)} \frac{x^2 y^3}{2x^2 + y^2} = 0$.

    But $f(0,0) = 1$, so f is discontinuous at $(0,0)$. For $(x,y) \neq (0,0)$, $f(x,y)$ is equal to a rational function and is therefore continuous. Therefore f is continuous on the set $\{(x,y) \mid (x,y) \neq (0,0)\}$.

49. $\lim_{(x,y) \to (0,0)} \frac{\sin(x^2 + y^2)}{x^2 + y^2} = \lim_{r \to 0^+} \frac{\sin(r^2)}{r^2}$ which is an indeterminate form of type $\frac{0}{0}$. Using

    l'Hospital's rule one gets $\lim_{r \to 0^+} \frac{\sin(r^2)}{r^2} \stackrel{H}{=} \lim_{r \to 0^+} \frac{2r \cos(r^2)}{2r} = \lim_{r \to 0^+} \cos(r^2) = 1$.

    [Or just use the fact that $\lim_{\theta \to 0} \frac{\sin \theta}{\theta} = 1$.]

51. (a) Let $\epsilon > 0$ be given. We need to find $\delta > 0$ such that $|x - a| < \epsilon$ whenever

    $0 < \sqrt{(x-a)^2 + (y-b)^2} < \delta$. But $|x - a| = \sqrt{(x-a)^2} \leq \sqrt{(x-a)^2 + (y-b)^2}$.

    Thus setting $\delta = \epsilon$ and letting $0 < \sqrt{(x-a)^2 + (y-b)^2} < \delta$, we have

    $|x - a| \leq \sqrt{(x-a)^2 + (y-b)^2} < \delta = \epsilon$. Hence, by 12.5, $\lim_{(x,y) \to (a,b)} x = a$.

    (b) The argument is the same as in (a) with the roles of x and y interchanged.

    (c) Let $\epsilon > 0$ be given and set $\delta = \epsilon$. Then $|f(x,y) - L| = |c - c| = 0$

    $\leq \sqrt{(x-a)^2 + (y-b)^2} < \delta = \epsilon$ whenever $0 < \sqrt{(x-a)^2 + (y-b)^2} < \delta$. Thus by

    definition 12.5 $\lim_{(x,y) \to (a,b)} c = c$.

53. Since $|\vec{x} - \vec{a}|^2 = |\vec{x}|^2 + |\vec{a}|^2 - 2|\vec{x}||\vec{a}|\cos\theta \geq |\vec{x}|^2 + |\vec{a}|^2 - 2|\vec{x}||\vec{a}| = (|\vec{x}| - |\vec{a}|)^2$, we have

    $||\vec{x}| - |\vec{a}|| \leq |\vec{x} - \vec{a}|$. Let $\epsilon > 0$ be given and set $\delta = \epsilon$. Then whenever $0 < |\vec{x} - \vec{a}| < \delta$,

    $||\vec{x}| - |\vec{a}|| \leq |\vec{x} - \vec{a}| < \delta = \epsilon$. Hence $\lim_{\vec{x} \to \vec{a}} |\vec{x}| = |\vec{a}|$ and $f(\vec{x}) = |\vec{x}|$ is continuous on R^n.

Section 12.3

EXERCISES 12.3

1. $f(x,y) = 16 - 4x^2 - y^2 \Rightarrow f_x(x,y) = -8x$ and $f_y(x,y) = -2y \Rightarrow f_x(1,2) = -8$ and $f_y(1,2) = -4$. The graph of f is the paraboloid $z = 16 - 4x^2 - y^2$ and the vertical plane $y = 2$ intersects it in the parabola $z = 12 - 4x^2$, $y = 2$, (*the curve C_1 in the first figure*). The slope of the tangent line to this parabola at $(1,2,8)$ is $f_x(1,2) = -8$. Similarly the plane $x = 1$ intersects the paraboloid in the parabola $z = 12 - y^2$, $x = 1$, (*the curve C_2 in the second figure*) and the slope of the tangent line at $(1,2,8)$ is $f_y(1,2) = -4$.

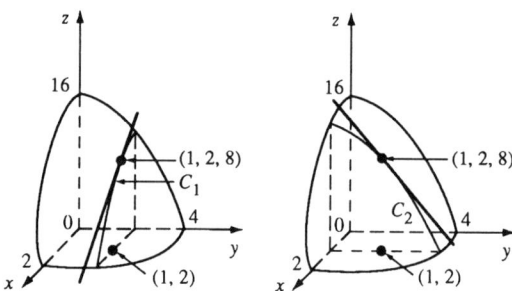

3. $f_x(x,y) = 3x^2 y^5$, $f_x(3,-1) = -27$.
5. $\partial f/\partial y = x(-1)e^{-y} + 3$, $(\partial f/\partial y)(1,0) = -1 + 3 = 2$.
7. $\partial z/\partial x = [3x^2(x^2+y^2) - (x^3+y^3)(2x)]/(x^2+y^2)^2$
   $= (x^4 + 3x^2 y^2 - 2xy^3)/(x^2+y^2)^2$
   $\partial z/\partial y = [3y^2(x^2+y^2) - (x^3+y^3)(2y)]/(x^2+y^2)^2$
   $= (3x^2 y^2 + y^4 - 2yx^3)/(x^2+y^2)^2$.
9. $\partial z/\partial x = (1/y) - (y/x^2)$.
11. $(\partial/\partial x)(xy + yz) = (\partial/\partial x)(xz)$ or $y + y(\partial z/\partial x) = z + x(\partial z/\partial x)$ or $(y-x)(\partial z/\partial x) = z - y$, so $\partial z/\partial x = (z-y)/(y-x)$
    $(\partial/\partial y)(xy + yz) = (\partial/\partial y)(xz)$ or $x + z + y(\partial z/\partial y) = x(\partial z/\partial y)$ or $(y-x)(\partial z/\partial y) = -(x+z)$, so $\partial z/\partial y = (x+z)/(x-y)$.
13. $(\partial/\partial x)(x^2+y^2-z^2) = (\partial/\partial x)[2x(y+z)]$ or $2x - 2z(\partial z/\partial x) = 2(y+z) + 2x(\partial z/\partial x)$ or $2(x+z)(\partial z/\partial x) = 2(x-y-z)$ so $\partial z/\partial x = (x-y-z)/(x+z)$.
    $(\partial/\partial y)(x^2+y^2-z^2) = (\partial/\partial y)[2x(y+z)]$ or $2y - 2z(\partial z/\partial y) = 2x[1 + (\partial z/\partial y)]$ or $2(x+z)(\partial z/\partial y) = 2(y-x)$ so $\partial z/\partial y = (y-x)/(x+z)$.
15. $\partial u/\partial x = y\sec(xy) + xy[\sec(xy)\tan(xy)](y) = y\sec(xy)[1 + xy\tan(xy)]$.
17. $f_y(x,y,z) = xz$ so $f_y(0,1,2) = 0$.

Section 12.3

19. $u_x = y + z$, $u_y = x + z$, $u_z = y + x$.
21. $f_x(x,y) = 3x^2y^5 - 4xy + 1$, $f_y(x,y) = 5x^3y^4 - 2x^2$.
23. $f_x(x,y) = 4x^3 + 2xy^2$, $f_y(x,y) = 2x^2y + 4y^3$.
25. $f_x(x,y) = [(1)(x+y) - (x-y)(1)]/(x+y)^2 = 2y/(x+y)^2$
    $f_y(x,y) = [(-1)(x+y) - (x-y)(1)]/(x+y)^2 = -2x/(x+y)^2$.
27. $f_x(x,y) = e^x \tan(x-y) + e^x(\sec^2(x-y)) = e^x[\tan(x-y) + \sec^2(x-y)]$
    $f_y(x,y) = e^x[\sec^2(x-y)](-1) = -e^x \sec^2(x-y)$.
29. $f_s(s,t) = (1/2)(2 - 3s^2 - 5t^2)^{-1/2}(-6s) = -3s/\sqrt{2 - 3s^2 - 5t^2}$
    $f_t(s,t) = (1/2)(2 - 3s^2 - 5t^2)^{-1/2}(-10t) = -5t/\sqrt{2 - 3s^2 - 5t^2}$.
31. $f_u(u,v) = (1/[1 + (u/v)^2])(1/v) = (1/v)[v^2/(u^2 + v^2)] = v/(u^2 + v^2)$
    $f_v(u,v) = (1/[1 + (u/v)^2])(-u/v^2) = (-u/v^2)[v^2/(u^2 + v^2)] = -u/(u^2 + v^2)$.
33. $g_x(x,y) = [y \sec^2(x^2y^3)](2xy^3) = 2xy^4 \sec^2(x^2y^3)$
    $g_y(x,y) = \tan(x^2y^3) + [y \sec^2(x^2y^3)](3x^2y^2)$
    $= \tan(x^2y^3) + 3x^2y^3 \sec^2(x^2y^3)$.
35. $\partial z/\partial x = [1/(x + \sqrt{x^2 + y^2})][1 + (1/2)(x^2 + y^2)^{-1/2}(2x)]$
    $= [(\sqrt{x^2 + y^2} + x)/\sqrt{x^2 + y^2}]/(x + \sqrt{x^2 + y^2}) = 1/\sqrt{x^2 + y^2}$
    $\partial z/\partial y = [1/(x + \sqrt{x^2 + y^2})](1/2)(x^2 + y^2)^{-1/2}(2y) = y/(x\sqrt{x^2 + y^2} + x^2 + y^2)$.
37. $\partial z/\partial x = (\cosh\sqrt{3x + 4y})(1/2)(3x + 4y)^{-1/2}(3) = (3\cosh\sqrt{3x + 4y})/(2\sqrt{3x + 4y})$
    $\partial z/\partial y = (\cosh\sqrt{3x + 4y})(1/2)(3x + 4y)^{-1/2}(4) = (2\cosh\sqrt{3x + 4y})/\sqrt{3x + 4y}$.
39. By the Fundamental Theorem of Calculus, Part I, $(d/dx)\int_a^x f(t)dt = f(x)$ for f continuous. Thus $f_x(x,y) = (\partial/\partial x)(\int_x^y e^{t^2}dt) = (\partial/\partial x)(-\int_y^x e^{t^2}dt) = -e^{x^2}$. And $f_y(x,y) = (\partial/\partial y)(\int_x^y e^{t^2}dt) = e^{y^2}$.
41. $f_x(x,y,z) = 2xyz^3 + y$, $f_y(x,y,z) = x^2z^3 + x$, $f_z(x,y,z) = 3x^2yz^2 - 1$.
43. $f_x(x,y,z) = yzx^{yz-1}$. By Theorem 6.33, $f_y(x,y,z) = x^{yz}\ln(x^z) = zx^{yz}\ln x$ and by symmetry $f_z(x,y,z) = yx^{yz}\ln x$.
45. $u_x = [z\cos(y/(x+z))](-y(x+z)^{-2}) = -yz\cos[y/(x+z)]/(x+z)^2$,
    $u_y = [z\cos(y/(x+z))](1/(x+z)) = z\cos[y/(x+z)]/(x+z)$,
    $u_z = \sin[y/(x+z)] + [z\cos(y/(x+z))](-y(x+z)^{-2})$
    $= \sin[y/(x+z)] - yz\cos[y/(x+z)]/(x+z)^2$.
47. $u_x = y^2z^3 \ln(x + 2y + 3z) = xy^2z^3(1/(x + 2y + 3z))$
    $= y^2z^3[\ln(x + 2y + 3z) + x/(x + 2y + 3z)]$
    $u_y = 2xyz^3 \ln(x + 2y + 3z) + xy^2z^3(1/(x + 2y + 3z))(2)$
    $= 2xyz^3[\ln(x + 2y + 3z) + y/(x + 2y + 3z)]$
    and by symmetry $u_z = 3xy^2z^2[\ln(x + 2y + 3z) + z/(x + 2y + 3z)]$.

Section 12.3

49. $f_x(x,y,z,t) = 1/(z-t)$, $f_y(x,y,z,t) = -1/(z-t)$,
$f_z(x,y,z,t) = (x-y)(-1)(z-t)^{-2} = (y-x)/(z-t)^2$
and $f_t(x,y,z,t) = (x-y)(-1)(z-t)^{-2}(-1) = (x-y)/(z-t)^2$.

51. For each $i = 1,...,n$, $u_{x_i} = (1/2)(x_1^2 + x_2^2 + ... + x_n^2)^{-1/2}(2x_i)$
$= x_i/\sqrt{x_1^2 + x_2^2 + ... + x_n^2}$.

53. $f_x(x,y) = \lim_{h \to 0} \dfrac{f(x+h,y) - f(x,y)}{h} = \lim_{h \to 0} \dfrac{(x+h)^2 - (x+h)y + 2y^2 - (x^2 - xy + 2y^2)}{h}$
$= \lim_{h \to 0} \dfrac{h(2x - y + h)}{h} = \lim_{h \to 0} (2x - y + h) = 2x - y$.
$f_y(x,y) = \lim_{h \to 0} \dfrac{f(x, y+h) - f(x,y)}{h} = \lim_{h \to 0} \dfrac{x^2 - x(y+h) + 2(y+h)^2 - (x^2 - xy + 2y^2)}{h}$
$= \lim_{h \to 0} \dfrac{h(4y - x + 2h)}{h} = \lim_{h \to 0} (4y - x + 2h) = 4y - x$.

55. $\partial z/\partial x = f'(x) \qquad \partial z/\partial y = g'(y)$.

57. Let $u = x + y$. Then $\partial z/\partial x = (df/du)(\partial u/\partial x) = df/d(x+y) = f'(x+y)$
and $\partial z/\partial y = (df/du)(\partial u/\partial y) = df/d(x+y) = f'(x+y)$.

59. Let $u = x/y$, then $\partial u/\partial x = 1/y$ and $\partial u/\partial y = -x/y^2$.
Hence $\partial z/\partial x = (df/du)(\partial u/\partial x) = [df/d(x/y)]/y = f'(x/y)/y$
and $\partial z/\partial y = [df/d(x/y)](-x/y^2) = -x[df/d(x/y)]/y^2 = -xf'(x/y)/y^2$.

61. $f_x = 2xy + \sqrt{y}$, $f_y = x^2 + x/(2\sqrt{y})$. Thus $f_{xx} = 2y$, $f_{xy} = 2x + 1/(2\sqrt{y})$, $f_{yx} = 2x + 1/(2\sqrt{y})$
and $f_{yy} = -x/(4y^{3/2})$.

63. $z_x = (3/2)(x^2 + y^2)^{1/2}(2x) = 3x(x^2 + y^2)^{1/2}$ and $z_y = 3y(x^2 + y^2)^{1/2}$. Thus
$z_{xx} = 3(x^2 + y^2)^{1/2} + 3x(x^2 + y^2)^{-1/2}(1/2)(2x) = [3(x^2 + y^2) + 3x^2]/\sqrt{x^2 + y^2}$
$= 3(2x^2 + y^2)/\sqrt{x^2 + y^2}$ and
$z_{xy} = 3x(1/2)(x^2 + y^2)^{-1/2}(2y) = 3xy/\sqrt{x^2 + y^2}$.
By symmetry $z_{yx} = 3xy/\sqrt{x^2 + y^2}$ and $z_{yy} = 3(x^2 + 2y^2)/\sqrt{x^2 + y^2}$.

65. $z_x = t(1/\sqrt{1 - (\sqrt{x})^2})(1/2)x^{-1/2} = t/(2\sqrt{x - x^2})$, $z_t = \sin^{-1}\sqrt{x}$. Thus
$z_{xx} = (t/2)(-1/2)(x - x^2)^{-3/2}(1 - 2x) = t(2x - 1)/[4(x - x^2)^{3/2}]$,
$z_{xt} = 1/(2\sqrt{x - x^2})$, $z_{tx} = (1/\sqrt{1 - (\sqrt{x})^2})(1/2)x^{-1/2} = 1/(2\sqrt{x - x^2})$, and
$z_{tt} = 0$.

67. $u_x = 5x^4y^4 - 6xy^3 + 4x$, $u_{xy} = 20x^4y^3 - 18xy^2$ and $u_y = 4x^5y^3 - 9x^2y^2$,
$u_{yx} = 20x^4y^3 - 18xy^2$. Thus $u_{xy} = u_{yx}$.

69. $u_x = (1/\sqrt{1 - (xy^2)^2})(y^2) = y^2\sqrt{1 - x^2y^4}$,
$u_{xy} = 2y(1 - x^2y^4)^{-1/2} + y^2(-1/2)(1 - x^2y^4)^{-3/2}(-4x^2y^3)$
$= [2y(1 - x^2y^4) + 2x^2y^5]/(1 - x^2y^4)^{3/2} = 2y/(1 - x^2y^4)^{3/2}$.

Section 12.3

And $u_y = (1/\sqrt{1-(xy^2)^2})(2xy) = 2xy/\sqrt{1-x^2y^4}$,

$u_{yx} = [2y\sqrt{1-x^2y^4} - 2xy(1/2)(1-x^2y^4)^{-1/2}(-2xy^4)]/(1-x^2y^4)$

$= (2y - 2x^2y^5 + 2x^2y^5)/(1-x^2y^4)^{3/2} = 2y/(1-x^2y^4)^{3/2}$.

Thus $u_{xy} = u_{yx}$.

71. $f_x = 2xy^3 - 8x^3y$, $f_{xx} = 2y^3 - 24x^2y$, $f_{xxx} = -48xy$.

73. $f_x = 5x^4 + 4x^3y^4z^3$, $f_{xy} = 16x^3y^3z^3$, and $f_{xyz} = 48x^3y^3z^2$.

75. $\partial z/\partial x = \sin y$, $\partial^2 z/\partial y \partial x = \cos y$, and $\partial^3 z/\partial^2 y \partial x = -\sin y$.

77. $\partial u/\partial z = [1/(x+2y^2+3z^3)](9z^2) = 9z^2(x+2y^2+3z^3)^{-1}$,

$\partial^2 u/\partial y \partial z = -9z^2(x+2y^2+3z^3)^{-2}(4y) = -36yz^2(x+2y^2+3z^3)^{-2}$, and

$\partial^3 u/\partial x \partial y \partial z = 72yz^2(x+2y^2+3z^3)^{-3}$.

79. $u_x = ke^{-\alpha^2 k^2 t}\cos kx$, $u_{xx} = -k^2 e^{-\alpha^2 k^2 t}\sin kx$, and

$u_t = -\alpha^2 k^2 e^{-\alpha^2 k^2 t}\sin kx$. Thus $\alpha^2 u_{xx} = u_t$.

81. $u_x = (-1/2)(x^2+y^2+z^2)^{-3/2}(2x) = -x(x^2+y^2+z^2)^{-3/2}$ and

$u_{xx} = -(x^2+y^2+z^2)^{-3/2} - x(-3/2)(x^2+y^2+z^2)^{-5/2}(2x)$

$= (2x^2 - y^2 - z^2)/(x^2+y^2+z^2)^{5/2}$.

By symmetry $u_{yy} = (2y^2 - x^2 - z^2)/(x^2+y^2+z^2)^{5/2}$

and $u_{zz} = (2z^2 - x^2 - y^2)/(x^2+y^2+z^2)^{5/2}$. Thus

$u_{xx} + u_{yy} + u_{zz} = (2x^2 - y^2 - z^2 + 2y^2 - x^2 - z^2 + 2z^2 - x^2 - y^2)/(x^2+y^2+z^2)^{5/2} = 0$.

83. Let $v = x + at$, $w = x - at$. Then $u_t = \partial[f(v) + g(w)]/\partial t$

$= (df(v)/dv)(\partial v/\partial t) + (dg(w)/dw)(\partial w/\partial t) = af'(v) - ag'(w)$ and

$u_{tt} = \partial[af'(v) - ag'(w)]/\partial t = a[af''(v) + ag''(w)] = a^2(f''(v) + g''(w))$. Similarly by using the chain rule we have $u_x = f'(v) + g'(w)$ and $u_{xx} = f''(v) + g''(w)$. Thus $u_{tt} = a^2 u_{xx}$.

85. $z_x = e^y + ye^x$, $z_{xx} = ye^x$, $\partial^3 z/\partial x^3 = ye^x$.

By symmetry $z_y = xe^y + e^x$, $z_{yy} = xe^y$ and $\partial^3 z/\partial y^3 = xe^y$.

Then $\partial^3 z/\partial x \partial y^2 = e^y$ and $\partial^3 z/\partial x^2 \partial y = e^x$.

Thus $z = xe^y + ye^x$ satisfies the given partial differential equation.

87. $f(x_1, \cdots, x_n) = (x_1^2 + \cdots + x_n^2)^{(2-n)/2} \Rightarrow \dfrac{\partial f}{\partial x_i} = (1 - n/2)2x_i(x_1^2 + \cdots + x_n^2)^{-n/2}$,

$1 \le i \le n \Rightarrow$

$\dfrac{\partial^2 f}{\partial x_i^2} = 2(1 - n/2)(x_1^2 + \cdots + x_n^2)^{-n/2} - (2n)(1 - n/2)(x_i^2)(x_1^2 + \cdots + x_n^2)^{-(2+n)/2}$

$1 \le i \le n$. Therefore, $\dfrac{\partial^2 f}{\partial x_1^2} + \cdots + \dfrac{\partial^2 f}{\partial x_n^2}$

$= \sum_{i=1}^{n}\left[(2-n)(x_1^2 + \cdots + x_n^2)^{-n/2} - n(2-n)(x_i^2)(x_1^2 + \cdots + x_n^2)^{-(2+n)/2}\right]$

$= (2-n)(x_1^2 + \cdots + x_n^2)^{-n/2} - n(2-n)(x_1^2 + \cdots + x_1^2)^{-(2+n)/2}(x_1^2 + \cdots + x_n^2)$

$= n(2-n)(x_1^2 + \cdots + x_n^2)^{-n/2} - n(2-n)(x_1^2 + \cdots + x_n^2)^{-n/2} = 0.$

89. Taking the partial of both sides w.r.t. R_1 gives by the chain rule
$(\partial R^{-1}/\partial R)(\partial R/\partial R_1) = \partial[(1/R_1) + (1/R_2) + (1/R_3)]/\partial R_1$ or $-R^{-2}(\partial R/\partial R_1) = -R_1^{-2}$.
Thus $\partial R/\partial R_1 = R^2/R_1^2$.

91. $\partial K/\partial m = \frac{1}{2}V^2$, $\partial K/\partial V = mV$, $\partial^2 K/\partial V^2 = m$. Thus $(\partial K/\partial m)(\partial^2 K/\partial V^2) = \frac{1}{2}V^2 m = K$.

93. $f_x(x,y) = x + 4y \Rightarrow f_{xy}(x,y) = 4$ and $f_y(x,y) = 3x - y \Rightarrow f_{yx}(x,y) = 3$. Since f_{xy} and f_{yx} are continuous everywhere but $f_{xy}(x,y) \neq f_{yx}(x,y)$, Clairaut's Theorem implies that such a function $f(x,y)$ does not exist.

95. By Clairaut's Theorem, $f_{xyy} = (f_{xy})_y = (f_{yx})_y = f_{yxy} = (f_y)_{xy} = (f_y)_{yx} = f_{yyx}$.

97. Let $g(x) = f(x,0) = x(x^2)^{-3/2}e^0 = x|x|^{-3}$. But we are using the point $(1,0)$, so near $(1,0)$, $g(x) = x^{-2}$. Then $g'(x) = -2x^{-3}$ and $g'(1) = -2$, so using (12.10) we have $f_x(1,0) = g'(1) = -2$.

99. (a) For $(x,y) \neq (0,0)$, $f_x(x,y) = [(3x^2 y - y^3)(x^2 + y^2) - (x^3 y - xy^3)(2x)]/(x^2 + y^2)^2$
$= (x^4 y + 4x^2 y^3 - y^5)/(x^2 + y^2)^2$
and by symmetry $f_y(x,y) = (x^5 - 4x^3 y^2 - xy^4)/(x^2 + y^2)^2$.

(b) $f_x(0,0) = \lim_{h \to 0}[f(h,0) - f(0,0)]/h = \lim_{h \to 0}[(0/h^2) - 0]/h = 0$ and
$f_y(0,0) = \lim_{h \to 0}[f(0,h) - f(0,0)]/h = 0$.

(c) Using 12.12, $f_{xy}(0,0) = \partial f_x/\partial y = \lim_{h \to 0}[f_x(0,h) - f_x(0,0)]/h$
$= \lim_{h \to 0}[(-h^5 - 0)/h^4]/h = -1$ while by 12.11,
$f_{yx}(0,0) = \partial f_y/\partial x = \lim_{h \to 0}[f_y(h,0) - f_y(0,0)]/h = \lim_{h \to 0}(h^5/h^4)/h = 1$.

(d) For $(x,y) \neq (0,0)$, $f_{xy}(x,y)$
$= [(x^4 + 12x^2 y^2)(x^2 + y^2)^2 - (x^4 y + 4x^2 y^3 - y^5)(2)(x^2 + y^2)(2y)]/(x^2 + y^2)^4$
$= (x^6 + 13x^4 y^2 + 12x^2 y^4 - 4x^4 y^2 - 16x^2 y^4 + 4y^6)/(x^2 + y^2)^3$
$= (x^6 + 9x^4 y^2 - 4x^2 y^4 + 4y^6)/(x^2 + y^2)^3$.
Now as $(x,y) \to (0,0)$ along the x-axis, $f_{xy}(x,y) \to 1$ while as $(x,y) \to (0,0)$ along the y-axis, $f_{xy}(x,y) \to 4$. Thus f_{xy} isn't continuous at $(0,0)$ and Clairaut's Theorem doesn't apply so there is no contradiction.

EXERCISES 12.4

1. $f_x(x,y) = 2x$, $f_y(x,y) = 8y$, $f_x(2,1) = 4$, $f_y(2,1) = 8$. Thus the equation of the tangent plane is $z - 8 = 4(x-2) + 8(y-1)$ or $4x + 8y - z = 8$.

Section 12.4

3. $f_x(x,y) = 2(x-1)$, $f_y(x,y) = 2(y+2)$, $f_x(2,0) = 2$, $f_y(2,0) = 4$ and the equation is
$z - 10 = 2(x-2) + 4y$ or $2x + 4y - z = -6$.

5. $f_x(x,y) = 2/(2x+y)$, $f_y(x,y) = 1/(2x+y)$, $f_x(-1,3) = 2$, $f_y(-1,3) = 1$. Thus the equation of the tangent plane is $z = 2(x+1) + (y-3)$ or $2x + y - z = 1$.

7. $f_x(-1,2) = 2$, $f_y(-1,2) = -1$ and the equation of the tangent plane is
$z + 2 = 2(x+1) + (-1)(y-2)$ or $2x - y - z = -2$.

9. $dz = \frac{\partial z}{\partial x} dx + \frac{\partial z}{\partial y} dy = 2xy^3 dx + 3x^2 y^2 dy$.

11. $dz = \frac{\partial z}{\partial x} dx + \frac{\partial z}{\partial y} dy = [-2/(x^2+y^2)^2](xdx + ydy)$.

13. $du = e^x(\cos xy - y \sin xy) dx - (xe^x \sin xy) dy$.

15. $dw = \frac{\partial w}{\partial x} dx + \frac{\partial w}{\partial y} dy + \frac{\partial w}{\partial z} dz = 2xy dx + (x^2 + 2yz) dy + y^2 dz$.

17. $dw = \frac{1}{2}[2x(x^2+y^2+z^2)^{-1/2} dx + 2y(x^2+y^2+z^2)^{-1/2} dy$
$+ 2z(x^2+y^2+z^2)^{-1/2} dz]/(x^2+y^2+z^2)^{1/2} = (xdx + ydy + zdz)/(x^2+y^2+z^2)$.

19. $\Delta x = 0.05$, $\Delta y = 0.1$, $z_x = 10x$, $z_y = 2y$. Thus when $x = 1$, $y = 2$,
$dz = (10)(0.05) + (4)(0.1) = 0.9$, while $\Delta z = f(1.05, 2.1) - f(1,2)$
$= 5(1.05)^2 + (2.1)^2 - 5 - 4 = 0.9225$.

21. $f(5,4) = \sqrt{25 - 16} = 3$ so set $(a,b) = (5,4)$ then $\Delta x = 0.01$, $\Delta y = 0.02$ and
$f_x = \frac{1}{2}(x^2+y^2)^{-1/2}(2x) = x/\sqrt{x^2-y^2}$, $f_y = -y/\sqrt{x^2-y^2}$. Thus
$f(5.01, 4.02) \approx f(5,4) + dz = 3 + [\frac{5}{3}(0.01) + (-\frac{4}{3})(0.02)] = 2.99$.

23. Since $f(7,2) = \ln(7-6) = 0$, set $(a,b) = (7,2)$. Then $\Delta x = -0.1$, $\Delta y = 0.06$,
$f_x = 1/(x-3y)$ and $f_y = -3/(x-3y)$.
Thus $f(6.9, 2.06) \approx f(7,2) + dz = 0 + (1)(-0.1) + (-3)(0.06) = -0.28$.

25. Since $f(1,1,3) = 81$, set $(a,b,c) = (1,1,3)$. Then $\Delta x = 0.05$,
$\Delta y = -0.1$, $\Delta z = 0.01$, $f_x = 2xy^3 z^4$, $f_y = 3x^2 y^2 z^4$ and $f_z = 4x^2 y^3 z^3$.
Thus $f(1.05, 0.9, 3.01) \approx f(1,1,3) + dw$
$= 81 + (162)(0.05) + (243)(-0.1) + (108)(0.01) = 65.88$.

27. Let $w = f(x,y,z) = x\sqrt{y - z^3}$, then $f(9,10,1) = 27$ so set $(a,b,c) = (9,10,1)$. Then
$\Delta x = -0.06$, $\Delta y = -0.01$, $\Delta z = 0.01$, $f_x = \sqrt{y - z^3}$,
$f_y = x/(2\sqrt{y-z^3})$ and $f_z = -3xz^2/(2\sqrt{y-z^3})$.
Thus $8.94\sqrt{9.99 - (1.01)^3} \approx 27 + (3)(-0.06) + \frac{9}{6}(-0.01) + (-\frac{27}{6})(0.01) = 26.76$.

29. Let $z = f(x,y) = \sqrt{x}\, e^y$, then $f(1,0) = 1$ so set $(a,b) = (1,0)$.
Thus $\sqrt{0.99}\, e^{0.02} \approx 1 + \frac{1}{2}(-0.01) + (1)(0.02) = 1.015$.

31. $dA = (\partial A/\partial x) dx + (\partial A/\partial y) dy = y dx + x dy$ and $|\Delta x| \leq 0.1$, $|\Delta y| \leq 0.1$.
Thus the maximum error in the area $\approx dA = 24(0.1) + 30(0.1) = 5.4$ cm^2.

Section 12.5

33. The volume of a can is $V = \pi r^2 h$ and $\Delta V \approx dV$ is an estimate to the amount of tin. Here $dV = 2\pi r h \, dr + \pi r^2 \, dh$ so $\Delta V \approx dV = 2\pi(48)(-0.04) + \pi(16)(-0.08) = -16.08$ cm^3. Thus the amount of tin is about 16 cm^3.

35. The area of the rectangle is $A = xy$ and $\Delta A \approx dA$ is an estimate of the area of paint in the stripe. Here $dA = y \, dx + x \, dy$, so $\Delta A \approx dA = (100)(\frac{1}{2}) + (200)(\frac{1}{2}) = 150$ ft^2. Thus there are approximately 150 ft^2 of paint in the stripe.

37. Taking the partial of both sides w.r.t. R_1 gives by the chain rule
$(\partial R^{-1}/\partial R)(\partial R/\partial R_1) = \partial[(1/R_1) + (1/R_2) + (1/R_3)]/\partial R_1 \Rightarrow -R^{-2}(\partial R/\partial R_1) = -R_1^{-2}$
$\Rightarrow \partial R/\partial R_1 = R^2/R_1^2$ and by symmetry $\partial R/\partial R_2 = R^2/R_2^2$, $\partial R/\partial R_3 = R^2/R_3^2$. When $R_1 = 25$, $R_2 = 40$ and $R_3 = 50$, $1/R = \frac{17}{200}$ or $R = \frac{200}{17}$ ohms. Since the possible error for each R_i is 0.5%, the maximum error of R is attained by setting $\Delta R_i = 0.005 R_i$. Then $\Delta R \approx dR = (\partial R/\partial R_1)\Delta R_1 + (\partial R/\partial R_2)\Delta R_2 + (\partial R/\partial R_3)\Delta R_3$
$= (0.005)R^2[(1/R_1) + (1/R_2) + (1/R_3)] = (0.005)R = \frac{1}{17} \approx 0.059$ ohms.

39. $\Delta z = f(a + \Delta x, b + \Delta y) - f(a, b) = (a + \Delta x)^2 + (b + \Delta y)^2 - (a^2 + b^2)$
$= a^2 + 2a\Delta x + (\Delta x)^2 + b^2 + 2b\Delta y + (\Delta y)^2 - a^2 - b^2$
$= 2a\Delta x + (\Delta x)^2 + 2b\Delta y + (\Delta y)^2$. But $f_x(a,b) = 2a$ and $f_y(a,b) = 2b$ and so
$\Delta z = f_x(a,b)\Delta x + f_y(a,b)\Delta y + \Delta x \Delta x + \Delta y \Delta y$ which is definition 12.26 if $\epsilon_1 = \Delta x$ and $\epsilon_2 = \Delta y$. Hence f is differentiable.

41. To show that f is continuous at (a,b) we need to show that $\lim_{(x,y) \to (a,b)} f(x,y) = f(a,b)$ or equivalently $\lim_{(\Delta x, \Delta y) \to (0,0)} f(a + \Delta x, b + \Delta y) = f(a,b)$. Since f is differentiable at (a,b),
$f(a + \Delta x, b + \Delta y) - f(a,b) = \Delta z = f_x(a,b)\Delta x + f_y(a,b)\Delta y + \epsilon_1 \Delta x + \epsilon_2 \Delta y$ where ϵ_1 and $\epsilon_2 \to 0$ as $(\Delta x, \Delta y) \to (0,0)$. Thus
$f(a + \Delta x, b + \Delta y) = f(a,b) + f_x(a,b)\Delta x + f_y(a,b)\Delta y + \epsilon_1 \Delta x + \epsilon_2 \Delta y$.
Taking the limit of both sides as $(\Delta x, \Delta y) \to (0,0)$ gives
$\lim_{(\Delta x, \Delta y) \to (0,0)} f(a + \Delta x, b + \Delta y) = f(a,b)$. Thus f is continuous at (a,b).

EXERCISES 12.5

1. $\frac{dz}{dt} = 2x\frac{dx}{dt} + 2y\frac{dy}{dt} = (2t^3)(3t^2) + 2(1 + t^2)(2t) = 6t^5 + 4t^3 + 4t$.

3. $\frac{dz}{dt} = (18x^2 - 3y)e^t + (-3x + 4y)(-\sin t)$
$= (18e^{2t} - 3\cos t)e^t + (3e^t - 4\cos t)(\sin t)$.

5. $\frac{dz}{dt} = \frac{1}{(x+y^2)}\frac{1}{2\sqrt{1+t}} + \frac{1}{(x+y^2)}2y\frac{1}{2\sqrt{t}} = \frac{1}{\sqrt{1+t}+1+\sqrt{t}}\left(\frac{1}{2\sqrt{1+t}} + \frac{1+\sqrt{t}}{\sqrt{t}}\right)$.

Section 12.5

7. $\frac{dw}{dt} = y^2z^3(\cos t) + 2xyz^3(-\sin t) + 3xy^2z^2(2e^{2t})$.

9. $\frac{\partial z}{\partial s} = (2x\sin y)(2s) + (x^2\cos y)(2t) = 4sx\sin y + 2tx^2\cos y$,

   $\frac{\partial z}{\partial t} = (2x\sin y)(2t) + (x^2\cos y)(2s) = 4xt\sin y + 2sx^2\cos y$.

11. $\frac{\partial z}{\partial s} = (2x - 6xy^3)(e^t) + (-9x^2y^2)(e^{-t}) = (2x - 6xy^3)e^t - 9x^2y^2e^{-t}$,

    $\frac{\partial z}{\partial t} = (2x - 6xy^3)(se^t) + (-9x^2y^2)(-se^{-t}) = (2x - 6xy^3)se^t + 9x^2y^2se^{-t}$.

13. $\frac{\partial z}{\partial s} = (z\ln 2)(2st) + z(-3\ln 2)(t^2) = (2^{x-3y}\ln 2)(2st - 3t^2)$,

    $\frac{\partial z}{\partial t} = (z\ln 2)(s^2) + z(-3\ln 2)(2st) = (2^{x-3y}\ln 2)(s^2 - 6st)$.

15. $\frac{\partial u}{\partial r} = \frac{\partial u}{\partial x}\frac{\partial x}{\partial r} + \frac{\partial u}{\partial y}\frac{\partial y}{\partial r}$, $\frac{\partial u}{\partial s} = \frac{\partial u}{\partial x}\frac{\partial x}{\partial s} + \frac{\partial u}{\partial y}\frac{\partial y}{\partial s}$, $\frac{\partial u}{\partial t} = \frac{\partial u}{\partial x}\frac{\partial x}{\partial t} + \frac{\partial u}{\partial y}\frac{\partial y}{\partial t}$.

17. $\frac{\partial v}{\partial x} = \frac{\partial v}{\partial p}\frac{\partial p}{\partial x} + \frac{\partial v}{\partial q}\frac{\partial q}{\partial x} + \frac{\partial v}{\partial r}\frac{\partial r}{\partial x}$, $\frac{\partial v}{\partial y} = \frac{\partial v}{\partial p}\frac{\partial p}{\partial y} + \frac{\partial v}{\partial q}\frac{\partial q}{\partial y} + \frac{\partial v}{\partial r}\frac{\partial r}{\partial y}$,

    $\frac{\partial v}{\partial z} = \frac{\partial v}{\partial p}\frac{\partial p}{\partial z} + \frac{\partial v}{\partial q}\frac{\partial q}{\partial z} + \frac{\partial v}{\partial r}\frac{\partial r}{\partial z}$.

19. $\frac{\partial w}{\partial s} = \frac{\partial w}{\partial x}\frac{\partial x}{\partial s} + \frac{\partial w}{\partial y}\frac{\partial y}{\partial s} + \frac{\partial w}{\partial z}\frac{\partial z}{\partial s} = 2xt + 2y\cos t + 2z\sin t$. When $s = 1$, $t = 0$, we have $x = 0$, $y = 1$ and $z = 0$, so $\frac{\partial w}{\partial s} = 2\cos 0 = 2$. Similarly

    $\frac{\partial w}{\partial t} = 2xs + 2y(-s\sin t) + 2z(s\cos t) = 0 + (-2)\sin 0 + 0 = 0$, when $s = 1$, $t = 0$.

21. $\frac{\partial z}{\partial t} = (y^2\sec^2 x)2tuv + (2y\tan x)v^2$, $\frac{\partial z}{\partial u} = (y^2\sec^2 x)t^2v + 2y\tan x$,

    $\frac{\partial z}{\partial v} = (y^2\sec^2 x)t^2u + (2y\tan x)2tv$. When $t = 2$, $u = 1$ and $v = 0$, we have $x = 0$, $y = 1$, so

    $\frac{\partial z}{\partial t} = 0$, $\frac{\partial z}{\partial u} = 0$, $\frac{\partial z}{\partial v} = 4$.

23. $\frac{\partial u}{\partial p} = \frac{1}{y+z} + \frac{(y+z)-(x+y)}{(y+z)^2} - \frac{x+y}{(y+z)^2} = \frac{(y+z)+(z-x)-(x+y)}{(y+z)^2} = 2\frac{z-x}{(y+z)^2} = 2\frac{-2t}{4p^2}$

    $= \frac{-t}{p^2}$, $\frac{\partial u}{\partial r} = \frac{1}{y+z} + \frac{z-x}{(y+z)^2}(-1) - \frac{x+y}{(y+z)^2} = 0$ and

    $\frac{\partial u}{\partial t} = \frac{1}{y+z} + \frac{z-x}{(y+z)^2} + \frac{x+y}{(y+z)^2} = 2\frac{y+z}{(y+z)^2} = \frac{2}{2p} = \frac{1}{p}$.

25. $\frac{\partial w}{\partial r} = [-\sin(x-y)]s^2t^3\sin\theta + [\sin(x-y)]2rst\cos\theta = st\sin(x-y)[2r\cos\theta - st^2\sin\theta]$,

    $\frac{\partial w}{\partial s} = [-\sin(x-y)]2rst^3\sin\theta + [\sin(x-y)]r^2t\cos\theta = [rt\sin(x-y)](r\cos\theta - 2st^2\sin\theta)$,

    $\frac{\partial w}{\partial t} = [-\sin(x-y)]3rs^2t^2\sin\theta + [\sin(x-y)]r^2s\cos\theta = [sr\sin(x-y)](r\cos\theta - 3st^2\sin\theta)$,

    $\frac{\partial w}{\partial \theta} = [-\sin(x-y)]rs^2t^3\cos\theta + [\sin(x-y)](-r^2st\sin\theta) = [-rst\sin(x-y)](st^2\cos\theta + r\sin\theta)$.

27. $F(x,y) = x^2 - xy + y^3 - 8 = 0$, so $\frac{dy}{dx} = -\frac{F_x}{F_y} = -\frac{(2x-y)}{-x+3y^2} = \frac{y-2x}{3y^2-x}$.

29. $F(x,y) = 2y^2 + \sqrt[3]{xy} - 3x^2 - 17$, so $\frac{dy}{dx} = -\frac{y/[3(xy)^{2/3}] - 6x}{4y + x/[3(xy)^{2/3}]} = \frac{18x - x^{-2/3}y^{1/3}}{12y + x^{1/3}y^{-2/3}}$.

Section 12.5

31. $\frac{\partial z}{\partial x} = -\frac{F_x}{F_z} = -\frac{y-z}{y-x} = \frac{z-y}{y-x}$, $\frac{\partial z}{\partial y} = -\frac{F_y}{F_z} = -\frac{x+z}{y-x} = \frac{x+z}{x-y}$.

33. $\frac{\partial z}{\partial x} = -\frac{2x - 2y - 2z}{-2z - 2x} = \frac{x - y - z}{z + x}$, $\frac{\partial z}{\partial y} = -\frac{2y - 2x}{-2z - 2x} = \frac{y - x}{z + x}$.

35. $\frac{\partial z}{\partial x} = -\frac{e^y + ze^x}{y + e^x}$, $\frac{\partial z}{\partial y} = -\frac{xe^y + z}{y + e^x}$.

37. $\frac{dr}{dt} = -1.2$, $\frac{dh}{dt} = 3$, $V = \pi r^2 h$ and $\frac{dV}{dt} = 2\pi r h \frac{dr}{dt} + \pi r^2 \frac{dh}{dt}$. Thus when $r = 80$ and $h = 150$,
$\frac{dV}{dt} = (-28800)\pi + (19200)\pi = -9600\pi$ cm^3/s.

39. $\frac{dP}{dt} = 0.05$, $\frac{dT}{dt} = 0.15$, $V = 8.31\frac{T}{P}$ and $\frac{dV}{dt} = \frac{8.31}{P}\frac{dT}{dt} - 8.31\frac{T}{P^2}\frac{dP}{dt}$. Thus when $P = 20$ and
$T = 320°$, $\frac{dV}{dt} = 8.31\left(\frac{0.15}{20} - \frac{(0.05)(320)}{400}\right) \approx -0.27$ L/s.

41. (a) Using the Chain Rule, $\frac{\partial z}{\partial r} = \frac{\partial z}{\partial x}\cos\theta + \frac{\partial z}{\partial y}\sin\theta$, $\frac{\partial z}{\partial \theta} = \frac{\partial z}{\partial x}(-r\sin\theta) + \frac{\partial z}{\partial y}r\cos\theta$.

   (b) $\left(\frac{\partial z}{\partial r}\right)^2 = \left(\frac{\partial z}{\partial x}\right)^2 \cos^2\theta + 2\frac{\partial z}{\partial x}\frac{\partial z}{\partial y}\cos\theta\sin\theta + \left(\frac{\partial z}{\partial y}\right)^2 \sin^2\theta$

   $\left(\frac{\partial z}{\partial \theta}\right)^2 = \left(\frac{\partial z}{\partial x}\right)^2 r^2 \sin^2\theta - 2\frac{\partial z}{\partial x}\frac{\partial z}{\partial y}r^2\cos\theta\sin\theta + \left(\frac{\partial z}{\partial y}\right)^2 r^2 \cos^2\theta$. Thus

   $\left(\frac{\partial z}{\partial r}\right)^2 + \frac{1}{r^2}\left(\frac{\partial z}{\partial \theta}\right)^2 = \left[\left(\frac{\partial z}{\partial x}\right)^2 + \left(\frac{\partial z}{\partial y}\right)^2\right](\cos^2\theta + \sin^2\theta) = \left(\frac{\partial z}{\partial x}\right)^2 + \left(\frac{\partial z}{\partial y}\right)^2$.

43. Let $u = x - y$, then $\frac{\partial z}{\partial x} = \frac{dz}{du}\frac{\partial u}{\partial x} = \frac{dz}{du}$ and $\frac{\partial z}{\partial y} = \frac{dz}{du}(-1)$. Thus $\frac{\partial z}{\partial x} + \frac{\partial z}{\partial y} = 0$.

45. Let $u = x + at$, $v = x - at$, then $z = f(u) + g(v)$, so $\frac{\partial z}{\partial u} = f'(u)$ and $\frac{\partial z}{\partial v} = g'(v)$. Thus
$\frac{\partial z}{\partial t} = \frac{\partial z}{\partial u}\frac{\partial u}{\partial t} + \frac{\partial z}{\partial v}\frac{\partial v}{\partial t} = af'(u) - ag'(v)$ and
$\frac{\partial^2 z}{\partial t^2} = a\frac{\partial}{\partial t}[f'(u) - g'(v)] = a\left(\frac{df'(u)}{du}\frac{\partial u}{\partial t} - \frac{dg'(v)}{dv}\frac{\partial v}{\partial t}\right) = a^2 f''(u) + a^2 g''(v)$. Similarly
$\frac{\partial z}{\partial x} = f'(u) + g'(v)$ and $\frac{\partial^2 z}{\partial x^2} = f''(u) + g''(v)$. Thus $\frac{\partial^2 z}{\partial t^2} = a^2 \frac{\partial^2 z}{\partial x^2}$.

47. $\frac{\partial z}{\partial s} = \frac{\partial z}{\partial x}2s + \frac{\partial z}{\partial y}2r$. Then $\frac{\partial^2 z}{\partial r \partial s} = \frac{\partial}{\partial r}\left(\frac{\partial z}{\partial x}2s\right) + \frac{\partial}{\partial r}\left(\frac{\partial z}{\partial y}2r\right)$

   $= \frac{\partial^2 z}{\partial x^2}\frac{\partial x}{\partial r}2s + \frac{\partial}{\partial y}\left(\frac{\partial z}{\partial x}\right)\frac{\partial y}{\partial r}2s + \frac{\partial z}{\partial x}\frac{\partial}{\partial r}(2s) + \frac{\partial^2 z}{\partial y^2}\frac{\partial y}{\partial r}2r + \frac{\partial}{\partial x}\left(\frac{\partial z}{\partial y}\right)\frac{\partial x}{\partial r}2r + \frac{\partial z}{\partial y}2$

   $= 4rs\frac{\partial^2 z}{\partial x^2} + \frac{\partial^2 z}{\partial y \partial x}4s^2 + 0 + 4rs\frac{\partial^2 z}{\partial y^2} + \frac{\partial^2 z}{\partial x \partial y}4r^2 + 2\frac{\partial z}{\partial y}$. And by the continuity of the partials

   $\frac{\partial^2 z}{\partial r \partial s} = 4rs\frac{\partial^2 z}{\partial x^2} + 4rs\frac{\partial^2 z}{\partial y^2} + (4r^2 + 4s^2)\frac{\partial^2 z}{\partial x \partial y} + 2\frac{\partial z}{\partial y}$.

49. $\frac{\partial z}{\partial r} = \frac{\partial z}{\partial x}\cos\theta + \frac{\partial z}{\partial y}\sin\theta$ and $\frac{\partial z}{\partial \theta} = -\frac{\partial z}{\partial x}r\sin\theta + \frac{\partial z}{\partial y}r\cos\theta$. Then

   $\frac{\partial^2 z}{\partial r^2} = \cos\theta\left(\frac{\partial^2 z}{\partial x^2}\cos\theta + \frac{\partial^2 z}{\partial y \partial x}\sin\theta\right) + \sin\theta\left(\frac{\partial^2 z}{\partial y^2}\sin\theta + \frac{\partial^2 z}{\partial x \partial y}\cos\theta\right)$

$$= \cos^2\theta \frac{\partial^2 z}{\partial x^2} + 2\cos\theta\sin\theta \frac{\partial^2 z}{\partial x \partial y} + \sin^2\theta \frac{\partial^2 z}{\partial y^2} \text{ and}$$

$$\frac{\partial^2 z}{\partial \theta^2} = -r\cos\theta \frac{\partial z}{\partial x} + (-r\sin\theta)\left(\frac{\partial^2 z}{\partial x^2}(-r\sin\theta) + \frac{\partial^2 z}{\partial y \partial x} r\cos\theta\right)$$

$$-r\sin\theta \frac{\partial z}{\partial y} + r\cos\theta \left(\frac{\partial^2 z}{\partial y^2} r\cos\theta + \frac{\partial^2 z}{\partial x \partial y}(-r\sin\theta)\right)$$

$$= -r\cos\theta \frac{\partial z}{\partial x} - r\sin\theta \frac{\partial z}{\partial y} + r^2\sin^2\theta \frac{\partial^2 z}{\partial x^2} - 2r^2\cos\theta\sin\theta \frac{\partial^2 z}{\partial x \partial y} + r^2\cos^2\theta \frac{\partial^2 z}{\partial y^2}. \text{ Thus}$$

$$\frac{\partial^2 z}{\partial r^2} + \frac{1}{r^2}\frac{\partial^2 z}{\partial \theta^2} + \frac{1}{r}\frac{\partial z}{\partial r} = (\cos^2\theta + \sin^2\theta)\frac{\partial^2 z}{\partial x^2} + (\sin^2\theta + \cos^2\theta)\frac{\partial^2 z}{\partial y^2} - \frac{1}{r}\cos\theta\frac{\partial z}{\partial x}$$

$$-\frac{1}{r}\sin\theta\frac{\partial z}{\partial y} + \frac{1}{r}\left(\cos\theta\frac{\partial z}{\partial x} + \sin\theta\frac{\partial z}{\partial y}\right) = \frac{\partial^2 z}{\partial x^2} + \frac{\partial^2 z}{\partial y^2} \text{ as desired.}$$

51. Differentiating both sides of $f(tx, ty) = t^n f(x,y)$ with respect to t using the Chain rule,

we get $\frac{\partial}{\partial t}f(tx, ty) = \frac{\partial}{\partial t}[t^n f(x,y)] \Leftrightarrow \frac{\partial}{\partial (tx)}f(tx,ty) \cdot \frac{\partial(tx)}{\partial t} + \frac{\partial}{\partial (ty)}f(tx,ty) \cdot \frac{\partial(ty)}{\partial t}$

$= x\frac{\partial}{\partial (tx)}f(tx,ty) + y\frac{\partial}{\partial (ty)}f(tx,ty) = nt^{n-1}f(x,y)$. Setting $t = 1$ gives

$x\frac{\partial}{\partial x}f(x,y) + y\frac{\partial}{\partial y}f(x,y) = nf(x,y)$.

53. Differentiating both sides of $f(tx, ty) = t^n f(x,y)$ with respect to x using the chain rule,

we get $\frac{\partial}{\partial x}f(tx,ty) = \frac{\partial}{\partial x}[t^n f(x,y)] \Leftrightarrow \frac{\partial}{\partial (tx)}f(tx,ty) \cdot \frac{\partial(tx)}{\partial x} + \frac{\partial}{\partial (ty)}f(tx,ty) \cdot \frac{\partial(ty)}{\partial x} = t^n \frac{\partial}{\partial x}f(x,y)$

$\Leftrightarrow tf_x(tx, ty) = t^n f_x(x,y)$. Thus $f_x(tx, ty) = t^{n-1} f_x(x,y)$.

EXERCISES 12.6

1. $D_{\vec{u}}f(x,y) = (2xy^3 + 8x^3 y)\cos(\pi/3) + (3x^2 y^2 + 2x^4)\sin(\pi/3)$. Thus
$D_{\vec{u}}f(1, -2) = (-16 - 16)(\frac{1}{2}) + (12 + 2)(\sqrt{3}/2) = 7\sqrt{3} - 16$.

3. $D_{\vec{u}}f(x,y) = (y^x \ln y)\cos(\pi/2) + (xy^{x-1})\sin(\pi/2) = xy^{x-1}$. Thus
$D_{\vec{u}}f(1, 2) = (1)(2)^{1-1} = 1$.

5. (a) $\nabla f(x,y) = (3x^2 - 8xy)\vec{i} + (2y - 4x^2)\vec{j}$
   (b) $\nabla f(0, -1) = -2\vec{j}$
   (c) $\nabla f(0, -1) \cdot \vec{u} = -\frac{8}{5}$.

7. (a) $\nabla f(x, y, z) = y^2 z^3 \vec{i} + 2xyz^3 \vec{j} + 3xy^2 z^2 \vec{k}$
   (b) $\nabla f(1, -2, 1) = 4\vec{i} - 4\vec{j} + 12\vec{k}$
   (c) $\nabla f(1, -2, 1) \cdot \vec{u} = (1/\sqrt{3})(20) = 20/\sqrt{3}$.

9. $\nabla f(x,y) = <\frac{1}{2}(x-y)^{-1/2}, -\frac{1}{2}(x-y)^{-1/2}>$, $\nabla f(5, 1) = <\frac{1}{4}, -\frac{1}{4}>$, a unit
vector in the direction of $\vec{v}$ is $\vec{u} = <\frac{12}{13}, \frac{5}{13}>$, so $D_{\vec{u}}f(5,1) = \nabla f(5,1) \cdot \vec{u} = \frac{12}{52} - \frac{5}{52} = \frac{7}{52}$.

Section 12.6

11. $\nabla g(x,y) = <e^{xy}(1+xy), x^2 e^{xy}>$, $\nabla g(-3,0) = <1,9>$, $\vec{u} = <2/\sqrt{13}, 3/\sqrt{13}>$

    and $D_{\vec{u}} g(-3,0) = \frac{2}{\sqrt{13}} + \frac{27}{\sqrt{13}} = \frac{29}{\sqrt{13}}$.

13. $\nabla f(x,y,z) = \frac{1}{2}(xyz)^{-1/2}<yz, xz, xy>$, $\nabla f(2,4,2) = <1, \frac{1}{2}, 1>$,

    $\vec{u} = <\frac{2}{3}, \frac{1}{3}, -\frac{2}{3}>$ and $D_{\vec{u}} f(2,4,2) = 1 \cdot \frac{2}{3} + \frac{1}{2} \cdot \frac{1}{3} + 1(-\frac{2}{3}) = \frac{1}{6}$.

15. $\nabla g(x,y,z) = <\tan^{-1}(y/z), xz/(y^2+z^2), -xy/(y^2+z^2)>$, $\nabla g(1,2,-2) =$
    $<-\pi/4, -1/4, -1/4>$, $\vec{u} = (1/\sqrt{3})<1,1,-1>$ and

    $D_{\vec{u}} g(1,2,-2) = \frac{(-\pi)(1)}{4\sqrt{3}} + \frac{(-1)(1)}{4\sqrt{3}} + \frac{(-1)(-1)}{4\sqrt{3}} = -\frac{\pi}{4\sqrt{3}}$.

17. $\nabla f(x,y) = <e^{-y}, 3 - xe^{-y}>$, $\nabla f(1,0) = <1, 2>$ is the direction and the maximum rate is $|\nabla f(1,0)| = \sqrt{5}$.

19. $\nabla f(x,y) = <x/\sqrt{x^2+2y}, 1/\sqrt{x^2+2y}>$. Thus the maximum rate of change is $|\nabla f(4,10)| = \sqrt{17}/6$ in the direction $<2/3, 1/6>$ or $<4, 1>$.

21. $\nabla f(x,y) = <-3\sin(3x+2y), -2\sin(3x+2y)>$, so the maximum rate is $|\nabla f(\pi/6, -\pi/8)| = \sqrt{13/2}$ in the direction $<-3\sqrt{2}/2, -\sqrt{2}>$ or in the direction $<-3, -2>$.

23. As in the proof of Theorem 12.48, $D_{\vec{u}} f = |\nabla f| \cos\theta$. Since the minimum value of $\cos\theta$ is -1 occurring when $\theta = \pi$, the minimum value of $D_{\vec{u}} f$ is $-|\nabla f|$ occurring when $\theta = \pi$, that is when $\vec{u}$ is in the opposite direction of ∇f (assuming $\nabla f \neq \vec{0}$).

25. $T = k/\sqrt{x^2+y^2+z^2}$ and $120 = T(1,2,2) = k/3$ so $k = 360$.

    (a) $\vec{u} = <1, -1, 1>/\sqrt{3}$, $D_{\vec{u}} T(1,2,2) = \nabla T(1,2,2) \cdot \vec{u}$
    $= [-360(x^2+y^2+z^2)^{-3/2}<x,y,z>]|_{(1,2,2)} \cdot \vec{u}$
    $= (-40/3)<1,2,2> \cdot (1/\sqrt{3})<1,-1,1> = -40/(3\sqrt{3})$.

    (b) From (a) $\nabla T = -360(x^2+y^2+z^2)^{-3/2}<x,y,z>$ and since $<x,y,z>$ is the position vector of the point (x,y,z), the vector $-<x,y,z>$, and thus ∇T, always points toward the origin.

27. $\nabla V(x,y,z) = <10x - 3y + yz, xz - 3x, xy>$, $\nabla V(3,4,5) = <38, 6, 12>$

    (a) $D_{\vec{u}} V(3,4,5) = <38, 6, 12> \cdot (1/\sqrt{3})<1, 1, -1> = 32/\sqrt{3}$.

    (b) $<38, 6, 12>$ or $<19, 3, 6>$.

    (c) $|\nabla V(x,y,z)| = \sqrt{(10x - 3y + yz)^2 + (xz - 3x)^2 + x^2 y^2}$ in general or at $(3,4,5)$ is $|\nabla V(3,4,5)| = \sqrt{1624} = 2\sqrt{406}$.

29. A unit vector in the direction of $\vec{AB}$ is $\vec{i}$ and a unit vector in the direction of $\vec{AC}$ is $\vec{j}$.

    Thus $D_{\vec{AB}} f(1,3) = f_x(1,3) = 3$ and $D_{\vec{AC}} f(1,3) = f_y(1,3) = 26$. Therefore,

Section 12.6

$\nabla f(1,3) = \langle f_x(1,3), f_y(1,3)\rangle = \langle 3, 26\rangle$ and by definition $D_{\overrightarrow{AD}} f(1,3) = \nabla f \cdot \vec{u}$ where u is a unit vector in the direction of $\overrightarrow{AD}$, which is $\langle \frac{5}{13}, \frac{12}{13}\rangle$. Therefore

$D_{\overrightarrow{AD}} f(1,3) = \langle 3, 26\rangle \cdot \langle \frac{5}{13}, \frac{12}{13}\rangle = 3(\frac{5}{13}) + 26(\frac{12}{13}) = \frac{327}{13}$.

31. $\nabla(au + bv) = <\partial(au+bv)/\partial x, \partial(au+bv)/\partial y>$
$= <a(\partial u/\partial x) + b(\partial v/\partial x), a(\partial u/\partial y) + b(\partial v/\partial y)>$
$= a<\partial u/\partial x, \partial u/\partial y> + b<\partial v/\partial x, \partial v/\partial y> = a\nabla u + b\nabla v$.

33. $\nabla(u/v) = <[v(\partial u/\partial x) - u(\partial v/\partial x)]/v^2, [v(\partial u/\partial y) - u(\partial v/\partial y)]/v^2>$
$= [v<\partial u/\partial x, \partial u/\partial y> - u<\partial v/\partial x, \partial v/\partial y>]/v^2 = (v\nabla u - u\nabla v)/v^2$.

35. (a) $F(x,y,z) = 4x^2 + y^2 + z^2$, $\nabla F(2,2,2) = <16, 4, 4>$ so the equation of the tangent plane is $16x + 4y + 4z = 48$ or $4x + y + z = 12$ and
(b) the normal line is given by $(x-2)/16 = (y-2)/4 = (z-2)/4$ or $(x-2)/4 = y - 2 = z - 2$.

37. $\nabla F(x,y,z) = <y+z, x+z, y+x>$, $\nabla F(1,1,1) = <2,2,2>$.
(a) $2x + 2y + 2z = 6$ or $x + y + z = 3$
(b) $x - 1 = y - 1 = z - 1$ or $x = y = z$.

39. $\nabla F(x,y,z) = <yz, xz, yx>$, $\nabla F(1,2,3) = <6, 3, 2>$.
(a) $6x + 3y + 2z = 18$
(b) $(x-1)/6 = (y-2)/3 = (z-3)/2$.

41. $F(x,y,z) = -z + xe^y \cos z$, $\nabla F(x,y,z) = <e^y \cos z, xe^y \cos z, -1 - xe^y \sin z>$
$\nabla F(1,0,0) = <1, 1, -1>$.
(a) $x + y - z = 1$
(b) $x - 1 = y = -z$.

43. $\nabla f(x,y) = <2x, 8y>$, $\nabla f(2,1) = <4, 8>$.
The tangent line has equation
$\nabla f(2,1) \cdot <x-2, y-1> = 0 \Rightarrow$
$4(x-2) + 8(y-1) = 0$ which simplifies to
$x + 2y = 4$.

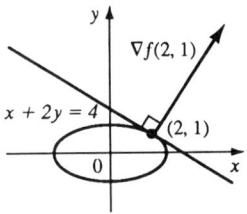

45. $\nabla F(x_0, y_0, z_0) = <2x_0/a^2, 2y_0/b^2, 2z_0/c^2>$. Thus the equation of the tangent plane at (x_0, y_0, z_0) is $(2x_0/a^2)x + (2y_0/b^2)y + (2z_0/c^2)z = 2[(x_0^2/a^2) + (y_0^2/b^2) + (z_0^2/c^2)]$
$= 2(1) = 2$ since (x_0, y_0, z_0) is a point on the ellipsoid. Hence
$(x_0/a^2)x + (y_0/b^2)y + (z_0/c^2)z = 1$ is the equation of the tangent plane.

47. $\nabla F(x_0, y_0, z_0) = <2x_0/a^2, 2y_0/b^2, -1/c>$, so the equation of the tangent plane is
$(2x_0/a^2)x + (2y_0/b^2)y - (1/c)z = (2x_0^2/a^2) + (2y_0^2/b^2) - (z_0/c)$

117

Section 12.6

or $(2x_0/a^2)x + (2y_0/b^2)y = (z/c) + 2[(x_0^2/a^2) + (y_0^2/b^2)] - (z_0/c)$.

But $z_0/c = (x_0^2/a^2) + (y_0^2/b^2)$, so the equation can be written as

$(2x_0/a^2)x + (2y_0/b^2)y = (z + z_0)/c$.

49. $\nabla f(x_0, y_0, z_0) = \langle 2x_0, -2y_0, 4z_0 \rangle$ and the given line has direction numbers $2, 4, 6$, so $\langle 2x_0, -2y_0, 4z_0 \rangle = k\langle 2, 4, 6 \rangle$ or $x_0 = k$, $y_0 = -2k$ and $z_0 = \frac{3}{2}k$. But $x_0^2 - y_0^2 + 2z_0^2 = 1$ or $(1 - 4 + \frac{9}{2})k^2 = 1$, so $k = \pm\sqrt{2/3} = \pm\sqrt{6}/3$ and there are two such points: $(\pm\sqrt{6}/3, \mp 2\sqrt{6}/3, \pm\sqrt{6}/2)$.

51. Let (x_0, y_0, z_0) be a point on the cone (other than $(0, 0, 0)$). Then the equation of the tangent plane to the cone at this point is $2x_0 x + 2y_0 y - 2z_0 z = 2(x_0^2 + y_0^2 - z_0^2)$. But $x_0^2 + y_0^2 = z_0^2$ so the tangent plane is given by $x_0 x + y_0 y - z_0 z = 0$, a plane which always contains the origin.

53. Let (x_0, y_0, z_0) be a point on the surface. Then the equation of the tangent plane to the point is $x/(2\sqrt{x_0}) + y/(2\sqrt{y_0}) + z/(2\sqrt{z_0}) = (\sqrt{x_0} + \sqrt{y_0} + \sqrt{z_0})/2$. But $\sqrt{x_0} + \sqrt{y_0} + \sqrt{z_0} = \sqrt{c}$, so the equation is $(x/\sqrt{x_0}) + (y/\sqrt{y_0}) + (z/\sqrt{z_0}) = \sqrt{c}$. The x-, y-, z- intercepts are $\sqrt{cx_0}$, $\sqrt{cy_0}$ and $\sqrt{cz_0}$ respectively. (*The x-intercept is found by setting $y = z = 0$ and solving the resulting equation for x. Similarly for the y- and z- intercepts.*) So the sum of the intercepts is $\sqrt{c}(\sqrt{x_0} + \sqrt{y_0} + \sqrt{z_0}) = c$, a constant.

55. Let $f(x, y, z) = y + z$ and $g(x, y, z) = x^2 + y^2$. Then the required tangent line is perpendicular to both ∇f and ∇g at $(1, 2, 1)$ and the vector $\vec{v} = \nabla f \times \nabla g$ will be parallel to the tangent line. We have:

$\nabla f(x, y, z) = \langle 0, 1, 1 \rangle \Rightarrow \nabla f(1, 2, 1) = \langle 0, 1, 1 \rangle$

$\nabla g(x, y, z) = \langle 2x, 2y, 0 \rangle \Rightarrow \nabla g(1, 2, 1) = \langle 2, 4, 0 \rangle$. Hence

$\vec{v} = \nabla f \times \nabla g = \begin{vmatrix} \vec{i} & \vec{j} & \vec{k} \\ 0 & 1 & 1 \\ 2 & 4 & 0 \end{vmatrix} = -4\vec{i} + 2\vec{j} - 2\vec{k}$

So parametric equations of the desired tangent line are: $x = 1 - 4t$, $y = 2 + 2t$, $z = 1 - 2t$.

57. The function $f(x, y) = (xy)^{1/3}$ is continuous on $\mathbf{R}^2$ by Theorem 12.7 since it is a composition of a polynomial and the cube root function, both of which are continuous.

$f_x(0, 0) = \lim_{h \to 0} \frac{f(0 + h, 0) - f(0, 0)}{h} = \lim_{h \to 0} \frac{(h \cdot 0)^{1/3} - 0}{h} = 0$

$f_y(0, 0) = \lim_{h \to 0} \frac{f(0, 0 + h) - f(0, 0)}{h} = \lim_{h \to 0} \frac{(0 \cdot h)^{1/3} - 0}{h} = 0$

Therefore, $f_x(0, 0)$ and $f_y(0, 0)$ do exist and are equal to 0. Now let $\vec{u}$ be any unit vector other than $\vec{i}$ and $\vec{j}$ (these correspond to f_x and f_y respectively). Then $\vec{u} = a\vec{i} + b\vec{j}$ where $a \neq 0$ and $b \neq 0$. Thus $D_{\vec{u}} f(0, 0) = \lim_{h \to 0} \frac{f(0 + ha, 0 + hb) - f(0, 0)}{h} = \lim_{h \to 0} \frac{\sqrt[3]{(ha)(hb)}}{h}$

$= \lim_{h \to 0} \frac{\sqrt[3]{ab}}{h^{1/3}}$ and this limit does not exist, so $D_{\vec{u}}f(0,0)$ does not exist.

EXERCISES 12.7

1. $f_x = 2x + 4$, $f_y = 2y - 6$, $f_{xx} = f_{yy} = 2$, $f_{xy} = 0$. Then $f_x = 0$ and $f_y = 0$ implies $(x,y) = (-2,3)$ and $D(-2,3) = 4 > 0$, so $f(-2,3) = -13$ is a local minimum.

3. $f_x = 4x + 2y + 2$, $f_y = 2y + 2x + 2$, $f_{xx} = 4$, $f_{yy} = 2$, $f_{xy} = 2$. Then $f_x = 0$ and $f_y = 0$ implies $2x = 0$ so the critical point is $(0,-1)$. $D(0,-1) = 8 - 4 > 0$, so $f(0,-1) = -1$ is a local minimum.

5. $f_x = 2x + 2xy$, $f_y = 2y + x^2$, $f_{xx} = 2 + 2y$, $f_{yy} = 2$, $f_{xy} = 2x$. Then $f_y = 0$ implies $y = -\frac{1}{2}x^2$, substituting into $f_x = 0$ gives $2x - x^3 = 0$ so $x = 0$ or $x = \pm\sqrt{2}$. Thus the critical points are $(0,0)$, $(\sqrt{2}, -1)$ and $(-\sqrt{2}, -1)$. Now $D(0,0) = 4$, $D(\sqrt{2},-1) = -8 = D(-\sqrt{2},-1)$, $f_{xx}(0,0) = 2$, $f_{xx}(\pm\sqrt{2},-1) = 0$. Thus $f(0,0) = 4$ is a local minimum and $(\pm\sqrt{2},-1)$ are saddle points.

7. $f_x = 3x^2 - 3y$, $f_y = 3y^2 - 3x$, $f_{xx} = 6x$, $f_{yy} = 6y$, $f_{xy} = -3$. Then $f_x = 0$ implies $x^2 = y$ and substituting into $f_y = 0$ gives $x = 0$ or $x = 1$. Thus the critical points are $(0,0)$ and $(1,1)$. Now $D(0,0) = -9 < 0$ so $(0,0)$ is a saddle point and $D(1,1) = 36 - 9 > 0$ while $f_{xx}(1,1) = 6$ so $f(1,1) = -1$ is a local minimum.

9. $f_x = y - 2$, $f_y = x - 1$, $f_{xx} = f_{yy} = 0$, $f_{xy} = 1$ and the only critical point is $(1,2)$. Now $D(1,2) = -1$ so $(1,2)$ is a saddle point and f has no local maxima or minima.

11. $f_x = y - x^{-2}$, $f_y = x + 8y^{-2}$, $f_{xx} = 2x^{-3}$, $f_{yy} = -16y^{-3}$ and $f_{xy} = 1$. Then $f_x = 0$ implies $y = x^{-2}$, substituting into $f_y = 0$ gives $x + 8x^4 = 0$ so $x = 0$ or $x = -\frac{1}{2}$ but $(0,y)$ is not in the domain of f. Thus the only critical point is $(-\frac{1}{2}, 4)$. Then $f_{xx}(-\frac{1}{2}, 4) = -16$ and $D(-\frac{1}{2}, 4) = 4 - 1 > 0$ so $f(-\frac{1}{2}, 4) = -6$ is a local maximum.

13. $f_x = e^x \cos y$, $f_y = -e^x \sin y$. Now $f_x = 0$ implies $\cos y = 0$ or $y = (\pi/2) + n\pi$ for $n \in \mathbb{Z}$. But $\sin[(\pi/2) + n\pi] \neq 0$, so there are no critical points.

15. $f_x = 6xy - 6x$, $f_y = 3x^2 + 3y^2 - 6y$. Then $f_x = 0$ implies $x = 0$ or $y = 1$ and when $x = 0$, $f_y = 0$ implies $y = 0$ or $y = 2$; when $y = 1$, $f_y = 0$ implies $x^2 = 1$ or $x = \pm 1$. Thus the critical points are $(0,0)$, $(0,2)$, $(\pm 1, 1)$. Now $f_{xx} = 6y - 6$, $f_{yy} = 6y - 6$ and $f_{xy} = 6x$, so $D(0,0) = D(0,2) = 36 > 0$ while $D(\pm 1, 1) = -36 < 0$ and $f_{xx}(0,0) = -6$, $f_{xx}(0,2) = 6$. Hence $(\pm 1, 1)$ are saddle points while $f(0,0) = 2$ is a local maximum and $f(0,2) = -2$ is a local minimum.

17. $f_x = \sin y$, $f_y = x\cos y$, $f_{xx} = 0$, $f_{yy} = -x\sin y$ and $f_{xy} = \cos y$. Then $f_x = 0$ iff $y = n\pi$, $n \in \mathbb{Z}$ and substituting into $f_y = 0$ requires $x = 0$ for each of these y values. Thus the critical points are $(0, n\pi)$, $n \in \mathbb{Z}$. But $D(0, n\pi) = -\cos^2(n\pi) < 0$ so each critical point is a saddle point.

19. $f_x = \cos x + \cos(x+y)$, $f_y = \cos y + \cos(x+y)$, $f_{xx} = -\sin x - \sin(x+y)$, $f_{yy} = -\sin y - \sin(x+y)$, $f_{xy} = -\sin(x+y)$. Setting $f_x = 0$ and $f_y = 0$ and subtracting gives $\cos x - \cos y = 0$ or $\cos x = \cos y$. Thus $x = y$ or $x = 2\pi - y$. If $x = y$, $f_x = 0$ becomes $\cos x + \cos 2x = 0$ or $2\cos^2 x + \cos x - 1 = 0$, a quadratic in $\cos x$. Thus $\cos x = -1$ or $1/2$ and $x = \pi$, $\pi/3$, or $5\pi/3$ ($0 \le x \le 2\pi$) yielding the critical points (π, π), $(\pi/3, \pi/3)$ and $(5\pi/3, 5\pi/3)$. Similarly if $x = 2\pi - y$, $f_x = 0$ becomes $(\cos x) + 1 = 0$ and the resulting critical point is (π, π). Now $D(x, y) = \sin x \sin y + \sin x \sin(x+y) + \sin y \sin(x+y)$. So $D(\pi, \pi) = 0$ and (12.56) doesn't apply. $D(\pi/3, \pi/3) = 9/4 > 0$ and $f_{xx}(\pi/3, \pi/3) < 0$ so $f(\pi/3, \pi/3) = 3\sqrt{3}/2$ is a local maximum while $D(5\pi/3, 5\pi/3) = 9/4 > 0$ and $f_{xx}(5\pi/3, 5\pi/3) > 0$, so $f(5\pi/3, 5\pi/3) = -3\sqrt{3}/2$ is a local minimum.

21. Since f is a polynomial it is continuous on D, so an absolute maximum and minimum exist. Here $f_x = -3$, $f_y = 4$ so there are no critical points inside D. Thus the absolute extrema must both occur on the boundary. Along L_1, $y = 0$ and $f(x, 0) = 5 - 3x$, a decreasing function in x so the maximum value is $f(0, 0) = 5$ and the minimum value is $f(4, 0) = -7$. Along L_2, $x = 4$ and $f(4, y) = -7 + 4y$, an increasing function in y, so the minimum value is $f(4, 0) = -7$ and the maximum value is $f(4, 5) = 13$. Along L_3, $y = 5x/4$ and $f(x, 5x/4) = 5 + 2x$, an increasing function in x, so the minimum value is $f(0, 0) = 5$ and the maximum value is $f(4, 5) = 13$. Thus the absolute minimum of f on D is $f(4, 0) = -7$ and the absolute maximum is $f(4, 5) = 13$.

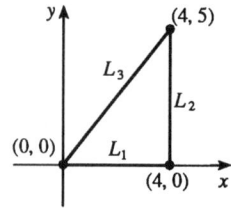

23. In Exercise 5, we found the critical points of f; only $(0, 0)$ with $f(0, 0) = 4$ is in D.
On L_1: $y = -1$, $f(x, -1) = 5$, a constant.
On L_2: $x = 1$, $f(1, y) = y^2 + y + 5$, a quadratic in y which attains its maximum at $(1, 1)$, $f(1, 1) = 7$ and its minimum at $(1, -1/2)$, $f(1, -1/2) = 17/4$.
On L_3: $f(x, 1) = 2x^2 + 5$ which attains its maximum at $(-1, 1)$ and $(1, 1)$ with $f(\pm 1, 1) = 7$ and its minimum at $(0, 1)$, $f(0, 1) = 5$. On L_4: $f(-1, y) = y^2 + y + 5$ with maximum at $(-1, 1)$, $f(-1, 1) = 7$ and minimum at $(-1, -1/2)$, $f(-1, -1/2) = 17/4$. Thus the absolute

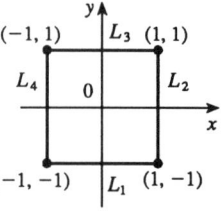

Section 12.7

maximum is attained at both $(\pm 1, 1)$ with $f(\pm 1, 1) = 7$ and the absolute minimum on D is attained at $(0,0)$ with $f(0,0) = 4$.

25. $f_x(x,y) = y - 1$ and $f_y(x,y) = x - 1$ and so the critical point is at $(1,1)$ where $f(1,1) = 0$ in D. Along L_1: $y = 4$, so
$f(x, 4) = 1 + 4x - x - 4 = 3x - 3$,
$-2 \le x \le 2$, which is an increasing function and has a maximum value when $x = 2$ where
$f(2,4) = 3$ and a minimum at $f(-2,4) = -9$.

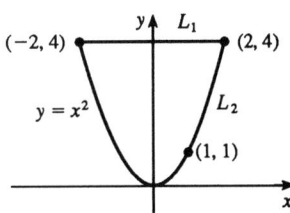

Along L_2: $y = x^2$, so let $g(x) = f(x, x^2) = x^3 - x^2 - x + 1$. Then $g'(x) = 3x^2 - 2x - 1 = 0$ $\Leftrightarrow x = -\frac{1}{3}$ or $x = 1$. $f(-\frac{1}{3}, \frac{1}{9}) = \frac{32}{27}$ and $f(1,1) = 0$. As a result, the absolute maximum and minimum values of f on D are $f(2,4) = 3$ and $f(-2,4) = -9$ respectively.

27. $f_x(x,y) = 6x^2$ and $f_y(x,y) = 4y^3$. And so $f_x = 0$ and $f_y = 0$ only occur when $x = y = 0$. Hence, the only critical point inside the disk is at $x = y = 0$ where $f(0,0) = 0$. Now on the circle $x^2 + y^2 = 1$, $y^2 = 1 - x^2$ so let $g(x) = f(x,y) = 2x^3 + (1-x^2)^2 = x^4 + 2x^3 - 2x^2 + 1$, $-1 \le x \le 1$. Then $g'(x) = 4x^3 + 6x^2 - 4x = 0 \Rightarrow x = 0, -2$, or $\frac{1}{2}$. $f(0, \pm 1) = g(0) = 1$, $f(\frac{1}{2}, \pm\sqrt{3}/2) = g(\frac{1}{2}) = \frac{13}{16}$, and $(-2, -3)$ is not in D. Checking the endpoints, we get $f(-1,0) = g(-1) = -2$ and $f(1,0) = g(1) = 2$. Thus the absolute maximum and minimum of f on D are $f(1,0) = 2$ and $f(-1,0) = -2$ respectively.
[Another method: On the boundary $x^2 + y^2 = 1$ we can write $x = \cos\theta$, $y = \sin\theta$, so $f(\cos\theta, \sin\theta) = 2\cos^3\theta + \sin^4\theta$, $0 \le \theta \le 2\pi$.]

29. Here $d = \sqrt{x^2 + y^2 + z^2}$ with $z = \frac{1}{3}(4 - x - 2y)$. So minimize
$d^2 = x^2 + y^2 + \frac{1}{9}(4 - x - 2y)^2 = f(x,y)$. Then $f_x = 2x - \frac{2}{9}(4 - x - 2y) = \frac{1}{9}(20x + 4y - 8)$, $f_y = \frac{1}{9}(26y + 4x - 16)$ so the only critical point is $(\frac{2}{7}, \frac{4}{7}, \frac{6}{7})$. Since the absolute minimum has to occur at a critical point, this is the closest point to the origin (or check $D(\frac{2}{7}, \frac{4}{7})$).

31. Again minimize $f(x,y) = (x-2)^2 + (y+2)^2 + (2x + \frac{4}{3}y - \frac{11}{3})^2$.
Then $f_x = 10x + \frac{16}{3}y - \frac{56}{3}$ and $f_y = \frac{50}{9}y + \frac{16}{3}x - \frac{52}{9}$.
Solving $50y + 48x = 52$ and $16y + 30x = 56$ simultaneously gives $x = \frac{164}{61}$, $y = -\frac{94}{61}$. Thus $d^2 = (\frac{42}{61})^2 + (\frac{28}{61})^2 + (-\frac{21}{61})^2$ or $d = 7/\sqrt{61}$.

33. Minimize $d^2 = x^2 + y^2 + z^2 = x^2 + y^2 + xy + 1$. Then $f_x = 2x + y$, $f_y = 2y + x$ so the critical point is $(0,0)$ and $D(0,0) = 4 - 1 > 0$ with $f_{xx}(0,0) = 2$ so this is a minimum. Thus $z^2 = 1$ or $z = \pm 1$ and the points on the surface are $(0, 0, \pm 1)$.

35. $x + y + z = 100$, so maximize $f(x,y) = xy(100 - x - y)$.
$f_x = 100y - 2xy - y^2$, $f_y = 100x - x^2 - 2xy$, $f_{xx} = -2y$, $f_{yy} = -2x$, $f_{xy} = 100 - 2x - 2y$.
Then $f_x = 0$ implies $y = 0$ or $y = 100 - 2x$. Substituting $y = 0$ into $f_y = 0$ gives $x = 0$ or $x = 100$ and substituting $y = 100 - 2x$ into $f_y = 0$ gives $3x^2 - 100x = 0$ so $x = 0$ or

121

100/3. Thus the critical points are $(0,0)$, $(100,0)$, $(0,100)$ and $(100/3, 100/3)$.

$D(0,0) = D(100,0) = D(0,100) = -10000$

while $D(100/3, 100/3) = 10000/3$ and $f_{xx}(100/3, 100/3) = -200/3 < 0$. Thus $(0,0)$, $(100,0)$ and $(0,100)$ are saddle points whereas $f(100/3, 100/3)$ is a local maximum.

Thus the numbers are $x = y = z = 100/3$.

37. Maximize $f(x,y) = xy(36 - 9x^2 - 36y^2)^{1/2}/2$ with (x,y,z) in first octant. Then

$f_x = [y(36 - 9x^2 - 36y^2)^{1/2}/2] + [-9x^2 y(36 - 9x^2 - 36y^2)^{-1/2}/2]$
$= (36y - 18x^2 y - 36y^3)/2(36 - 9x^2 - 36y^2)^{1/2}$ and

$f_y = (36x - 9x^3 - 72xy^2)/2(36 - 9x^2 - 36y^2)^{1/2}$. Setting $f_x = 0$ gives $y = 0$ or $y^2 = (2 - x^2)/2$ but $y > 0$, so only the latter solution applies. Substituting this y into $f_y = 0$ gives $x^2 = 4/3$ or $x = 2/\sqrt{3}$, $y = 1/\sqrt{3}$ and then $z^2 = (36 - 12 - 12)/4 = 3$. That this gives a maximum volume follows from the geometry. This maximum volume is

$V = (2x)(2y)(2z) = 8(2/\sqrt{3})(1/\sqrt{3})(\sqrt{3}) = 16/\sqrt{3}$.

39. Maximize $f(x,y) = \frac{xy}{3}(6 - x - 2y)$, then the maximum volume is $V = xyz$.

$f_x = \frac{1}{3}(6y - 2xy - y^2) = \frac{y}{3}(6 - 2x - 2y)$ and $f_y = \frac{x}{3}(6 - x - 4y)$. Setting $f_x = 0$ and $f_y = 0$ gives the critical point $(2,1)$ which geometrically must yield a maximum. Thus the volume of the largest such rectangle is $V = (2)(1)(\frac{2}{3}) = \frac{4}{3}$.

41. Let the dimensions be x, y, and z; then $4x + 4y + 4z = c$ and the volume is $V = xyz$
$= xy[(c/4) - x - y] = (c/4)xy - x^2 y - xy^2$, $x > 0$, $y > 0$. Then $V_x = (c/4)y - 2xy - y^2$ and $V_y = (c/4)x - x^2 - 2xy$, so $V_x = 0 = V_y$ when $2x+y = c/4$ and $x+2y = c/4$. Solving, we get $x = c/12$, $y = c/12$ and $z = (c/4) - x - y = c/12$. From the geometrical nature of the problem, this critical point must give an absolute maximum. Thus the box is a cube with edge length $= c/12$.

43. Let the dimensions be x, y and z, then minimize $xy + 2(xz + yz)$ if $xyz = 32{,}000$ m^3.
Then $f(x,y) = xy + [64{,}000(x+y)/xy] = xy + 64{,}000(x^{-1} + y^{-1})$, $f_x = y - 64{,}000 x^{-2}$, $f_y = x - 64{,}000 y^{-2}$. And $f_x = 0$ implies $y = 64{,}000/x^2$; substituting into $f_y = 0$ implies $x^3 = 64{,}000$ or $x = 40$ and then $y = 40$. Now $D(x,y) = [(2)(64{,}000)]^2 x^{-3} y^{-3} - 1 > 0$ for $(40, 40)$ and $f_{xx}(40, 40) > 0$ so this is indeed a minimum. Thus the dimensions of the box are $x = y = 40$ cm, $z = 20$ cm.

45. Note: Here the variables are m and b and $f(m,b) = \sum_{i=1}^{n} [y_i - (mx_i + b)]^2$.

Then $f_m = \sum_{i=1}^{n} -2x_i[y_i - (mx_i + b)] = 0$ implies $\sum_{i=1}^{n}(x_i y_i - mx_i^2 - bx_i) = 0$ or

$\sum_{i=1}^{n} x_i y_i = m \sum_{i=1}^{n} x_i^2 + b \sum_{i=1}^{n} x_i$ and $f_b = \sum_{i=1}^{n} -2[y_i - (mx_i + b)] = 0$ implies

$\sum_{i=1}^{n} y_i = m \sum_{i=1}^{n} x_i + \sum_{i=1}^{n} b = m(\sum_{i=1}^{n} x_i) + nb$. Thus we have the two desired

Section 12.8

equations. Now $f_{mm} = \sum\limits_{i=1}^{n} 2x_i^2$, $f_{bb} = \sum\limits_{i=1}^{n} 2 = 2n$ and $f_{mb} = \sum\limits_{i=1}^{n} 2x_i$. And $f_{mm}(m,b) > 0$ always and $D(m,b) = 4n(\sum\limits_{i=1}^{n} x_i^2) - 4(\sum\limits_{i=1}^{n} x_i)^2$

$= 4[n(\sum\limits_{i=1}^{n} x_i^2) - (\sum\limits_{i=1}^{n} x_i)^2] > 0$ always so the solutions of these two

equations do indeed minimize $\sum\limits_{i=1}^{n} d_i^2$.

47.

	x_i	y_i	x_i^2	$x_i y_i$
	69	138	4761	9522
	65	127	4225	8255
	71	178	5041	12638
	73	185	5329	13505
	68	141	4624	9588
	63	122	3969	7686
	70	158	4900	11060
	67	135	4489	9045
	69	145	4761	10005
	70	162	4900	11340
$\sum\limits_{i=1}^{n}$	685	1491	46999	102644

From the table $685m + 10b = 1491$ and $46999m + 685b = 102644$. Solving simultaneously gives $m = 1021/153 \approx 6.67$ and $b = 149.1 - 68.5(1021/153)$ ≈ -308.01. Thus a 6 ft. boy is predicted to weigh about $(6.67)(72) - 308.01$ or 172.2 lbs.

EXERCISES 12.8

1. $\nabla f = <2x, -2y>$, $\lambda \nabla g = <2\lambda x, 2\lambda y>$. Then $2x = 2\lambda x$ implies $x = 0$ or $\lambda = 1$. If $x = 0$, then $g(x,y) = x^2 + y^2 = 1$ implies $y = \pm 1$ and if $\lambda = 1$, then $-2y = 2\lambda y$ implies $y = 0$ and thus $x = \pm 1$. Thus the possible points for the extrema of f are $(\pm 1, 0)$, $(0, \pm 1)$. But $f(\pm 1, 0) = 1$ while $f(0, \pm 1) = -1$ so the maximum value of f on $x^2 + y^2 = 1$ is $f(\pm 1, 0) = 1$ and the minimum value is $f(0, \pm 1) = -1$.

3. $\nabla f = <y, x>$, $\lambda \nabla g = <18\lambda x, 2\lambda y>$, $g(x,y) = 4$. Then $y = 18\lambda x$ implies $(x,y) = (0,0)$ or $\lambda = y/18x$ and $x = 2\lambda y$ implies $(x,y) = (0,0)$ or $\lambda = x/2y$. Thus $(x,y) = (0,0)$ or $y/18x = x/2y$ implies $y^2 = 9x^2$. Now $(x,y) = (0,0)$ doesn't satisfy $g(x,y) = 4$ and when $y^2 = 9x^2$, $g(x,y) = 4$ implies $x^2 = 2/9$ or $x = \pm\sqrt{2}/3$. Hence the possible points are $(\pm\sqrt{2}/3, \sqrt{2})$, $(\pm\sqrt{2}/3, -\sqrt{2})$ and the maximum value of f on the ellipse is $f(\sqrt{2}/3, \sqrt{2}) = f(-\sqrt{2}/3, -\sqrt{2}) = 2/3$ while the minimum value is $f(-\sqrt{2}/3, \sqrt{2}) = f(\sqrt{2}/3, -\sqrt{2}) = -2/3$.

5. $\nabla f = <1, 3, 5>$, $\lambda \nabla g = <2\lambda x, 2\lambda y, 2\lambda z>$. Then $\nabla f = \lambda \nabla g$ implies $\lambda = 1/(2x) = 3/(2y) = 5/(2z)$ so $x = z/5$, $y = 3z/5$. Then $g(x,y,z) = 1$ implies $(z^2/25) + (9z^2/25) + z^2 = 1$ or $z = \pm\sqrt{5/7}$. Thus the possible points are $(\pm 1/\sqrt{35}, \pm 3/\sqrt{35}, \pm 5/\sqrt{35})$ with the maximum being $f(1/\sqrt{35}, 3/\sqrt{35}, 5/\sqrt{35}) = \sqrt{35}$ and

Section 12.8

minimum being $f(-1/\sqrt{35}, -3/\sqrt{35}, -5/\sqrt{35}) = -\sqrt{35}$.

7. $\nabla f = <yz, xz, xy>$, $\lambda \nabla g = <2\lambda x, 4\lambda y, 6\lambda z>$. Then $\nabla f = \lambda \nabla g$ implies $\lambda = yz/(2x) = xz/(4y) = xy/(6z)$ or $x^2 = 2y^2$ and $z^2 = 2y^2/3$. Thus $g(x,y,z) = 6$ implies $6y^2 = 6$ or $y = \pm 1$. Then the possible points are $(\sqrt{2}, \pm 1, \sqrt{2/3})$, $(\sqrt{2}, \pm 1, -\sqrt{2/3})$, $(-\sqrt{2}, \pm 1, \sqrt{2/3})$, $(-\sqrt{2}, \pm 1, -\sqrt{2/3})$. And the maximum value of f on the ellipsoid is $2/\sqrt{3}$ occurring when all coordinates are positive or exactly two are negative and the minimum is $-2/\sqrt{3}$ occurring when 1 or 3 of the coordinates are negative.

9. $\nabla f = <2x, 2y, 2z>$, $\lambda \nabla g = <4\lambda x^3, 4\lambda y^3, 4\lambda z^3>$. (1) If $x \neq 0$, $y \neq 0$ and $z \neq 0$, then $\nabla f = \lambda \nabla g$ implies $\lambda = 1/(2x^2) = 1/(2y^2) = 1/(2z^2)$ or $x^2 = y^2 = z^2$ and $3x^4 = 1$ or $x = \pm 1/\sqrt[4]{3}$ giving the points $(\pm 1/\sqrt[4]{3}, 1/\sqrt[4]{3}, 1/\sqrt[4]{3})$, $(\pm 1/\sqrt[4]{3}, -1/\sqrt[4]{3}, 1/\sqrt[4]{3})$, $(\pm 1/\sqrt[4]{3}, 1/\sqrt[4]{3}, -1/\sqrt[4]{3})$, $(\pm 1/\sqrt[4]{3}, -1/\sqrt[4]{3}, -1/\sqrt[4]{3})$ all with an f value of $\sqrt{3}$.
    (2) If one of the variables equals zero and the other two are not zero, then the squares of the two nonzero coordinates are equal with common value $1/\sqrt{2}$ and corresponding f value of $\sqrt{2}$.
    (3) If exactly two of the variables are zero, then the third variable has value ± 1 with the corresponding f value of 1. Thus on $x^4 + y^4 + z^4 = 1$, the maximum value of f is $\sqrt{3}$ and the minimum value is 1.

11. $<1,1,1,1> = <2\lambda x, 2\lambda y, 2\lambda z, 2\lambda t>$ so $\lambda = 1/(2x) = 1/(2y) = 1/(2z) = 1/(2t)$ and $x = y = z = t$. But $x^2 + y^2 + z^2 + t^2 = 1$, so the possible points are $(\pm 1/2, \pm 1/2, \pm 1/2, \pm 1/2)$. Thus the maximum value of f is $f(1/2, 1/2, 1/2, 1/2) = 2$ and the minimum value is $f(-1/2, -1/2, -1/2, -1/2) = -2$.

13. $\nabla f = <1, 2, 0>$, $\lambda \nabla g = <\lambda, \lambda, \lambda>$ and $\mu \nabla h = <0, 2\mu y, 2\mu z>$. Then $1 = \lambda$, $2 = \lambda + 2\mu y$ and $0 = \lambda + 2\mu z$ so $\mu y = 1/2 = -\mu z$ or $y = 1/(2\mu)$, $z = -1/(2\mu)$. Thus $g(x,y,z) = 1$ implies $x = 1$ and $h(x,y,z) = 4$ implies $\mu = \pm 1/(2\sqrt{2})$. Then the possible points are $(1, \pm\sqrt{2}, \mp\sqrt{2})$ and the maximum value is $f(1, \sqrt{2}, -\sqrt{2}) = 1 + 2\sqrt{2}$ and the minimum value is $f(1, -\sqrt{2}, \sqrt{2}) = 1 - 2\sqrt{2}$.

15. $\nabla f = <y, x+z, y>$, $\lambda \nabla g = <\lambda y, \lambda x, 0>$, $\mu \nabla h = <0, 2\mu y, 2\mu z>$. Then $y = \lambda y$ implies $\lambda = 1$ ($y \neq 0$ since $g(x,y,z) = 1$), $x + z = \lambda x + 2\mu y$ and $y = 2\mu z$. Thus $\mu = z/(2y) = y/(2z)$ or $y^2 = z^2$, and so $h(x,y,z) = 1$ implies $y = \pm 1/\sqrt{2}$, $z = \pm 1/\sqrt{2}$. Then $g(x,y,z) = 1$ implies $x = \pm \sqrt{2}$ and the possible points are $(\pm\sqrt{2}, \pm 1/\sqrt{2}, 1/\sqrt{2}), (\pm\sqrt{2}, \pm 1/\sqrt{2}, -1/\sqrt{2})$. Hence the maximum of f subject to the constraints is $f(\pm\sqrt{2}, \pm 1/\sqrt{2}, \pm 1/\sqrt{2}) = 3/2$ and the minimum is $f(\pm\sqrt{2}, \pm 1/\sqrt{2}, \mp 1/\sqrt{2}) = 1/2$.
    (*Note: since $xy = 1$ is one of the constraints we could have solved the problem by solving $f(y,z) = yz + 1$ subject to $y^2 + z^2 = 1$.*)

17. $\nabla Q = <\alpha K x^{\alpha - 1} y^{1-\alpha}, (1-\alpha) K x^\alpha y^{-\alpha}>$, $\lambda \nabla g = <\lambda m, \lambda n>$. Then $\alpha K (y/x)^{1-\alpha} = \lambda m$ and $(1-\alpha) K (x/y)^\alpha = \lambda n$ and $mx + ny = p$, so $\alpha K (y/x)^{1-\alpha}/m = (1-\alpha) K (x/y)^\alpha / n$ or $n\alpha/[m(1-\alpha)] = (x/y)^\alpha (x/y)^{1-\alpha}$ or $x = yn\alpha/[m(1-\alpha)]$. Substituting into $g(x,y) = p$

124

Section 12.8

gives $y = p(1-\alpha)/n$ and $x = p\alpha/m$ for the maximum production.

19. $\nabla f(x,y) = <y,x>$, $\lambda \nabla g = <2\lambda, 2\lambda>$, $g(x,y) = 2x+2y = p$. Then $\lambda = y/2 = x/2$ implies $x = y$ and the rectangle with maximum area is a square with side length $p/4$.

21. $f(x,y,z) = x^2 + y^2 + z^2$, $g(x,y,z) = x + 2y + 3z = 4$. Then
$\nabla f = <2x, 2y, 2z> = \lambda \nabla g = <\lambda, 2\lambda, 3\lambda> \Rightarrow x = \lambda/2$, $y = \lambda$, $z = 3\lambda/2$ and $(\lambda/2) + 2\lambda + (9\lambda/2) = 4 \Rightarrow \lambda = 4/7$. Hence the point closest to the origin is $(2/7, 4/7, 6/7)$.

23. $\nabla f = <2(x-2), 2(y+2), 2(z-3)> = \lambda \nabla g = <6\lambda, 4\lambda, -3\lambda>$, so $x = 3\lambda + 2$, $y = 2\lambda - 2$, $z = (-3/2)\lambda + 3$ and $(18\lambda + 12) + (8\lambda - 8) + (9/2)\lambda - 9 = 2$ implies $\lambda = 14/61$. Thus the shortest distance is $\sqrt{(42/61)^2 + (28/61)^2 + (-21/61)^2} = 7/\sqrt{61}$.

25. $f(x,y,z) = x^2 + y^2 + z^2$, $\nabla f = <2x, 2y, 2z> = \lambda \nabla g = <-\lambda y, -\lambda x, 2\lambda z>$. Then $2z = 2\lambda z$ implies $z = 0$ or $\lambda = 1$. If $z = 0$ then $g(x,y,z) = 1$ implies $xy = -1$ or $x = -1/y$. Thus $2x = -\lambda y$ and $2y = -\lambda x$ imply $\lambda = 2/y^2 = 2y^2$ or $y = \pm 1$, $x = \pm 1$. If $\lambda = 1$, then $2x = -y$ and $2y = -x$ imply $x = y = 0$, so $z = \pm 1$. Hence the possible points are $(\pm 1, \mp 1, 0)$, $(0, 0, \pm 1)$ and the minimum value of f is $f(0, 0, \pm 1) = 1$, so the points closest to the origin are $(0, 0, \pm 1)$.

27. $\nabla f = <yz, xz, xy> = \lambda \nabla g = <\lambda, \lambda, \lambda>$, $g(x,y,z) = x + y + z = 100$. Then $\lambda = yz = xz = xy$ implies $x = y = z = 100/3$.

29. If the dimensions are $2x$, $2y$ and $2z$ then $f(x,y,z) = 8xyz$, and
$\nabla f = <8yz, 8xz, 8xy> = \lambda \nabla g = <18\lambda x, 72\lambda y, 8\lambda z>$, $g(x,y,z) = 36$. Thus $18\lambda x = 8yz$, $72\lambda y = 8xz$, $8\lambda z = 8xy$ so $x^2 = 4y^2$, $z^2 = 9y^2$ and $36y^2 + 36y^2 + 36y^2 = 36$ or $y = 1/\sqrt{3}$ ($y > 0$). Thus the volume of the largest such rectangle is $8(1/\sqrt{3})(2/\sqrt{3})(3/\sqrt{3}) = 16\sqrt{3}$.

31. $\nabla f = <yz, xz, xy> = \lambda \nabla g = <\lambda, 2\lambda, 3\lambda>$, $g(x,y,z) = 6$. Then $\lambda = yz = xz/2 = xy/3$ implies $x = 2y$, $z = 2y/3$. But $2y + 2y + 2y = 6$ so $y = 1$, $x = 2$, $z = 2/3$ and the volume is $V = 4/3$.

33. $\nabla f = <yz, xz, xy>$, $g(x,y,z) = 4(x+y+z) = c$, $\lambda \nabla g = <4\lambda, 4\lambda, 4\lambda>$. Thus $4\lambda = yz = xz = xy$ or $x = y = z = c/12$ are the dimensions.

35. $\nabla f = <2z + y, 2z + x, 2(x+y)> = \lambda \nabla g = <\lambda yz, \lambda xz, \lambda xy>$, $g(x,y,z) = 32,000$ cm^3. Then (1) $\lambda yz = 2z + y$, (2) $\lambda xz = 2z + x$, (3) $\lambda xy = 2(x+y)$. And (1) − (2) implies $\lambda z(y-x) = y - x$, so $x = y$ or $\lambda = 1/z$. If $\lambda = 1/z$ then (1) implies $z = 0$ which can't be, so $x = y$. But twice (2) minus (3) together with $x = y$ implies $\lambda y(2x - y) = (4z + 2y) - 4y$ or $\lambda y(2z - y) = 2(2z - y)$ so $z = y/2$ or $\lambda = 2/y$. If $\lambda = 2/y$ then (3) implies $y = 0$ which can't be. Thus $x = y = 2z$ and $y^3/2 = 32,000$ or $y = 40$ and the dimensions which minimize the volume are $x = y = 40$ cm, $z = 20$ cm.

37. We need to find the extrema of $x^2 + y^2 + z^2$ subject to the two constraints $g(x,y,z) = x + y + 2z = 2$ and $h(x,y,z) = x^2 + y^2 - z = 0$. $\nabla f = <2x, 2y, 2z>$,

125

$\lambda \nabla g = <\lambda, \lambda, 2\lambda>$ and $\mu \nabla h = <2\mu x, 2\mu y, -\mu>$. Thus we need (1) $2x = \lambda + 2\mu x$, (2) $2y = \lambda + 2\mu y$, (3) $2z = 2\lambda - \mu$, (4) $x + y + 2z = 2$, and (5) $x^2 + y^2 - z = 0$. From (1) and (2) $2(x - y) = 2\mu(x - y)$. so if $x \neq y$, $\mu = 1$. Putting this in (3) says $2z = 2\lambda - 1$ or $\lambda = \frac{2z+1}{2}$, but putting $\mu = 1$ into (1) says $\lambda = 0$. Hence $\frac{2z+1}{2} = 0$ or $z = -\frac{1}{2}$. Then (4) and (5) become $x + y - 3 = 0$ and $x^2 + y^2 + \frac{1}{2} = 0$. The last equation cannot be true, so this case gives no solution. It must be then that $x = y$. Then (4) and (5) become $2x + 2z = 2$ and $2x^2 - z = 0$ which imply $z = 1 - x$ and $z = 2x^2$. Thus $2x^2 = 1 - x$ or $2x^2 + x - 1 = (2x - 1)(x + 1) = 0$ so $x = \frac{1}{2}$ or $x = -1$. The two points to check are $(\frac{1}{2}, \frac{1}{2}, \frac{1}{2})$ and $(-1, -1, 2)$; $f(\frac{1}{2}, \frac{1}{2}, \frac{1}{2}) = \frac{3}{4}$ and $f(-1, -1, 2) = 6$. Thus $(\frac{1}{2}, \frac{1}{2}, \frac{1}{2})$ is the point on the ellipse nearest the origin and $(-1, -1, 2)$ is the one farthest from the origin.

REVIEW EXERCISES FOR CHAPTER 12

1. True. $f_y(a, b) = \lim_{h \to 0} \frac{f(a, b+h) - f(a, b)}{h}$ from (12.12). Let $h = y - b$. As $h \to 0$, $y \to b$. Then by substituting, we get $f_y(a, b) = \lim_{y \to b} \frac{f(a, y) - f(a, b)}{y - b}$.

3. False. $f_{xy} = \frac{\partial^2 f}{\partial y \partial x}$.

5. False. See Example 3 in Section 12.2.

7. True. If f has a local minimum and f is differentiable at (a, b) then by (12.55) $f_x(a, b) = 0$ and $f_y(a, b) = 0$ so $\nabla f(a, b) = <f_x(a, b), f_y(a, b)> = <0, 0> = \vec{0}$.

9. False. $\nabla f(x, y) = <0, \frac{1}{y}>$.

11. $x \neq 1$ and $x + y + 1 > 0$, so
    $D = \{(x, y) | y > -x - 1, x \neq 1\}$.

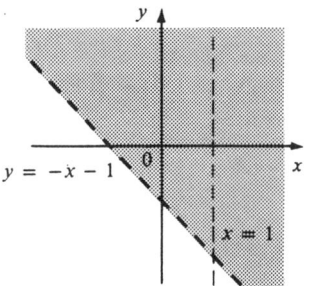

Chapter 12 Review

13. $D = \{(x,y) \mid -1 \leq x \leq 1, -\infty < y < \infty\}$.

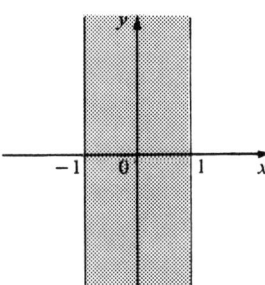

15. $z = f(x,y) = 1 - x^2 - y^2$, a paraboloid with vertex $(0,0,1)$.

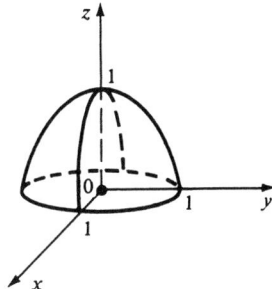

17. Let $k = e^{-c} = e^{-(x^2+y^2)}$ be the level curves, then $-\ln k = c = x^2 + y^2$, so we have a family of concentric circles.

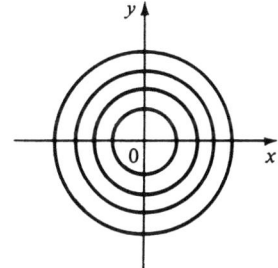

19. Since $0 \leq x^2y^2/(x^2+2y^2) \leq x^2y^2/(x^2+y^2) \leq (x^2+y^2)(x^2+y^2)/(x^2+y^2) = x^2 + y^2$, given $\epsilon > 0$, let $\delta = \sqrt{\epsilon}$. Then whenever $0 < \sqrt{x^2+y^2} < \delta$,

$|x^2y^2/(x^2+2y^2) - 0| = x^2y^2/(x^2+2y^2) \leq x^2 + y^2 < \delta^2 = \epsilon$. Hence $\lim_{(x,y)\to(0,0)} x^2y^2/(x^2+2y^2) = 0$. [Or use the Squeeze Theorem.]

21. $f_x = 12x^3 - \sqrt{y}$, $f_y = -\frac{1}{2}xy^{-1/2}$.

23. $f_s = 2e^{2s}\cos \pi t$, $f_t = -\pi e^{2s} \sin \pi t$.

25. $f_x = y^z$, $f_y = xzy^{z-1}$, and $\ln f = \ln xy^z = \ln x + z \ln y$ so $f_z = xy^z \ln y$.

27. $f_x = 2xy^3 - 8x^3$, $f_y = 3x^2y^2 + 2y$, $f_{xx} = 2y^3 - 24x^2$, $f_{yy} = 6x^2y + 2$, and $f_{xy} = f_{yx} = 6xy^2$.

Chapter 12 Review

29. $f_x = y^2z^3$, $f_y = 2xyz^3$, $f_z = 3xy^2z^2$, $f_{xx} = 0$, $f_{yy} = 2xz^3$, $f_{zz} = 6xy^2z$,
    $f_{xy} = f_{yx} = 2yz^3$, $f_{xz} = f_{zx} = 3y^2z^2$, and $f_{yz} = f_{zy} = 6xyz^2$.

31. $u_x = yx^{y-1}$, $u_y = x^y \ln x$ and $(x/y)u_x + (\ln x)^{-1}u_y = x^y + x^y = 2u$.

33. $z_x(0,1) = 0$, $z_y(0,1) = 6$ and the equation of the tangent plane is $z - 5 = 6(y - 1)$
    or $z - 6y = -1$.

35. $z_x(1,0) = 1$, $z_y(1,0) = 1$, and the tangent plane is $z - 1 = x - 1 + y$ or $x + y - z = 0$.

37. $F(x,y,z) = x^2 + 2y^2 - 3z^2$, $F_x = 2x$, $F_y = 4y$, $F_z = -6z$; $F_x(3,2,-1) = 6$,
    $F_y(3,2,-1) = 8$, $F_z(3,2,-1) = 6$. So the equation of the tangent plane is
    $6(x-3) + 8(y-2) + 6(z+1) = 0$ or $3x + 4y + 3z = 14$.

39. $F(x,y,z) = x^2 + y^2 + z^2$, $\nabla F(x_0, y_0, z_0) = <2x_0, 2y_0, 2z_0> = k<2, 1, -3>$ or
    $x_0 = k$, $y_0 = k/2$ and $z_0 = -3k/2$. But $x_0^2 + y_0^2 + z_0^2 = 1$, so $7k^2/2 = 1$ and
    $k = \pm\sqrt{2/7}$. Hence there are two such points: $(\pm\sqrt{2/7}, \pm 1/\sqrt{14}, \mp 3/\sqrt{14})$.

41. Let $w = f(x,y,z) = x^3\sqrt{y^2 + z^2}$, then $f(2,3,4) = 8(5) = 40$ so set
    $(a,b,c) = (2,3,4)$. Then $\Delta x = -0.02$, $\Delta y = 0.01$, $\Delta z = -0.03$,
    $f_x = 3x^2\sqrt{y^2 + z^2}$, $f_y = yx^3/\sqrt{y^2 + z^2}$, $f_z = zx^3/\sqrt{y^2 + z^2}$. Thus
    $(1.98)^3\sqrt{(3.01)^2 + (3.97)^2} \approx 40 + (60)(-0.02) + (24/5)(0.01) + (32/5)(-0.03) = 38.656$.

43. $dw/dt = [1/(2\sqrt{x})](2e^{2t}) + (2y/z)(3t^2 + 4) + (-y^2/z^2)(2t)$
    $= e^t + (2y/z)(3t^2 + 4) - 2t(y^2/z^2)$.

45. $\partial z/\partial x = 2xf'(x^2 - y^2)$, $\partial z/\partial y = 1 - 2yf'(x^2 - y^2)$ (where $f' \equiv df/d(x^2 - y^2)$).
    Then $y(\partial z/\partial x) + x(\partial z/\partial y) = 2xyf'(x^2 - y^2) + x - 2xyf'(x^2 - y^2) = x$.

47. $\dfrac{\partial z}{\partial x} = \dfrac{\partial z}{\partial u}y + \dfrac{\partial z}{\partial v}\dfrac{-y}{x^2}$ and $\dfrac{\partial^2 z}{\partial x^2} = y\dfrac{\partial}{\partial x}\left(\dfrac{\partial z}{\partial u}\right) + \dfrac{2y}{x^3}\dfrac{\partial z}{\partial v} + \dfrac{-y}{x^2}\dfrac{\partial}{\partial x}\left(\dfrac{\partial z}{\partial v}\right)$

    $= \dfrac{2y}{x^3}\dfrac{\partial z}{\partial v} + y\left(\dfrac{\partial^2 z}{\partial u^2}y + \dfrac{\partial^2 z}{\partial v\partial u}\dfrac{-y}{x^2}\right) + \dfrac{-y}{x^2}\left(\dfrac{\partial^2 z}{\partial v^2}\dfrac{-y}{x^2} + \dfrac{\partial^2 z}{\partial u\partial v}y\right)$

    $= \dfrac{2y}{x^3}\dfrac{\partial z}{\partial v} + y^2\dfrac{\partial^2 z}{\partial u^2} - \dfrac{2y^2}{x^2}\dfrac{\partial^2 z}{\partial u\partial v} + \dfrac{y^2}{x^4}\dfrac{\partial^2 z}{\partial v^2}$. Also $\dfrac{\partial z}{\partial y} = x\dfrac{\partial z}{\partial u} + \dfrac{1}{x}\dfrac{\partial z}{\partial v}$ and

    $\dfrac{\partial^2 z}{\partial y^2} = x\dfrac{\partial}{\partial y}\left(\dfrac{\partial z}{\partial u}\right) + \dfrac{1}{x}\dfrac{\partial}{\partial y}\left(\dfrac{\partial z}{\partial v}\right) = x\left(\dfrac{\partial^2 z}{\partial u^2}x + \dfrac{\partial^2 z}{\partial v\partial u}\dfrac{1}{x}\right) + \dfrac{1}{x}\left(\dfrac{\partial^2 z}{\partial v^2}\dfrac{1}{x} + \dfrac{\partial^2 z}{\partial u\partial v}x\right)$

    $= x^2\dfrac{\partial^2 z}{\partial u^2} + 2\dfrac{\partial^2 z}{\partial u\partial v} + \dfrac{1}{x^2}\dfrac{\partial^2 z}{\partial v^2}$. Thus

    $x^2\dfrac{\partial^2 z}{\partial x^2} - y^2\dfrac{\partial^2 z}{\partial y^2} = \dfrac{2y}{x}\dfrac{\partial z}{\partial v} + x^2y^2\dfrac{\partial^2 z}{\partial u^2} - 2y^2\dfrac{\partial^2 z}{\partial u\partial v} + \dfrac{y^2}{x^2}\dfrac{\partial^2 z}{\partial v^2}$

    $\quad - x^2y^2\dfrac{\partial^2 z}{\partial u^2} - 2y^2\dfrac{\partial^2 z}{\partial u\partial v} - \dfrac{y^2}{x^2}\dfrac{\partial^2 z}{\partial v^2}$

    $= \dfrac{2y}{x}\dfrac{\partial z}{\partial v} - 4y^2\dfrac{\partial^2 z}{\partial u\partial v} = 2v\dfrac{\partial z}{\partial v} - 4uv\dfrac{\partial^2 z}{\partial u\partial v}$ since $y = xv = \dfrac{uv}{y}$ or $y^2 = uv$.

49. $\nabla f = <z^2\sqrt{y}e^{x\sqrt{y}}, xz^2e^{x\sqrt{y}}/(2\sqrt{y}), 2ze^{x\sqrt{y}}> = ze^{x\sqrt{y}}<z\sqrt{y}, xz/(2\sqrt{y}), 2>$.

51. $\nabla f = <1/\sqrt{x}, -2y>$, $\nabla f(1,5) = <1, -10>$, $\vec{u} = <3, -4>/5$. Then $D_{\vec{u}}f(1,5) = 43/5$.

Chapter 12 Review

53. $\nabla f = <2xy, x^2 + 1/(2\sqrt{y})>$, $|\nabla f(2,1)| = |<4, 9/2>|$. Thus the maximum rate of change of f at $(2,1)$ is $\sqrt{145}/2$ in the direction $<4, 9/2>$.

55. $\dfrac{x^2+y^2}{(x-1)^2+y^2} \to \infty$ as $(x,y) \to (1,0)$ and so $\lim\limits_{(x,y)\to(1,0)} \tan^{-1}\left[\dfrac{x^2+y^2}{(x-1)^2+y^2}\right] = \pi/2$

57. $f_x = 2x - y + 9$, $f_y = -x + 2y - 6$, $f_{xx} = 2 = f_{yy}$, $f_{xy} = -1$. Then $f_x = 0$ and $f_y = 0$ imply $y = 1$, $x = -4$. Thus the only critical point is $(-4, 1)$ and $f_{xx}(-4, 1) > 0$, $D(-4, 1) = 3 > 0$ so $f(-4, 1) = -9$ is a local minimum.

59. $f_x = 3y - 2xy - y^2$, $f_y = 3x - x^2 - 2xy$, $f_{xx} = -2y$, $f_{yy} = -2x$, $f_{xy} = 3 - 2x - 2y$. Then $f_x = 0$ implies $y(3 - 2x - y) = 0$ so $y = 0$ or $y = 3 - 2x$. Substituting into $f_y = 0$ implies $x(3 - x) = 0$ or $3x(-1 + x) = 0$. Hence the critical points are $(0,0)$, $(3,0)$, $(0,3)$ and $(1,1)$. $D(0,0) = D(3,0) = D(0,3) = -9 < 0$ so $(0,0)$, $(3,0)$ and $(0,3)$ are saddle points. $D(1,1) = 3 > 0$ and $f_{xx}(1,1) = -2 < 0$, so $f(1,1) = 1$ is a local maximum.

61. First solve inside of D. Here
$f_x = 4y^2 - 2xy^2 - y^3$,
$f_y = 8xy - 2x^2y - 3xy^2$. Then $f_x = 0$ implies
$y = 0$ or $y = 4 - 2x$, but $y = 0$ isn't inside D.
Substituting $y = 4 - 2x$ into $f_y = 0$ implies
$x = 0$, $x = 2$ or $x = 1$, but $x = 0$ isn't inside

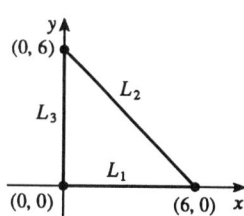

D, and when $x = 2$, $y = 0$ but $(2,0)$ isn't inside D. Thus the only critical point inside D is $(1,2)$ and $f(1,2) = 4$. Secondly we consider the boundary of D. On L_1, $f(x,0) = 0$ and so $f \equiv 0$ on L_1. On L_2, $x = -y + 6$ and $f(-y+6, y) = y^2(6-y)(-2) = -2(6y^2 - y^3)$ which has critical points at $y = 0$ and $y = 4$. Then $f(6,0) = 0$ while $f(2,4) = -64$.
On L_3, $f(0,y) = 0$, so $f \equiv 0$ on L_3. Thus on D the absolute maximum of f is $f(1,2) = 4$ while the absolute minimum is $f(2,4) = -64$.

63. $\nabla f = <2xy, x^2> = \lambda \nabla g = <2\lambda x, 2\lambda y>$, $g(x,y) = 1$. Then $2xy = 2\lambda x$ and $x^2 = 2\lambda y$ imply $\lambda = x^2/(2y)$ and $\lambda = y$ if $x \neq 0$ and $y \neq 0$. Hence $x^2 = 2y^2$. Then $g(x,y) = 1$ implies $3y^2 = 1$ so $y = \pm 1/\sqrt{3}$ and $x = \pm\sqrt{2/3}$. (Note if $x = 0$ then $x^2 = 2\lambda y$ implies $y = 0$ and $f(0,0) = 0$.) Thus the possible points are $(\pm\sqrt{2/3}, \pm 1/\sqrt{3})$ and the absolute maxima are $f(\pm\sqrt{2/3}, 1/\sqrt{3}) = 2/(3\sqrt{3})$ while the absolute minima are $f(\pm\sqrt{2/3}, -1/\sqrt{3}) = -2/(3\sqrt{3})$.

65. $\nabla f = <1,1,1> = \lambda \nabla g = <-\lambda x^{-2}, -\lambda y^{-2}, -\lambda z^{-2}>$, $g(x,y,z) = 1$.
Thus $\lambda = -x^2 = -y^2 = -z^2$ or $y = \pm x$, $z = \pm x$. Substituting into $g(x,y,z) = 1$ gives
(1) $3/x = 1$ so $x = 3$, or (2) $1/x = 1$ so $x = 1$, or (3) $-1/x = 1$ so $x = -1$ with the associated points (1) $(3,3,3)$, (2) $(1,1,-1)$, or $(1,-1,1)$, (3) $(-1,1,1)$. Thus the absolute maximum is $f(3,3,3) = 9$ and the absolute minimum is
$f(1,1,-1) = f(1,-1,1) = f(-1,1,1) = 1$.

Chapter 12 Review

67. $f(x, y, z) = x^2 + y^2 + z^2$, $g(x, y, z) = xy^2z^3 = 2$,
$\nabla f = \langle 2x, 2y, 2z \rangle = \lambda \nabla g = \langle \lambda y^2z^3, 2\lambda xyz^3, 3\lambda xy^2z^2 \rangle$. Since $g(x, y, z) = 2$, $x \neq 0$, $y \neq 0$ and $z \neq 0$, so (1) $2x = \lambda y^2z^3$, (2) $1 = \lambda xz^3$, (3) $2 = 3\lambda xy^2z$. Then (2) and (3) imply $1/(xz^3) = 2/(3xy^2z)$ or $y^2 = (2/3)z^2$ so $y = \pm z\sqrt{2/3}$. Similarly (1) and (3) imply $2x/(y^2z^3) = 2/(3xy^2z)$ or $3x^2 = z^2$ so $x = \pm z/\sqrt{3}$. But $xy^2z^3 = 2$ so x and z must have the same sign, i.e., $x = z/\sqrt{3}$. Thus $g(x, y, z) = 2$ implies $(z/\sqrt{3})(2z^2/3)z^3 = 2$ or $z = \pm 3^{1/4}$ and the possible points are $(\pm 3^{-1/4}, \sqrt{2}3^{-1/4}, \pm 3^{1/4})$, $(\pm 3^{-1/4}, -\sqrt{2}3^{-1/4}, \pm 3^{1/4})$. However at each of these points f takes on the same value, $2\sqrt{3}$. But $(2, 1, 1)$ also satisfies $g(x, y, z) = 2$ and $f(2, 1, 1) = 6 > 2\sqrt{3}$. Thus f has an absolute minimum value of $2\sqrt{3}$ and no absolute maximum subject to the constraint $g(x, y, z) = 2$.

<u>Alternate solution</u>

$g(x, y, z) = xy^2z^3 = 2$ implies $y^2 = 2/(xz^3)$, so minimize $f(x, z) = x^2 + 2/(xz^3) + z^2$.
Then $f_x = 2x - 2/(x^2z^3)$, $f_z = -6/(xz^4) + 2z$, $f_{xx} = 2 + 4/(x^3z^3)$, $f_{zz} = 24/(xz^5) + 2$ and $f_{xz} = 6/(x^2z^4)$. Now $f_x = 0$ implies $2x^3z^3 - 2 = 0$ or $z = 1/x$. Substituting into $f_y = 0$ implies $-6x^3 + 2x^{-1} = 0$ or $x = \sqrt[4]{1/3}$, so the two critical points are $(\pm\sqrt[4]{1/3}, \pm\sqrt[4]{3})$. Then $D(\pm\sqrt[4]{1/3}, \pm\sqrt[4]{3}) = (2+4)(2+24/3) - [6/\sqrt{3}]^2 > 0$ and $f_{xx}(\pm\sqrt[4]{1/3}, \pm\sqrt[4]{3}) = 6 > 0$, so each point is a minimum. Finally $y^2 = 2/(xz^3)$, so the four points closest to the origin are $(\pm\sqrt[4]{1/3}, (\sqrt{2}/\sqrt[4]{3}), \pm\sqrt[4]{3})$, $(\pm\sqrt[4]{1/3}, (-\sqrt{2}/\sqrt[4]{3}), \pm\sqrt[4]{3})$.

69.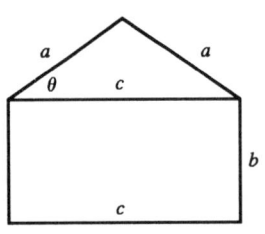
The area of the triangle is $\frac{1}{2}ca\sin\theta$ and the area of the rectangle is bc. Thus, the area of the whole object is $f(a, b, c) = \frac{1}{2}ca\sin\theta + bc$. The perimeter of the object is $g(a, b, c) = 2a + 2b + c = P$. To simplify $\sin\theta$ in terms of a, b, and c notice that $a^2\sin^2\theta + (\frac{1}{2}c)^2 = a^2$
$\Rightarrow \sin\theta = \frac{1}{2a}\sqrt{4a^2 - c^2}$. Thus $f(a, b, c) = \frac{c}{4}\sqrt{4a^2 - c^2} + bc$.

(Instead of using θ, we could just have used the Pythagorean Theorem.) As a result, by Lagrange's method, we must find a, b, c, and λ by solving $\nabla f = \lambda \nabla g$ which gives the equations: (1) $ca(4a^2 - c^2)^{-1/2} = 2\lambda$, (2) $c = 2\lambda$,

(3) $\frac{1}{4}(4a^2 - c^2)^{1/2} - \frac{c^2}{4}(4a^2 - c^2)^{-1/2} + b = \lambda$, and (4) $2a + 2b + c = P$. From (2), $\lambda = c/2$ and so (1) produces $ca(4a^2 - c^2)^{-1/2} = c \Rightarrow (4a^2 - c^2)^{1/2} = a \Rightarrow 4a^2 - c^2 = a^2 \Rightarrow c = \sqrt{3}a$ (5). Similarly, since $(4a^2 - c^2)^{1/2} = a$ and $\lambda = c/2$, (3) gives $\frac{1}{4}a - \frac{c^2}{4a} + b = c/2 \Rightarrow$ from (5) $\frac{a}{4} - \frac{3a}{4} + b = \sqrt{3}a/2 \Rightarrow -\frac{a}{2} - \sqrt{3}a/2 = -b \Rightarrow b = \frac{a}{2}(1 + \sqrt{3})$ (6). Inserting (5) and (6) into (4) we get: $2a + a(1 + \sqrt{3}) + \sqrt{3}a = P \Rightarrow 3a + 2\sqrt{3}a = P \Rightarrow$
$a = P/(3 + 2\sqrt{3}) = P(2\sqrt{3} - 3)/3$ and thus $b = P(2\sqrt{3} - 3)(1 + \sqrt{3})/6 = P(3 - \sqrt{3})/6$ and $c = P(2 - \sqrt{3})$.

PROBLEMS PLUS (after Chapter 12)

1. Since 3-dimensional situations are often difficult to visualize and work with, let us first try to find an analogous problem in 2 dimensions. The analogue of a cube is a square and the analogue of a sphere is a circle. Thus a similar problem in 2 dimensions is the following: if 5 circles with the same radius r are contained in a square of side 1 m so that the circles touch each other and 4 of the circles touch 2 sides of the square, find r.

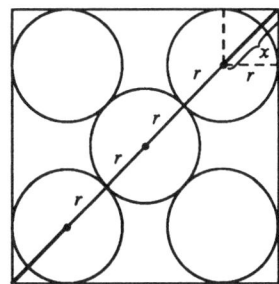

The diagonal of the square is $\sqrt{2}$. The diagonal is also $4r + 2x$. But x is the diagonal of a smaller square of side r. Therefore $x = \sqrt{2}r \Rightarrow \sqrt{2} = 4r + 2x = 4r + 2\sqrt{2}r$
$= (4 + 2\sqrt{2})r \Rightarrow r = \dfrac{\sqrt{2}}{4 + 2\sqrt{2}}$.

Let us use these ideas to solve the original 3-dimensional problem. The diagonal of the cube is $\sqrt{1^2 + 1^2 + 1^2} = \sqrt{3}$. The diagonal of the cube is also $4r + 2x$ where x is the diagonal of a smaller cube with edge r. Therefore $x = \sqrt{r^2 + r^2 + r^2} = \sqrt{3}r$
$\Rightarrow \sqrt{3} = 4r + 2x = 4r + 2\sqrt{3}r = (4 + 2\sqrt{3})r$. Therefore $r = \dfrac{\sqrt{3}}{4 + 2\sqrt{3}} = \dfrac{2\sqrt{3} - 3}{2}$. The radius of each ball is $\sqrt{3} - \dfrac{3}{2}$ m.

3. The areas of the smaller rectangles are $A_1 = xy$,
$A_2 = (L-x)y$, $A_3 = (L-x)(W-y)$,
$A_4 = x(W-y)$. Let $f(x,y) = A_1^2 + A_2^2 + A_3^2 + A_4^2$
$= x^2y^2 + (L-x)^2y^2 + (L-x)^2(W-y)^2 + x^2(W-y)^2$
$= [x^2 + (L-x)^2][y^2 + (W-y)^2]$, $0 \le x \le L$,
$0 \le y \le W$. Then we need to find the maximum and minimum values of $f(x,y)$. Here $f_x(x,y) = [2x - 2(L-x)][y^2 + (W-y)^2] = 0 \Rightarrow 4x - 2L = 0$ or $x = L/2$, and $f_y(x,y) = [x^2 + (L-x)^2][2y - 2(W-y)] = 0 \Rightarrow 4y - 2W = 0$ or $y = W/2$. Also $f_{xx} = 4[y^2 + (W-y)^2]$, $f_{yy} = 4[x^2 + (L-x)^2]$, and $f_{xy} = (4x - 2L)(4y - 2W)$. Then $D = 16[y^2 + (W-y)^2][x^2 + (L-x)^2] - (4x - 2L)^2(4y - 2W)^2$. Thus when $x = L/2$ and

$y = W/2$, $D > 0$ and $f_{xx} = 2W^2 > 0$. Thus a minimum of f occurs at $(L/2, W/2)$ and this minimum value is $f(L/2, W/2) = \frac{1}{4}L^2W^2$. There are no other critical points, so the maximum must occur on the boundary. Now along the width of the rectangle let
$g(y) = f(0, y) = f(L, y) = L^2[y^2 + (W - y)^2]$, $0 \le y \le W$. Then
$g'(y) = L^2[2y - 2(W - y)] = 0 \Leftrightarrow y = W/2$. And $g(W/2) = \frac{1}{2}L^2W^2$. Checking the endpoints we get $g(0) = g(W) = L^2W^2$. Along the length of the rectangle let
$h(x) = f(x, 0) = f(x, W) = W^2[x^2 + (L - x)^2]$, $0 \le x \le L$. By symmetry
$h'(x) = 0 \Leftrightarrow x = L/2$ and $h(L/2) = \frac{1}{2}L^2W^2$. At the endpoints we have
$h(0) = h(L) = L^2W^2$. Therefore L^2W^2 is the maximum value of f. This maximum value of f occurs when the "cutting" lines correspond to sides of the rectangle.

5. Let $g(x, y) = xf(y/x)$. Then $g_x(x, y) = f(y/x) + xf'(y/x)(-y/x^2) = f(y/x) - (y/x)f'(y/x)$ and $g_y(x, y) = xf'(y/x)(1/x) = f'(y/x)$. Thus the tangent plane at (x_0, y_0, z_0) on the surface has equation $z - x_0 f(y_0/x_0) = [f(y_0/x_0) - y_0 x_0^{-1}f'(y_0/x_0)](x - x_0) + f'(y_0/x_0)(y - y_0)$
$\Rightarrow [f(y_0/x_0) - y_0 x_0^{-1}f'(y_0/x_0)]x + [f'(y_0/x_0)]y - z = 0$. But any plane whose equation is of the form $ax + by + cz = 0$ *(which all of these tangent planes are)* passes through the origin. Thus the origin is the common point of intersection.

7. (a) $x = r\cos\theta$, $y = r\sin\theta$, $z = z$. Then $\dfrac{\partial u}{\partial r} = \dfrac{\partial u}{\partial x}\dfrac{\partial x}{\partial r} + \dfrac{\partial u}{\partial y}\dfrac{\partial y}{\partial r} + \dfrac{\partial u}{\partial z}\dfrac{\partial z}{\partial r}$

$= \dfrac{\partial u}{\partial x}\cos\theta + \dfrac{\partial u}{\partial y}\sin\theta$ and

$\dfrac{\partial^2 u}{\partial r^2} = \cos\theta\left[\dfrac{\partial^2 u}{\partial x^2}\dfrac{\partial x}{\partial r} + \dfrac{\partial^2 u}{\partial y\partial x}\dfrac{\partial y}{\partial r} + \dfrac{\partial^2 u}{\partial z\partial x}\dfrac{\partial z}{\partial r}\right]$

$+ \sin\theta\left[\dfrac{\partial^2 u}{\partial y^2}\dfrac{\partial y}{\partial r} + \dfrac{\partial^2 u}{\partial x\partial y}\dfrac{\partial x}{\partial r} + \dfrac{\partial^2 u}{\partial z\partial y}\dfrac{\partial z}{\partial r}\right]$

$= \dfrac{\partial^2 u}{\partial x^2}\cos^2\theta + \dfrac{\partial^2 u}{\partial y^2}\sin^2\theta + 2\dfrac{\partial^2 u}{\partial y\partial x}\cos\theta\sin\theta$. Similarly $\dfrac{\partial u}{\partial \theta} = -\dfrac{\partial u}{\partial x}r\sin\theta + \dfrac{\partial u}{\partial y}r\cos\theta$, and

$\dfrac{\partial^2 u}{\partial \theta^2} = \dfrac{\partial^2 u}{\partial x^2}r^2\sin^2\theta + \dfrac{\partial^2 u}{\partial y^2}r^2\cos^2\theta - 2\dfrac{\partial^2 u}{\partial y\partial x}r^2\sin\theta\cos\theta - \dfrac{\partial u}{\partial x}r\cos\theta - \dfrac{\partial u}{\partial y}r\sin\theta$.

Thus $\dfrac{\partial^2 u}{\partial r^2} + \dfrac{1}{r}\dfrac{\partial u}{\partial r} + \dfrac{1}{r^2}\dfrac{\partial^2 u}{\partial \theta^2} + \dfrac{\partial^2 u}{\partial z^2}$

$= \dfrac{\partial^2 u}{\partial x^2}\cos^2\theta + \dfrac{\partial^2 u}{\partial y^2}\sin^2\theta + 2\dfrac{\partial^2 u}{\partial y\partial x}\cos\theta\sin\theta + \dfrac{\partial u}{\partial x}\dfrac{\cos\theta}{r} + \dfrac{\partial u}{\partial y}\dfrac{\sin\theta}{r}$

$+ \dfrac{\partial^2 u}{\partial x^2}\sin^2\theta + \dfrac{\partial^2 u}{\partial y^2}\cos^2\theta - 2\dfrac{\partial^2 u}{\partial y\partial x}\sin\theta\cos\theta - \dfrac{\partial u}{\partial x}\dfrac{\cos\theta}{r} - \dfrac{\partial u}{\partial y}\dfrac{\sin\theta}{r} + \dfrac{\partial^2 u}{\partial z^2}$

$= \dfrac{\partial^2 u}{\partial x^2} + \dfrac{\partial^2 u}{\partial y^2} + \dfrac{\partial^2 u}{\partial z^2}$.

(b) $x = \rho \sin\phi \cos\theta$, $y = \rho \sin\phi \sin\theta$, $z = \rho \cos\phi$. Then $\dfrac{\partial u}{\partial \rho} = \dfrac{\partial u}{\partial x}\dfrac{\partial x}{\partial \rho} + \dfrac{\partial u}{\partial y}\dfrac{\partial y}{\partial \rho} + \dfrac{\partial u}{\partial z}\dfrac{\partial z}{\partial \rho}$

$= \dfrac{\partial u}{\partial x}\sin\phi\cos\theta + \dfrac{\partial u}{\partial y}\sin\phi\sin\theta + \dfrac{\partial u}{\partial z}\cos\phi$, and

$\dfrac{\partial^2 u}{\partial \rho^2} = \sin\phi\cos\theta\left[\dfrac{\partial^2 u}{\partial x^2}\dfrac{\partial x}{\partial \rho} + \dfrac{\partial^2 u}{\partial y\partial x}\dfrac{\partial y}{\partial \rho} + \dfrac{\partial^2 u}{\partial z\partial x}\dfrac{\partial z}{\partial \rho}\right]$

$\quad + \sin\phi\sin\theta\left[\dfrac{\partial^2 u}{\partial y^2}\dfrac{\partial y}{\partial \rho} + \dfrac{\partial^2 u}{\partial x\partial y}\dfrac{\partial x}{\partial \rho} + \dfrac{\partial^2 u}{\partial z\partial y}\dfrac{\partial z}{\partial \rho}\right]$

$\quad + \cos\phi\left[\dfrac{\partial^2 u}{\partial z^2}\dfrac{\partial z}{\partial \rho} + \dfrac{\partial^2 u}{\partial x\partial z}\dfrac{\partial x}{\partial \rho} + \dfrac{\partial^2 u}{\partial y\partial z}\dfrac{\partial y}{\partial \rho}\right]$

$= 2\dfrac{\partial^2 u}{\partial y\partial x}\sin^2\phi\sin\theta\cos\theta + 2\dfrac{\partial^2 u}{\partial z\partial x}\sin\phi\cos\phi\cos\theta + 2\dfrac{\partial^2 u}{\partial y\partial z}\sin\phi\cos\phi\sin\theta$

$\quad + \dfrac{\partial^2 u}{\partial x^2}\sin^2\phi\cos^2\theta + \dfrac{\partial^2 u}{\partial y^2}\sin^2\phi\sin^2\theta + \dfrac{\partial^2 u}{\partial z^2}\cos^2\phi$. Similarly

$\dfrac{\partial u}{\partial \phi} = \dfrac{\partial u}{\partial x}\rho\cos\phi\cos\theta + \dfrac{\partial u}{\partial y}\rho\cos\phi\sin\theta - \dfrac{\partial u}{\partial z}\rho\sin\phi$, and

$\dfrac{\partial^2 u}{\partial \phi^2} = 2\dfrac{\partial^2 u}{\partial y\partial x}\rho^2\cos^2\phi\sin\theta\cos\theta - 2\dfrac{\partial^2 u}{\partial x\partial z}\rho^2\sin\phi\cos\phi\cos\theta - 2\dfrac{\partial^2 u}{\partial y\partial z}\rho^2\sin\phi\cos\phi\sin\theta$

$\quad + \dfrac{\partial^2 u}{\partial x^2}\rho^2\cos^2\phi\cos^2\theta + \dfrac{\partial^2 u}{\partial y^2}\rho^2\cos^2\phi\sin^2\theta + \dfrac{\partial^2 u}{\partial z^2}\rho^2\sin^2\phi$

$\quad - \dfrac{\partial u}{\partial x}\rho\sin\phi\cos\theta - \dfrac{\partial u}{\partial y}\rho\sin\phi\sin\theta - \dfrac{\partial u}{\partial z}\rho\cos\phi$. And

$\dfrac{\partial u}{\partial \theta} = -\dfrac{\partial u}{\partial x}\rho\sin\phi\sin\theta + \dfrac{\partial u}{\partial y}\rho\sin\phi\cos\theta$, while

$\dfrac{\partial^2 u}{\partial \theta^2} = -2\dfrac{\partial^2 u}{\partial y\partial x}\rho^2\sin^2\phi\cos\theta\sin\theta + \dfrac{\partial^2 u}{\partial x^2}\rho^2\sin^2\phi\sin^2\theta + \dfrac{\partial^2 u}{\partial y^2}\rho^2\sin^2\phi\cos^2\theta$

$\quad - \dfrac{\partial u}{\partial x}\rho\sin\phi\cos\theta - \dfrac{\partial u}{\partial y}\rho\sin\phi\sin\theta$. Therefore

$\dfrac{\partial^2 u}{\partial \rho^2} + \dfrac{2}{\rho}\dfrac{\partial u}{\partial \rho} + \dfrac{\cot\phi}{\rho^2}\dfrac{\partial u}{\partial \phi} + \dfrac{1}{\rho^2}\dfrac{\partial^2 u}{\partial \phi^2} + \dfrac{1}{\rho^2\sin^2\phi}\dfrac{\partial^2 u}{\partial \theta^2}$

$= \dfrac{\partial^2 u}{\partial x^2}[(\sin^2\phi\cos^2\theta) + (\cos^2\phi\cos^2\theta) + \sin^2\theta]$

$\quad + \dfrac{\partial^2 u}{\partial y^2}[(\sin^2\phi\sin^2\theta) + (\cos^2\phi\sin^2\theta) + \cos^2\theta] + \dfrac{\partial^2 u}{\partial z^2}[\cos^2\phi + \sin^2\phi]$

$\quad + \dfrac{\partial u}{\partial x}\left[\dfrac{2\sin^2\phi\cos\theta + \cos^2\phi\cos\theta - \sin^2\phi\cos\theta - \cos\theta}{\rho\sin\phi}\right]$

$\quad + \dfrac{\partial u}{\partial y}\left[\dfrac{2\sin^2\phi\sin\theta + \cos^2\phi\sin\theta - \sin^2\phi\sin\theta - \sin\theta}{\rho\sin\phi}\right]$. But

$2\sin^2\phi\cos\theta + \cos^2\phi\cos\theta - \sin^2\phi\cos\theta - \cos\theta$
$= (2\sin^2\phi + \cos^2\phi - \sin^2\phi - 1)\cos\theta = 0$ and similarly the coefficient of $\dfrac{\partial u}{\partial y}$ is zero.

Problems Plus

Also $\sin^2\phi\cos^2\theta + \cos^2\phi\cos^2\theta + \sin^2\theta = \cos^2\theta(\sin^2\phi + \cos^2\phi) + \sin^2\theta = 1$, and similarly the coefficient of $\dfrac{\partial^2 u}{\partial y^2}$ is 1. So Laplace's Equation in spherical coordinates is as stated.

9. At $(x_1, y_1, 0)$ the equations of the tangent planes to $z = f(x, y)$ and $z = g(x, y)$ are P_1: $z - f(x_1, y_1) = f_x(x_1, y_1)(x - x_1) + f_y(x_1, y_1)(y - y_1)$ and P_2: $z - g(x_1, y_1) = g_x(x_1, y_1)(x - x_1) + g_y(x_1, y_1)(y - y_1)$, respectively. P_1 intersects the xy-plane in the line given by $f_x(x_1, y_1)(x - x_1) + f_y(x_1, y_1)(y - y_1) = -f(x_1, y_1)$, $z = 0$; and P_2 intersects the xy-plane in the line given by
$g_x(x_1, y_1)(x - x_1) + g_y(x_1, y_1)(y - y_1) = -g(x_1, y_1)$, $z = 0$. The point $(x_2, y_2, 0)$ is the point of intersection of these two lines, since $(x_2, y_2, 0)$ is the point where the line of intersection of the two tangent planes intersects the xy-plane. Thus (x_2, y_2) is the solution of the simultaneous equations
$f_x(x_1, y_1)(x_2 - x_1) + f_y(x_1, y_1)(y_2 - y_1) = -f(x_1, y_1)$ and
$g_x(x_1, y_1)(x_2 - x_1) + g_y(x_1, y_1)(y_2 - y_1) = -g(x_1, y_1)$. For simplicity, rewrite $f_x(x_1, y_1)$ as f_x and similarly for f_y, g_x, g_y, f and g and solve the equations
$(f_x)(x_2 - x_1) + (f_y)(y_2 - y_1) = -f$, and $(g_x)(x_2 - x_1) + (g_y)(y_2 - y_1) = -g$
simultaneously for $(x_2 - x_1)$ and $(y_2 - y_1)$. Then
$$y_2 - y_1 = \frac{gf_x - fg_x}{g_x f_y - f_x g_y} \text{ or } y_2 = y_1 - \frac{gf_x - fg_x}{f_x g_y - g_x f_y} \text{ and}$$
$$(f_x)(x_2 - x_1) + \frac{(f_y)(gf_x - fg_x)}{g_x f_y - f_x g_y} = -f \text{ so}$$
$$(x_2 - x_1) = \frac{-f - [(f_y)(gf_x - fg_x)/(g_x f_y - f_x g_y)]}{f_x} = \frac{fg_y - f_y g}{g_x f_y - f_x g_y}. \text{ Hence}$$
$$x_2 = x_1 - \frac{fg_y - f_y g}{f_x g_y - g_x f_y}.$$

11. Since we are minimizing the area of the ellipse, and the circle lies above the x-axis, the ellipse will intersect the circle for only one value of y. This y-value must satisfy both the equation of the circle and the ellipse. Now
$\dfrac{x^2}{a^2} + \dfrac{y^2}{b^2} = 1 \Rightarrow x^2 = \dfrac{a^2}{b^2}(b^2 - y^2)$.
Substituting into the equation of the circle gives
$\dfrac{a^2}{b^2}(b^2 - y^2) + y^2 - 2y = 0 \Rightarrow \left(\dfrac{b^2 - a^2}{b^2}\right)y^2 - 2y + a^2 = 0$.

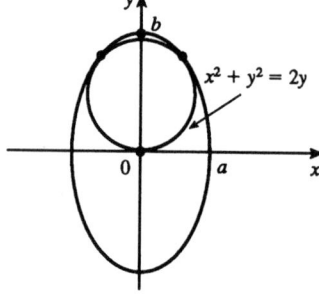

In order for there to be only one solution to this quadratic equation, the discriminant must be 0, so $4 - 4a^2(b^2 - a^2)/b^2 = 0 \Rightarrow b^2 - a^2 b^2 + a^4 = 0$. The area of the ellipse is $A(a, b) = \pi ab$, and we minimize this function subject to the constraint $g(a, b) = b^2 - a^2 b^2 + a^4 = 0$. Now

$\nabla A = \lambda \nabla g \Leftrightarrow \pi b = \lambda(4a^3 - 2ab^2),\ \pi a = \lambda(2b - 2ba^2) \Rightarrow$ (1) $\lambda = \dfrac{\pi b}{2a(2a^2 - b^2)}$,

(2) $\lambda = \dfrac{\pi a}{2b(1 - a^2)}$, (3) $b^2 - a^2 b^2 + a^4 = 0$. Comparing (1) and (2) gives

$\dfrac{\pi b}{2a(2a^2 - b^2)} = \dfrac{\pi a}{2b(1 - a^2)} \Rightarrow 2\pi b^2 = 4\pi a^4 \Leftrightarrow a^2 = \dfrac{b}{\sqrt{2}}$. Substitute this into (3) to get

$b = \dfrac{3}{\sqrt{2}} \Rightarrow a = \sqrt{\dfrac{3}{2}}$.

13. (a) (i) If $r = a$, $z = b\theta$ in cylindrical coordinates then $x = a\cos\theta$, $y = a\sin\theta$, $z = b\theta$.

So the position of the particle is $\vec{r}(\theta) = (a\cos\theta)\vec{i} + (a\sin\theta)\vec{j} + b\theta\vec{k}$, and

the velocity is $\vec{v}(\theta) = \vec{r}'(\theta) = -a\sin\theta \dfrac{d\theta}{dt}\vec{i} + a\cos\theta \dfrac{d\theta}{dt}\vec{j} + b\dfrac{d\theta}{dt}\vec{k}$. So

$|\vec{v}| = \sqrt{2gz} \Rightarrow a^2 \sin^2\theta \left(\dfrac{d\theta}{dt}\right)^2 + a^2 \cos^2\theta \left(\dfrac{d\theta}{dt}\right)^2 + b^2 \left(\dfrac{d\theta}{dt}\right)^2 = 2gb\theta$

$\Rightarrow (a^2 + b^2)\left(\dfrac{d\theta}{dt}\right)^2 = 2gb\theta$ so that $\dfrac{d\theta}{dt} = \sqrt{\dfrac{2gb}{a^2 + b^2}}\sqrt{\theta}$.

(ii) From (i) we have $\dfrac{d\theta}{\sqrt{\theta}} = \sqrt{\dfrac{2gb}{a^2 + b^2}}dt$. Integrating gives $2\sqrt{\theta} = \sqrt{\dfrac{2gb}{a^2 + b^2}}t + C$, but

$\theta(0) = 0 \Rightarrow C = 0$. Now squaring both sides gives $4\theta = \dfrac{2gb}{a^2 + b^2}t^2$, i.e.,

$\theta = \dfrac{gb}{2(a^2 + b^2)}t^2 = kt^2$ where $k = \dfrac{gb}{2(a^2 + b^2)}$. So the parametric equations are

$x = a\cos kt^2$, $y = a\sin kt^2$, $z = bkt^2$.

(b) (i) As in (a) we let the position of the particle be

$\vec{r}(\theta) = (a\theta \cos\theta)\vec{i} + (a\theta \sin\theta)\vec{j} + b\theta\vec{k}$. Then

$\vec{v}(\theta) = (a\cos\theta - a\theta \sin\theta)\dfrac{d\theta}{dt}\vec{i} + (a\sin\theta + a\theta \cos\theta)\dfrac{d\theta}{dt}\vec{j} + b\dfrac{d\theta}{dt}\vec{k}$. So

$|\vec{v}| = \sqrt{2gz} \Rightarrow (a\cos\theta - a\theta\sin\theta)^2 \left(\dfrac{d\theta}{dt}\right)^2 + (a\sin\theta + a\theta\cos\theta)^2 \left(\dfrac{d\theta}{dt}\right)^2 + b^2\left(\dfrac{d\theta}{dt}\right)^2$

$= (a^2 + a^2\theta^2 + b^2)\left(\dfrac{d\theta}{dt}\right)^2 = 2gb\theta$. Hence $\dfrac{d\theta}{dt} = \sqrt{\dfrac{2gb\theta}{a^2 + a^2\theta^2 + b^2}}$.

(ii) If $\vec{r}(\theta)$ is as in (i), then the distance traveled as a function of θ is the arclength

function $s(\theta) = \displaystyle\int_0^\theta |\vec{r}'(u)|du = \int_0^\theta \sqrt{(a^2 + b^2) + a^2 u^2}\,du = [\text{setting } t = au]$

$\dfrac{1}{a}\displaystyle\int_0^{a\theta} \sqrt{(a^2 + b^2) + t^2}\,dt = \dfrac{1}{a}\left[\dfrac{t}{2}\sqrt{(a^2 + b^2) + t^2} + \dfrac{a^2 + b^2}{2}\ln\left|t + \sqrt{(a^2 + b^2) + t^2}\right|\right]_0^{a\theta}$

$= \dfrac{1}{2a}\left\{a\theta\sqrt{a^2 + b^2 + a^2\theta^2} + (a^2 + b^2)\ln\left(a\theta + \sqrt{a^2 + b^2 + a^2\theta^2}\right) - (a^2 + b^2)\ln\sqrt{a^2 + b^2}\right\}$.

Section 13.1

CHAPTER 13

EXERCISES 13.1

1. (a) $\sum_{i=1}^{2}\sum_{j=1}^{2} f(x_{ij}^*, y_{ij}^*) \Delta A_{ij} = (1/2)[f(0, 3/2) + f(0, 2) + f(1, 3/2) + f(1, 2)]$
   $= (1/2)[(-27/4) + (-12) + (1 - 27/4) + (1 - 12)] = (1/2)(-71/2) = -17.75.$
   (b) $(1/2)[f(1, 3/2) + f(1, 2) + f(2, 3/2) + f(2, 2)]$
   $= (1/2)[(-23/4) + (-11) + (-19/4) + (-10)] = (1/2)(-63/2) = -15.75.$
   (c) $(1/2)[f(0, 1) + f(0, 3/2) + f(1, 1) + f(1, 3/2)] = (1/2)[-3 - (27/4) - 2 - (23/4)] = -8.75.$
   (d) $(1/2)[f(1, 1) + f(1, 3/2) + f(2, 1) + f(2, 3/2)] = (1/2)[-2 - (23/4) - 1 - (19/4)] = -6.75.$

3.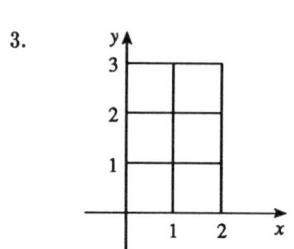
$\Delta A_{ij} = 1$ for $i = 1, 2$, $j = 1, 2, 3$. $\iint_R (x^2 + 4y) dA$
$\approx (1)[f(1, 1) + f(1, 2) + f(1, 3) + f(2, 1) + f(2, 2) + f(2, 3)]$
$= [5 + 9 + 13 + 8 + 12 + 16] = 63.$ $\|P\| = \sqrt{1 + 1} = \sqrt{2}.$

5.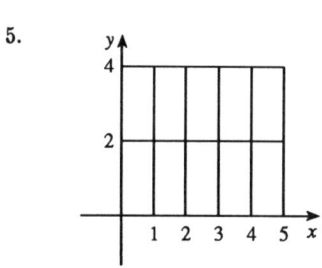
$\Delta A_{ij} = 2$ for $i = 1, 2, 3, 4, 5$, $j = 1, 2$. $\iint_R (xy - y^2) dA$
$\approx (2)[f(1/2, 1) + f(1/2, 3) + f(3/2, 1) + f(3/2, 3) + f(5/2, 1)$
$+ f(5/2, 3) + f(7/2, 1) + f(7/2, 3) + f(9/2, 1) + f(9/2, 3)]$
$= (2)[(-1/2) + (-15/2) + (1/2) + (-9/2) + (3/2)$
$+ (-3/2) + (5/2) + (3/2) + (7/2) + (9/2)] = 0$ and
$\|P\| = \sqrt{1 + 4} = \sqrt{5}.$

7.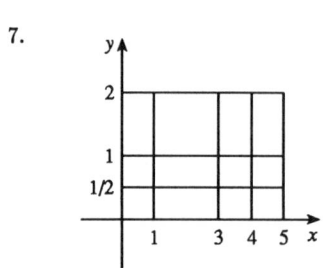
$\iint_R (x^2 - y^2) dA \approx (1/2)f(0, 1/2) + (1/2)f(0, 1) + (1)f(0, 2)$
$+ (1)f(1, 1/2) + (1)f(1, 1) + (2)f(1, 2) + (1/2)f(3, 1/2)$
$+ (1/2)f(3, 1) + (1)f(3, 2) + (1/2)f(4, 1/2) + (1/2)f(4, 1)$
$+ (1)f(4, 2) = (1/2)(-1/4) + (1/2)(-1) + (1)(-4)$
$+ (1)(3/4) + (1)(0) + 2(-3) + (1/2)(35/4) + (1/2)(8)$
$+ (1)(5) + (1/2)(63/4) + (1/2)(15) + (1)(12)$
$= (1/2)(185/4) + (55/4) - 6 = 247/8.$ And the length of the
longest diagonal is $\|P\| = \sqrt{1 + 4} = \sqrt{5}.$

Section 13.2

9. $z = f(x, y) = 4 - 2y \geq 0$ for $0 \leq y \leq 1$.
Thus the integral represents the volume of
that part of the rectangular solid
$[0, 1] \times [0, 1] \times [0, 4]$ which lies below the
plane $z = 4 - 2y$. So
$$\iint_R (4 - y)\, dA = (1)(1)(2) + \tfrac{1}{2}(1)(1)(2) = 3.$$

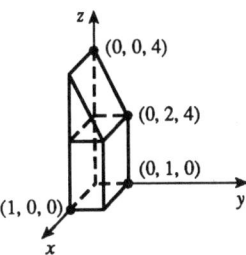

11. For any partition, $\iint_R k\, dA \approx \sum_i \sum_j f(x_{ij}^*, y_{ij}^*)\Delta A_{ij}$ but $f(x_{ij}^*, y_{ij}^*) = k$ always and $\sum_i \sum_j \Delta A_{ij}$ = area of $R = (b-a)(c-d)$. Thus for every partition $\sum_i \sum_j f(x_{ij}^*, y_{ij}^*)\Delta A_{ij}$
$= k \sum_i \sum_j \Delta A_{ij} = k(b-a)(c-d)$ and so as $\|P\| \to 0$ the limit is $k(b-a)(c-d)$.

EXERCISES 13.2

1. $\int_0^2 x^2 y^3\, dy = x^2(y^4/4)\Big|_0^2 = 4x^2$, $\int_0^1 x^2 y^3\, dx = y^3(x^3/3)\Big|_0^1 = y^3/3$.

3. $\int_0^2 xe^{x+y}\, dy = xe^{x+y}\Big|_0^2 = x(e^{x+2} - e^x) = xe^x(e^2 - 1)$, $\int_0^1 xe^{x+y}\, dx = e^y \int_0^1 xe^x\, dx$
$= e^y(xe^x - e^x)\Big|_0^1 = e^y$.

5. $\int_0^4 \int_0^2 x\sqrt{y}\, dx\, dy = \int_0^4 \sqrt{y}(x^2/2\Big|_0^2)\, dy = \int_0^4 2\sqrt{y}\, dy = 4y^{3/2}/3\Big|_0^4 = 32/3$.

7. $\int_{-1}^1 \int_0^1 (x^3 y^2 + 3xy^2)\, dy\, dx = \int_{-1}^1 [(x^3 y^4/4 + xy^3)\Big|_{y=0}^{y=1}]\, dx = \int_{-1}^1 [(x^3/4) + x]\, dx$
$= (x^4/16 + x^2/2)\Big|_{-1}^1 = 0$. Alternate solution: Applying Fubini's Theorem the integral
equals $\int_0^1 \int_{-1}^1 (x^3 y^2 + 3xy^2)\, dx\, dy = \int_0^1 [y^2(x^4/4) + 3y^2(x^2/2)]\Big|_{x=-1}^{x=1}\, dy = \int_0^1 0\, dy = 0$.

9. $\int_0^3 \int_0^1 \sqrt{x+y}\, dx\, dy = \int_0^3 (2/3)(x+y)^{3/2}\Big|_{x=0}^{x=1}\, dy = (2/3)\int_0^3 [(1+y)^{3/2} - y^{3/2}]\, dy$
$= (2/3)[(2/5)(1+y)^{5/2} - (2/5)y^{5/2}]\Big|_0^3 = (4/15)[32 - 3^{5/2} - 1] = (4/15)(31 - 9\sqrt{3})$.

11. $\int_0^{\pi/4} \int_0^3 \sin x\, dy\, dx = 3\int_0^{\pi/4} \sin x\, dx = 3(-\cos x)\Big|_0^{\pi/4} = 3(1 - (1/\sqrt{2}))$.

13. $\int_0^{\ln 2} \int_0^{\ln 5} e^{2x-y}\, dx\, dy = \left(\int_0^{\ln 5} e^{2x}\, dx\right)\left(\int_0^{\ln 2} e^{-y}\, dy\right)$
$= \left(\tfrac{1}{2}e^{2x}\Big|_0^{\ln 5}\right)\left(-e^{-y}\Big|_0^{\ln 2}\right) = \left(\tfrac{25}{2} - \tfrac{1}{2}\right)\left(-\tfrac{1}{2} + 1\right) = 6$

Section 13.2

15. $\int_1^2 \int_0^3 (2y^2 - 3xy^3) dy\, dx = \int_1^2 (2y^3/3 - 3xy^4/4|_{y=0}^{y=3}) dx = \int_1^2 (18 - 243x/4) dx$
$= 18x - 243x^2/8|_1^2 = -585/8$.

17. $\int_0^{\pi/6} \int_1^4 x\sin y\, dx\, dy = (\int_0^{\pi/6} \sin y\, dy)(\int_1^4 x\, dx) = (1 - (\sqrt{3}/2))(15/2) = 15(2 - \sqrt{3})/4$.

19. $\int_0^{\pi/6} \int_0^{\pi/3} x\sin(x+y) dy\, dx = \int_0^{\pi/6} (-x\cos(x+y)|_0^{\pi/3}) dx$
$= \int_0^{\pi/6} [(x\cos x) - x(\cos(x + \pi/3))] dx$
$= x[\sin x - \sin(x + \pi/3)]|_0^{\pi/6} - \int_0^{\pi/6} [\sin x - \sin(x + \pi/3)] dx$
$= (\pi/6)[(1/2) - 1] - [-\cos x + \cos(x + \pi/3)]|_0^{\pi/6} = -\pi/12 - [-\sqrt{3}/2 + 0 - (-1 + 1/2)]$
$= (\sqrt{3} - 1)/2 - (\pi/12)$.

21. $\int_0^1 \int_1^2 [1/(x+y)] dx\, dy = \int_0^1 \ln(x+y)|_1^2 dy = \int_0^1 [\ln(2+y) - \ln(1+y)] dy$
$= \{[(2+y)\ln(2+y) - (2+y)] - [(1+y)\ln(1+y) - (1+y)]\}|_0^1$
$= (3\ln 3) - 3 - (2\ln 2) + 2 - [(2\ln 2 - 2) - (0 - 1)] = 3\ln 3 - 4\ln 2 = \ln(27/16)$.

23. $z = f(x,y) = 4 - x - 2y \geq 0$ for
$0 \leq x \leq 1$ and $0 \leq y \leq 1$. So the
solid is the region in the first octant
which lies below the plane
$z = 4 - x - 2y$ and above $[0,1] \times [0,1]$.

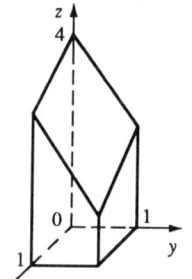

25. $V = \int_1^4 \int_{-1}^0 (2x + 5y + 1) dx\, dy = \int_1^4 (x^2 + 5xy + x|_{-1}^0) dy = \int_1^4 5y\, dy = \frac{5}{2}y^2|_1^4 = 75/2$.

27. $V = \int_{-2}^2 \int_{-1}^1 [1 - (x^2/4) - (y^2/9)] dx\, dy = 4\int_0^2 \int_0^1 (1 - x^2/4 - y^2/9) dx\, dy$
$= 4\int_0^2 [x - x^3/12 - y^2 x/9|_0^1] dy = 4\int_0^2 ((11/12) - y^2/9) dy = 4((11/12)y - y^3/27)|_0^2$
$= 4(83/54) = 166/27$.

29. Here we need the volume of the solid lying under the surface $z = x\sqrt{x^2 + y}$ and above the square $R = [0,1] \times [0,1]$ in the xy-plane.
$V = \int_0^1 \int_0^1 x\sqrt{x^2 + y}\, dx\, dy = \int_0^1 (1/3)(x^2 + y)^{3/2}|_0^1 dy$
$= (1/3)\int_0^1 ((1+y)^{3/2} - y^{3/2}) dy = (1/3)(2/5)[(1+y)^{5/2} - y^{5/2}]|_0^1 = (4/15)(2\sqrt{2} - 1)$.

31. Here we need the volume of the solid lying under the surface $z = 6 - xy$ and above the rectangle $R = [-2, 2] \times [0, 3]$ in the xy-plane.
$V = \int_{-2}^2 \int_0^3 (6 - xy) dy\, dx = \int_{-2}^2 (6y - xy^2/2|_0^3) dx = \int_{-2}^2 (18 - (9/2)x) dx$

138

Section 13.3

$= (18x - (9/4)x^2)\big|_{-2}^{2} = 72.$

33. $A(R) = 2 \cdot 5 = 10$, so $f_{ave} = \frac{1}{A(R)} \iint_R f(x, y) \, dA$

$= \frac{1}{10} \int_0^5 \int_{-1}^1 x^2 y \, dx \, dy = \frac{1}{10} \int_0^5 \left(\frac{yx^3}{3} \bigg|_{-1}^1 \right) dy = \frac{1}{10} \int_0^5 \frac{2y}{3} \, dy = \frac{1}{10} \left[\frac{y^2}{3} \right]_0^5 = \frac{5}{6}.$

EXERCISES 13.3

1. $\int_0^1 \int_0^y x \, dx \, dy = \int_0^1 (x^2/2) \big|_0^y dy = \int_0^1 (y^2/2) dy = 1/6.$

3. $\int_0^2 \int_{\sqrt{x}}^3 (x^2 + y) dy \, dx = \int_0^2 (x^2 y + (y^2/2) \big|_{\sqrt{x}}^3) dx = \int_0^2 [3x^2 + (9/2) - x^{5/2} - (x/2)] dx$
$= x^3 + (9/2)x - (2/7)x^{7/2} - (x^2/4) \big|_0^2 = 16(1 - (\sqrt{2}/7)).$

5. $\int_0^1 \int_0^x \sin(x^2) dy \, dx = \int_0^1 x \sin(x^2) dx = (1/2)(-\cos(x^2)) \big|_0^1 = (1 - \cos 1)/2.$

7. $\int_0^1 \int_{x^2}^{\sqrt{x}} xy \, dy \, dx = \int_0^1 (xy^2/2 \big|_{x^2}^{\sqrt{x}}) dx = (1/2) \int_0^1 (x^2 - x^5) dx = (1/2)(x^3/3 - x^6/6) \big|_0^1 = 1/12.$

9. $\int_{\pi/6}^{\pi/4} \int_{\sin x}^{\cos x} (3x + y) dy \, dx = \int_{\pi/6}^{\pi/4} (3xy + y^2/2 \big|_{\sin x}^{\cos x}) dx$
$= \int_{\pi/6}^{\pi/4} [3x(\cos x - \sin x) + (\cos^2 x)/2 - (\sin^2 x)/2] dx$
$= 3x(\sin x + \cos x) \big|_{\pi/6}^{\pi/4} - 3 \int_{\pi/6}^{\pi/4} (\sin x + \cos x) dx + (\sin 2x)/4 \big|_{\pi/6}^{\pi/4}$
$= 3(\pi/4)\sqrt{2} - (\pi/2)(1 + \sqrt{3})/2 + 3[0 + (1 - \sqrt{3})/2] + (1/4)(1 - (\sqrt{3}/2))$
$= \pi(3\sqrt{2} - 1 - \sqrt{3})/4 + (14 - 13\sqrt{3})/8.$

11. $\int_0^1 \int_{-y}^{1+y} (y - xy^2) dx \, dy = \int_0^1 (xy - x^2 y^2/2 \big|_{-y}^{1+y}) dy$
$= \int_0^1 (-y^3 + (3y^2/2) + y) dy = -y^4/4 + y^3/2 + y^2/2 \big|_0^1 = 3/4.$

13. $\int_1^2 \int_y^{y^3} e^{x/y} dx \, dy = \int_1^2 (y e^{x/y} \big|_y^{y^3}) dy = \int_1^2 (y e^{y^2} - ey) dy = e^{y^2}/2 - ey^2/2 \big|_1^2$
$= (e^4 - 4e)/2.$

15. $\int_0^1 \int_0^{x^2} x \cos y \, dy \, dx = \int_0^1 (x \sin y \big|_0^{x^2}) dx = \int_0^1 x \sin x^2 dx = (-\cos x^2)/2 \big|_0^1 = (1 - \cos 1)/2.$

139

Section 13.3

17.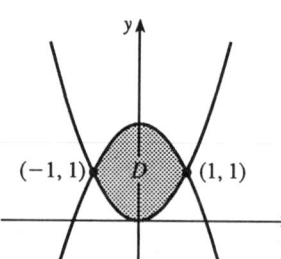

$$\int_{-1}^{1}\int_{x^2}^{2-x^2}(x^2+y)dy\,dx = 2\int_{0}^{1}\int_{x^2}^{2-x^2}(x^2+y)dy\,dx$$
$$= 2\int_{0}^{1}(x^2 y + y^2/2\big|_{x^2}^{2-x^2})dx = 2\int_{0}^{1}(-2x^4+2)dx$$
$$= 4(-x^5/5+x)\big|_{0}^{1} = 16/5.$$

19.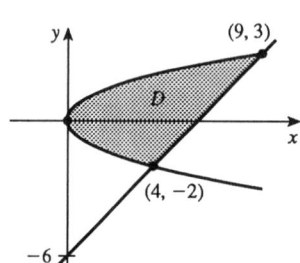

$$\int_{-2}^{3}\int_{y^2}^{y+6} 4y^3 dx\,dy = \int_{-2}^{3}(4y^4+24y^3-4y^5)dy$$
$$= 4y^5/5+6y^4-2y^6/3\big|_{-2}^{3} = 3^4(12/5)-16(26/15) = 500/3.$$

21.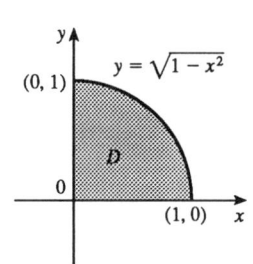

$$\int_{0}^{1}\int_{0}^{\sqrt{1-x^2}} xy\,dy\,dx = \int_{0}^{1}(xy^2/2\big|_{0}^{\sqrt{1-x^2}})dx$$
$$= \int_{0}^{1}(x-x^3)/2\,dx = (1/2)(x^2/2-x^4/4)\big|_{0}^{1} = 1/8.$$

23.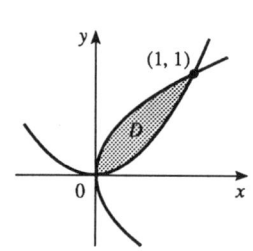

$$V = \int_{0}^{1}\int_{x^2}^{\sqrt{x}}(x^2+y^2)dy\,dx$$
$$= \int_{0}^{1}(x^{5/2}-x^4+x^{3/2}/3-x^6/3)dx$$
$$= (2/7)x^{7/2}-x^5/5+2x^{5/2}/15-x^7/21\big|_{0}^{1} = 18/105 = 6/35.$$

25.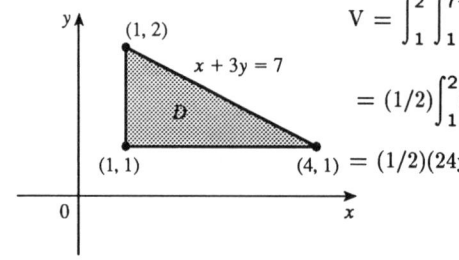

$$V = \int_{1}^{2}\int_{1}^{7-3y} xy\,dx\,dy = \int_{1}^{2}(yx^2/2\big|_{1}^{7-3y})dy$$
$$= (1/2)\int_{1}^{2}(48y-42y^2+9y^3)dy$$
$$= (1/2)(24y^2-14y^3+9y^4/4)\big|_{1}^{2} = 31/8.$$

Section 13.3

27.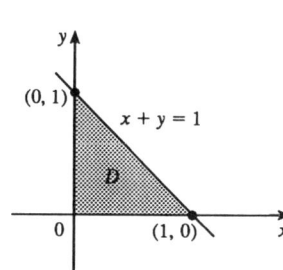

$$V = \int_0^1 \int_0^{1-x}(x^2+y^2+4)dy\,dx$$
$$= \int_0^1 (x^2y + y^3/3 + 4y|_0^{1-x})dx$$
$$= \int_0^1 [x^2 - x^3 + (1-x)^3/3 + 4(1-x)]dx$$
$$= x^3/3 - x^4/4 - (1-x)^4/12 - 2(1-x)^2\Big|_0^1 = 13/6.$$

29.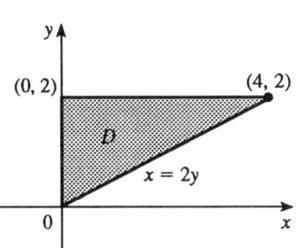

$$V = \int_0^2 \int_0^{2y}\sqrt{4-y^2}dx\,dy = \int_0^2 2y\sqrt{4-y^2}dy$$
$$= -(2/3)(4-y^2)^{3/2}\Big|_0^2 = 0 + 16/3 = 16/3.$$

31.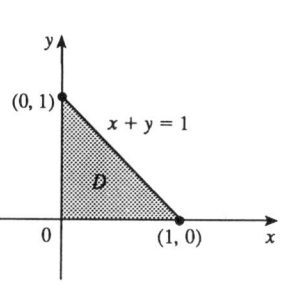

$$V = \int_0^1 \int_0^{1-x}(1-x-y)dy\,dx = \int_0^1 [(1-x)^2 - (1-x)^2/2]dx$$
$$= \int_0^1 (1-x)^2/2\,dx = -(1-x)^3/6\Big|_0^1 = 1/6.$$

33.

$$V = \int_0^1 \int_0^{\sqrt{1-x^2}} y\,dy\,dx = \int_0^1 (1/2)(1-x^2)dx$$
$$= (1/2)(x - x^3/3)\Big|_0^1 = 1/3.$$

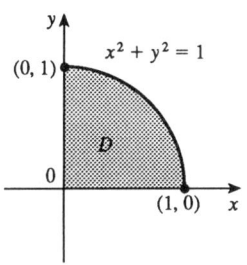

35. $\int_0^1 \int_y^1 f(x,y)dx\,dy.$

[Figure: triangle with vertices at origin, (1,0), (1,1); bounded by y = x, x = 1, y = 0]

141

37.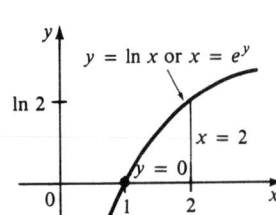
$$\int_0^{\ln 2} \int_{e^y}^2 f(x,y)\,dx\,dy.$$

39.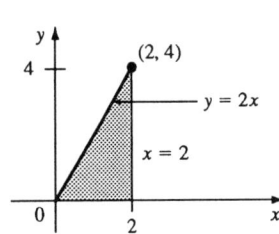
$$\int_0^2 \int_0^{2x} f(x,y)\,dy\,dx.$$

41.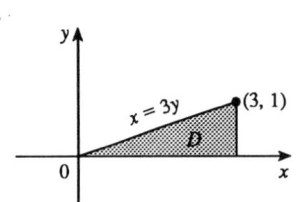
$$\int_0^1 \int_{3y}^3 e^{x^2}\,dx\,dy = \int_0^3 \int_0^{x/3} e^{x^2}\,dy\,dx$$
$$= \int_0^3 (x/3)e^{x^2}\,dx = e^{x^2}/6\Big|_0^3 = (e^9-1)/6.$$

43.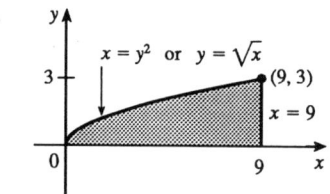
$$\int_0^3 \int_{y^2}^9 y\cos x^2\,dx\,dy = \int_0^9 \int_0^{\sqrt{x}} y\cos x^2\,dy\,dx$$
$$= \int_0^9 \cos x^2 \frac{y^2}{2}\Big|_0^{\sqrt{x}}\,dx = \int_0^9 \frac{x}{2}\cos x^2\,dx = \tfrac{1}{4}\sin x^2\Big|_0^9$$
$$= \tfrac{1}{4}\sin 81.$$

45.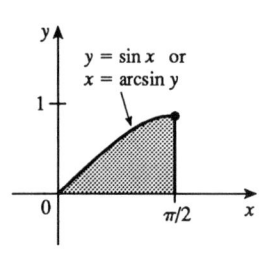
$$\int_0^1 \int_{\arcsin y}^{\pi/2} \cos x\sqrt{1+\cos^2 x}\,dx\,dy$$
$$= \int_0^{\pi/2} \int_0^{\sin x} \cos x\sqrt{1+\cos^2 x}\,dy\,dx$$
$$= \int_0^{\pi/2} \cos x\sqrt{1+\cos^2 x}\,y\Big|_0^{\sin x}\,dx$$
$$= \int_0^{\pi/2} \cos x\sqrt{1+\cos^2 x}\sin x\,dx$$
$$\left[\text{Let } u = \cos x,\ du = -\sin x\,dx,\ dx = du/(-\sin x)\right] = \int_1^0 -u\sqrt{1+u^2}\,du$$

Section 13.4

$= -\frac{1}{3}(1+u^2)^{3/2}\Big|_1^0 = \frac{\sqrt{8}-1}{3} = \frac{1}{3}(2\sqrt{2}-1).$

47. $D = \{(x,y) | 0 \le x \le 1, -x+1 \le y \le 1\} \cup \{(x,y) | -1 \le x \le 0, x+1 \le y \le 1\}$
$\cup \{(x,y) | 0 \le x \le 1, -1 \le y \le x-1\} \cup \{(x,y) | -1 \le x \le 0, -1 \le y \le -x-1\},$
all type 1.

$\iint_D x^2\, dA = \int_0^1 \int_{1-x}^1 x^2\, dy\, dx + \int_{-1}^0 \int_{x+1}^1 x^2\, dy\, dx + \int_0^1 \int_{-1}^{x-1} x^2\, dy\, dx + \int_{-1}^0 \int_{-1}^{-x-1} x^2\, dy\, dx$

$= 4\int_0^1 \int_{1-x}^1 x^2\, dy\, dx$ (by symmetry of the regions and since $f(x,y) = x^2 \ge 0$)

49. For $D = [0,1] \times [0,1]$, $0 \le \sqrt{x^3+y^3} \le \sqrt{2}$ and $A(D) = 1$, so $0 \le \iint_D \sqrt{x^3+y^3}\, dA \le \sqrt{2}.$

51. Since $m \le f(x,y) \le M$, $\iint_D m\, dA \le \iint_D f(x,y)\, dA \le \iint_D M\, dA$ by (13.21) $\Rightarrow$

$m\iint_D 1\, dA \le \iint_D f(x,y)\, dA \le M\iint_D 1\, dA$ by (13.20) $\Rightarrow mA(D) \le \iint_D f(x,y)\, dA \le MA(D)$ by (13.23).

53. $\iint_D (x^2\tan x + y^3 + 4)\, dA = \iint_D x^2\tan x\, dA + \iint_D y^3\, dA + \iint_D 4\, dA.$ But $x^2\tan x$ is an odd function and D is symmetric with respect to the y-axis, so $\iint_D x^2\tan x\, dA = 0.$

Similarly, y^3 is an odd function and D is symmetric with respect to the x-axis, so

$\iint_D y^3\, dA = 0.$ Thus, $\iint_D (x^2\tan x + y^3 + 4)\, dA = 4\iint_D dA = 4(\text{area of } D) = 16\pi.$

EXERCISES 13.4

1. $\iint_R x\, dA = \int_0^{2\pi} \int_0^5 r^2\cos\theta\, dr\, d\theta = \left(\int_0^{2\pi} \cos\theta\, d\theta\right)\left(\int_0^5 r^2\, dr\right) = 0.$

3. $\iint_R xy\, dA = \int_0^{\pi/2} \int_2^5 r^3\cos\theta\sin\theta\, dr\, d\theta = \left(\int_0^{\pi/2} (\sin 2\theta)/2\, d\theta\right)\left(\int_2^5 r^3\, dr\right)$
$= (1/2)(5^4 - 2^4)/4 = 609/8.$

5. The circle $r = 1$ intersects the cardioid $r = 1 + \sin\theta$ when $1 = 1 + \sin\theta \Rightarrow \theta = 0$ or

$\theta = \pi$, so $\iint_D \frac{1}{\sqrt{x^2+y^2}}\, dA = \int_0^\pi \int_1^{1+\sin\theta} \left(\frac{1}{r}\right) r\, dr\, d\theta = \int_0^\pi r\Big|_1^{1+\sin\theta}\, d\theta = \int_0^\pi \sin\theta\, d\theta$
$= -\cos\theta\Big|_0^\pi = 2.$

7. $\int_0^{2\pi} \int_\theta^{2\theta} r^3\, dr\, d\theta = \int_0^{2\pi} (15\theta^4/4)\, d\theta = (3/4)\theta^5\Big|_0^{2\pi} = 24\pi^5.$

143

Section 13.4

9. $A = \int_{-\pi/6}^{\pi/6} \int_0^{\cos 3\theta} r\, dr\, d\theta = \int_{-\pi/6}^{\pi/6} \frac{r^2}{2}\Big|_0^{\cos 3\theta} d\theta = \frac{1}{2}\int_{-\pi/6}^{\pi/6} \cos^2 3\theta\, d\theta$

$= \int_0^{\pi/6} \frac{1+\cos 6\theta}{2} d\theta = \frac{1}{2}\Big[\theta + \frac{1}{6}\sin 6\theta\Big]_0^{\pi/6} = \frac{\pi}{12}.$

11. By symmetry, the two loops of the lemniscate are equal in area, so

$A = 2\int_{-\pi/4}^{\pi/4} \int_0^{2\sqrt{\cos 2\theta}} r\, dr\, d\theta = \int_{-\pi/4}^{\pi/4} 4\cos 2\theta\, d\theta = 8\int_0^{\pi/4} \cos 2\theta\, d\theta = 4\sin 2\theta\Big|_0^{\pi/4} = 4.$

13. $3\cos\theta = 1+\cos\theta$ implies $\cos\theta = \frac{1}{2}$, so $\theta = \pm\frac{\pi}{3}$. Then by symmetry

$A = 2\int_0^{\pi/3}\int_{1+\cos\theta}^{3\cos\theta} r\, dr\, d\theta = 2\int_0^{\pi/3}\left(\frac{r^2}{2}\Big|_{1+\cos\theta}^{3\cos\theta}\right)d\theta = \int_0^{\pi/3}(9\cos^2\theta - 1 - 2\cos\theta - \cos^2\theta)\,d\theta$

$= \int_0^{\pi/3}\left(8(\frac{1+\cos 2\theta}{2}) - 2\cos\theta - 1\right)d\theta = \Big[4\theta + 2\sin 2\theta - 2\sin\theta - \theta\Big]_0^{\pi/3} = \pi.$

15. $V = \iint\limits_{x^2+y^2 \leq 9} (x^2+y^2)\,dA = \int_0^{2\pi}\int_0^3 r^3\, dr\, d\theta = 2\pi(81/4) = 81\pi/2.$

17.

$V = \iint\limits_{x^2+(y-1/2)^2 \leq 1/4} (12-6x-4y)\,dA$

$x^2+(y-\frac{1}{2})^2 = 1/4$, $(0, \frac{1}{2})$

$= \int_0^{\pi}\int_0^{\sin\theta}(12r - 6r^2\cos\theta - 4r^2\sin\theta)\,dr\,d\theta$

$= \int_0^{\pi}(6\sin^2\theta - 2\sin^3\theta\cos\theta - (4/3)\sin^4\theta)\,d\theta$

$= \Big[(5/2)(\theta - \sin\theta\cos\theta) - (\sin^4\theta)/2 + (\sin^2\theta\cos\theta)/3\Big]_0^{\pi}$

$= 5\pi/2$ (which we already knew since the volume of this cylinder is $\pi(1/4)(12+8)/2 = 5\pi/2$).

19. The cone $z = \sqrt{x^2+y^2}$ intersects the sphere $x^2+y^2+z^2 = 1$ when $x^2+y^2+(\sqrt{x^2+y^2})^2 = 1$ or $x^2+y^2 = \frac{1}{2}$.

So $V = \iint\limits_{x^2+y^2 \leq 1/2}(\sqrt{1-x^2-y^2} - \sqrt{x^2+y^2})\,dA = \int_0^{2\pi}\int_0^{1/\sqrt{2}}(\sqrt{1-r^2} - r)r\,dr\,d\theta$

$= (2\pi/3)[-(1-r^2)^{3/2} - r^3]\Big|_0^{1/\sqrt{2}} = (2\pi/3)(-1/\sqrt{2}+1) = \pi(2-\sqrt{2})/3.$

21. The given solid is the region inside the cylinder $x^2+y^2 = 4$ between the surfaces $z = \sqrt{64-4x^2-4y^2}$ and $z = -\sqrt{64-4x^2-4y^2}$. So

$V = \iint\limits_{x^2+y^2 \leq 4}(\sqrt{64-4x^2-4y^2} - (-\sqrt{64-4x^2-4y^2}))\,dA$

$= \iint\limits_{x^2+y^2 \leq 4} 2\sqrt{64-4x^2-4y^2}\,dA = 4\int_0^{2\pi}\int_0^2 \sqrt{16-r^2}\,r\,dr\,d\theta = 8\pi(-(16-r^2)^{3/2}/3)\Big|_0^2$

Section 13.4

$= (8\pi/3)(64 - 12^{3/2}) = (8\pi/3)(64 - 24\sqrt{3})$.

23. $V = 2\iint\limits_{x^2+y^2 \leq a^2} \sqrt{a^2 - x^2 - y^2}\, dA = 2\int_0^{2\pi}\int_0^a \sqrt{a^2 - r^2}\, r\, dr\, d\theta$

    $= 4\pi(-(a^2 - r^2)^{3/2}/3)\big|_0^a = 4\pi(a^3/3)$.

25. $\int_0^{\pi/2}\int_0^1 re^{r^2}\, dr\, d\theta = (\pi/2)(e^{r^2}/2\big|_0^1) = \pi(e-1)/4$.

    [Figure: quarter disk $x^2 + y^2 = 1$, region D]

27. $\int_0^\pi \int_0^2 (r^4 \cos^2\theta \sin^2\theta) r\, dr\, d\theta = \int_0^\pi \int_0^2 (r^5 \sin^2 2\theta)/4\, dr\, d\theta$

    $= (8/3)\int_0^\pi \sin^2 2\theta\, d\theta = (8/12)(2\theta - \sin 2\theta \cos 2\theta)\big|_0^\pi = 4\pi/3$.

    [Figure: half disk $x^2 + y^2 = 4$, region D]

29. $\int_0^3 \int_0^{\sqrt{9-x^2}} \arctan(y/x)\, dy\, dx = \int_0^{\pi/2}\int_0^3 \arctan(\tan\theta)\, r\, dr\, d\theta$

    $= \int_0^{\pi/2}\int_0^3 \theta\, r\, dr\, d\theta = \int_0^{\pi/2} \theta\left(\frac{r^2}{2}\Big|_0^3\right) d\theta = \int_0^{\pi/2} \frac{9}{2}\theta\, d\theta = \frac{9}{4}\theta^2\Big|_0^{\pi/2} = \frac{9}{16}\pi^2$.

31. (a) $\iint\limits_{D_a} e^{-(x^2+y^2)}\, dA = \int_0^{2\pi}\int_0^a re^{-r^2}\, dr\, d\theta = 2\pi(-e^{-r^2}/2)\big|_0^a = \pi(1 - e^{-a^2})$ for

    each a. Then $\lim\limits_{a \to \infty} \pi(1 - e^{-a^2}) = \pi$ since $e^{-a^2} \to 0$ as $a \to \infty$. Hence

    $\int_{-\infty}^\infty \int_{-\infty}^\infty e^{-(x^2+y^2)}\, dA = \pi$.

    (b) $\iint\limits_{S_a} e^{-(x^2+y^2)}\, dA = \int_{-a}^a \int_{-a}^a e^{-x^2}\cdot e^{-y^2}\, dx\, dy$

    $= (\int_{-a}^a e^{-x^2}\, dx)(\int_{-a}^a e^{-y^2}\, dy)$ for each a. Then from (a) $\pi = \iint\limits_{\mathbb{R}^2} e^{-(x^2+y^2)}\, dA$, so

Section 13.5

$$\pi = \lim_{a \to \infty} \iint_{S_a} e^{-(x^2+y^2)} dA = \lim_{a \to \infty} (\int_{-a}^{a} e^{-x^2} dx)(\int_{-a}^{a} e^{-y^2} dy)$$

$$= (\int_{-\infty}^{\infty} e^{-x^2} dx)(\int_{-\infty}^{\infty} e^{-y^2} dy). \text{ To evaluate } \lim_{a \to \infty} (\int_{-a}^{a} e^{-x^2} dx)(\int_{-a}^{a} e^{-y^2} dy), \text{ we are}$$

using the fact that these integrals are bounded. This is true since on $[-1, 1]$,

$0 < e^{-x^2} \leq 1$ while on $(-\infty, -1)$, $0 < e^{-x^2} \leq e^x$ and on $(1, \infty)$, $0 < e^{-x^2} < e^{-x}$.

Hence $0 \leq \int_{-\infty}^{\infty} e^{-x^2} dx \leq \int_{-\infty}^{-1} e^x dx + \int_{-1}^{1} dx + \int_{1}^{\infty} e^{-x} dx = 2(e^{-1} + 1)$.

(c) Since $(\int_{-\infty}^{\infty} e^{-x^2} dx)(\int_{-\infty}^{\infty} e^{-y^2} dy) = \pi$ and y can be replaced by x $(\int_{-\infty}^{\infty} e^{-x^2} dx)^2 = \pi$

implies $\int_{-\infty}^{\infty} e^{-x^2} dx = \pm\sqrt{\pi}$. But $e^{-x^2} \geq 0$ for all x so $\int_{-\infty}^{\infty} e^{-x^2} dx = \sqrt{\pi}$.

(d) Letting $t = \sqrt{2}x$, $\int_{-\infty}^{\infty} e^{-x^2} dx = \int_{-\infty}^{\infty} (e^{-t^2/2})/\sqrt{2}\, dt$, so that $\sqrt{\pi} = (1/\sqrt{2})\int_{-\infty}^{\infty} e^{-t^2/2} dt$

or $\int_{-\infty}^{\infty} e^{-t^2/2} dt = \sqrt{2\pi}$.

EXERCISES 13.5

1. $Q = \iint_D (x^2 + 3y^2) dA = \int_0^2 \int_1^2 (x^2 + 3y^2) dy\, dx = \int_0^2 (x^2 + 7) dx = 14 + (8/3) = 50/3$ C.

3. $m = \int_{-1}^{1} \int_0^1 x^2 dy\, dx = \int_{-1}^{1} x^2 dx = 2/3$, $\bar{x} = (3/2)\int_{-1}^{1} \int_0^1 x^3 dy\, dx = 0$,

   $\bar{y} = (3/2)\int_{-1}^{1} \int_0^1 x^2 y\, dy\, dx = (3/2)\int_{-1}^{1} (x^2/2) dx = (3/2)(2/6) = 1/2$. Hence $(\bar{x}, \bar{y}) = (0, 1/2)$.

5. $m = \int_0^2 \int_{x/2}^{3-x} (x+y) dy\, dx = \int_0^2 (x(3-3x/2) + (3-x)^2/2 - x^2/8) dx$

   $= \int_0^2 [-(9x^2/8) + (9/2)] dx = 6$, $M_y = \int_0^2 \int_{x/2}^{3-x} (x^2 + xy) dy\, dx = \int_0^2 (x^2 y + xy^2/2|_{x/2}^{3-x}) dx$

   $= \int_0^2 (9x/2 - 9x^3/8)\, dx = 9/2$, and $M_x = \int_0^2 \int_{x/2}^{3-y} (xy + y^2) dy\, dx = \int_0^2 (9 - 9x/2) dx = 9$.

   Hence $m = 6$, $(\bar{x}, \bar{y}) = (3/4, 3/2)$.

7. $m = \int_0^1 \int_{x^2}^{1} xy\, dy\, dx = \int_0^1 (x/2 - x^5/2) dx = 1/4 - 1/12 = 1/6$,

   $M_y = \int_0^1 \int_{x^2}^{1} x^2 y\, dy\, dx = \int_0^1 (x^2/2 - x^6/2) dx = 1/6 - 1/14 = 2/21$ and

   $M_x = \int_0^1 \int_{x^2}^{1} xy^2 dy\, dx = \int_0^1 (x/3 - x^7/3) dx = 1/6 - 1/24 = 1/8$.

   Hence $m = 1/6$, $(\bar{x}, \bar{y}) = (4/7, 3/4)$.

Section 13.5

9.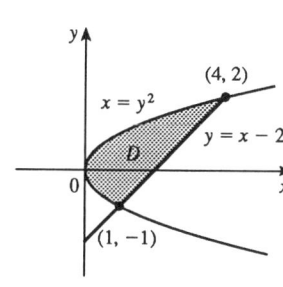

$$m = \int_{-1}^{2}\int_{y^2}^{y+2} 3\,dx\,dy = \int_{-1}^{2}(3y+6-3y^2)dy = 27/2,$$

$$M_y = \int_{-1}^{2}\int_{y^2}^{y+2} 3x\,dx\,dy = \int_{-1}^{2}(3/2)((y+2)^2 - y^4)dy$$

$$= (1/2)(y+2)^3 - 3y^5/10\big|_{-1}^{2} = 108/5 \text{ and}$$

$$M_x = \int_{-1}^{2}\int_{y^2}^{y+2} 3y\,dx\,dy = \int_{-1}^{2}(3y^2 + 6y - 3y^3)dy$$

$$= y^3 + 3y^2 - 3y^4/4\big|_{-1}^{2} = 27/4.$$

Hence $m = 27/2$, $(\bar{x}, \bar{y}) = (8/5, 1/2)$.

11. $m = \int_0^{\pi}\int_0^{\sin x} y\,dy\,dx = \int_0^{\pi} \tfrac{1}{2}\sin^2 x\,dx = \tfrac{1}{4}x - \tfrac{1}{8}\sin 2x\big|_0^{\pi} = \tfrac{1}{4}\pi,$

$M_y = \int_0^{\pi}\int_0^{\sin x} xy\,dy\,dx = \int_0^{\pi} \tfrac{1}{2}x\sin^2 x\,dx = \tfrac{1}{8}x^2 - \tfrac{1}{8}x\sin 2x - \tfrac{1}{16}\cos 2x\big|_0^{\pi} = \tfrac{1}{8}\pi^2,$ and

$M_x = \int_0^{\pi}\int_0^{\sin x} y^2 dy\,dx = \int_0^{\pi} \tfrac{1}{3}\sin^3 x\,dx = \tfrac{1}{3}[-\cos x + \tfrac{1}{3}\cos^3 x]_0^{\pi} = 4/9.$

Hence $m = \pi/4$, $(\bar{x}, \bar{y}) = (\pi/2, 16/(9\pi))$.

13. $\rho(x,y) = ky = kr\sin\theta$, $m = \int_0^{\pi/2}\int_0^1 kr^2\sin\theta\,dr\,d\theta = (k/3)\int_0^{\pi/2}\sin\theta\,d\theta$

$= (-k/3)\cos\theta\big|_0^{\pi/2} = k/3$, $M_y = \int_0^{\pi/2}\int_0^1 kr^3\sin\theta\cos\theta\,dr\,d\theta$

$= (k/4)\int_0^{\pi/2}\sin\theta\cos\theta\,d\theta = (k/8)(-\cos 2\theta\big|_0^{\pi/2}) = k/8,$

$M_x = \int_0^{\pi/2}\int_0^1 kr^3\sin^2\theta\,dr\,d\theta = k/4\int_0^{\pi/2}\sin^2\theta\,d\theta = (k/8)(\theta + \sin 2\theta)\big|_0^{\pi/2} = k\pi/16.$

Hence $(\bar{x}, \bar{y}) = (3/8, 3\pi/16)$.

15. Placing the vertex opposite the hypotenuse at $(0,0)$, $\rho(x,y) = k(x^2+y^2)$. Then

$$m = \int_0^a\int_0^{a-x} k(x^2+y^2)dy\,dx = k\int_0^a[ax^2 - x^3 + (a-x)^3/3]dx$$

$= k[ax^3/3 - x^4/4 - (a-x)^4/12]\big|_0^a = ka^4/6.$ By symmetry,

$M_y = M_x = \int_0^a\int_0^{a-x} ky(x^2+y^2)dy\,dx = k\int_0^a[(a-x)^2 x^2/2 + (a-x)^4/4]dx$

$= k[a^2x^3/6 - ax^4/4 + x^5/10 - (a-x)^5/20]\big|_0^a = ka^5/15.$ Hence $(\bar{x}, \bar{y}) = (2a/5, 2a/5).$

17. $I_x = \int_0^1\int_{x^2}^1 y^2(xy)dy\,dx = \int_0^1(x - x^9)/4\,dx = (1/8) - (1/40) = 1/10,$

$I_y = \int_0^1\int_{x^2}^1 x^3 y\,dy\,dx = \int_0^1 (x^3 - x^7)/2\,dx = (1/8) - (1/16) = 1/16,$

$I_0 = I_x + I_y = 13/80.$

19. $I_x = \int_{-1}^2\int_{y^2}^{y+2} 3y^2 dx\,dy = \int_{-1}^2(3y^3 + 6y^2 - 3y^4)dy = 3y^4/4 + 2y^3 - 3y^5/5\big|_{-1}^2 = 189/20,$

147

Section 13.6

$$I_y = \int_{-1}^{2} \int_{y^2}^{y+2} 3x^2 dx\, dy = \int_{-1}^{2} [(y+2)^3 - y^6] dy = (y+2)^4/4 - y^7/7 \Big|_{-1}^{2} = 1269/28, \text{ and}$$

$$I_o = I_x + I_y = 1917/35.$$

21. $I_x = \int_0^a \int_0^a \rho y^2 dx\, dy = \rho a(a^3/3) = \rho a^4/3 = I_y$ by symmetry, and $m = \rho a^2$ since the lamina is homogeneous. Hence $\bar{x} = \bar{y} = [(\rho a^4/3)/(\rho a^2)]^{1/2} = a/\sqrt{3}$.

23. Since $m = \iint_D \rho(x,y) dA = \rho \iint_D dA = \rho(\text{area of } D) = \rho A(D)$,

$$\bar{x} = \frac{1}{\rho A(D)} \iint_D x\rho\, dA = \frac{1}{A} \int_a^b \int_{g(x)}^{f(x)} x\, dy\, dx = \frac{1}{A} \int_a^b x[f(x) - g(x)] dx \text{ and}$$

$$\bar{y} = \frac{1}{\rho A(D)} \iint_D y\rho\, dA = \frac{1}{A} \int_a^b \int_{g(x)}^{f(x)} y\, dy\, dx = \frac{1}{A} \int_a^b [y^2/2]_{g(x)}^{f(x)} dx = \frac{1}{A} \int_a^b \tfrac{1}{2}[f(x)^2 - g(x)^2] dx.$$

EXERCISES 13.6

1. Here $z = f(x,y) = 4 - x - 2y$ with $0 \le x^2 + y^2 \le 4$. Thus by (13.40)

$$A(S) = \iint_D \sqrt{(-1)^2 + (-2)^2 + 1}\, dA = \sqrt{6} \iint_{x^2+y^2 \le 4} dA = \sqrt{6}\pi(2)^2 = 4\sqrt{6}\pi.$$

3. $y^2 + z^2 = 9 \Rightarrow z = \sqrt{9 - y^2}$. $f_x = 0$, $f_y = -y(9-y^2)^{-1/2} \Rightarrow$

$$A(S) = \int_0^4 \int_0^2 \sqrt{(-y(9-y^2)^{-1/2})^2 + 1}\, dy\, dx = \int_0^4 \int_0^2 \sqrt{\frac{y^2}{9-y^2} + 1}\, dy\, dx = \int_0^4 \int_0^2 \frac{3\, dy}{\sqrt{9-y^2}}\, dx$$

$$= 3\int_0^4 \sin^{-1}\tfrac{y}{3}\Big|_0^2 dx = 3\Big[(\sin^{-1}\tfrac{2}{3})x\Big]_0^4 = 12\sin^{-1}\tfrac{2}{3}.$$

5. $z = f(x,y) = y^2 - x^2$ with $1 \le x^2 + y^2 \le 4$. $f_x = -2x$, $f_y = 2y \Rightarrow$

$$A(S) = \iint_D \sqrt{1 + 4x^2 + 4y^2}\, dA = \int_0^{2\pi}\int_1^2 \sqrt{1+4r^2}\, r\, dr\, d\theta = \tfrac{4\pi}{24}(1+4r^2)^{3/2}\Big|_1^2 = \tfrac{\pi}{6}(17\sqrt{17} - 5\sqrt{5}).$$

7. $z = f(x,y) = xy$ with $0 \le x^2 + y^2 \le 1$, so $f_x = y$, $f_y = x \Rightarrow$

$$A(S) = \iint_D \sqrt{y^2 + x^2 + 1}\, dA = \int_0^{2\pi}\int_0^1 \sqrt{r^2 + 1}\, r\, dr\, d\theta = \int_0^{2\pi} \tfrac{1}{3}(r^2+1)^{3/2}\Big|_0^1 d\theta$$

$$= \int_0^{2\pi} \tfrac{1}{3}(2\sqrt{2} - 1)\, d\theta = \tfrac{2\pi}{3}(2\sqrt{2} - 1).$$

Section 13.7

9. $z = \sqrt{a^2 - x^2 - y^2}$, $z_x = -x(a^2 - x^2 - y^2)^{-1/2}$,

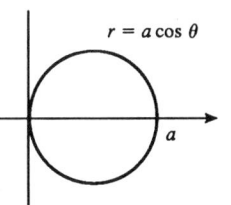

$r = a\cos\theta$

$z_y = -y(a^2 - x^2 - y^2)^{-1/2}$,

$A(S) = \iint_D \sqrt{[(x^2 + y^2)/(a^2 - x^2 - y^2)] + 1}\,dA$

$= \int_{-\pi/2}^{\pi/2} \int_0^{a\cos\theta} \sqrt{[r^2/(a^2 - r^2)] + 1}\,r\,dr\,d\theta$

$= \int_{-\pi/2}^{\pi/2} \int_0^{a\cos\theta} \frac{ar}{\sqrt{a^2 - r^2}}\,dr\,d\theta = \int_{-\pi/2}^{\pi/2} \left(-a\sqrt{a^2 - r^2}\,\Big|_0^{a\cos\theta}\right) d\theta$

$= \int_{-\pi/2}^{\pi/2} -a\left(\sqrt{a^2 - a^2\cos^2\theta} - a\right) d\theta = 2a^2 \int_0^{\pi/2} \left(1 - \sqrt{1 - \cos^2\theta}\right) d\theta$

$= 2a^2 \int_0^{\pi/2} d\theta - 2a^2 \int_0^{\pi/2} \sqrt{\sin^2\theta}\,d\theta = a^2\pi - 2a^2 \int_0^{\pi/2} \sin\theta\,d\theta = a^2(\pi - 2)$.

11. $z = f(x, y) = x^2 + y^2$ and $D = \{(x, y)\,|\,0 \leq x, y \leq 1\}$, so

$A(S) = \int_0^1 \int_0^1 \sqrt{(2x)^2 + (2y)^2 + 1}\,dy\,dx = \int_0^1 \int_0^1 \sqrt{4x^2 + 4y^2 + 1}\,dy\,dx$.

13. Here $z = f(x, y) = ax + by + c$, $f_x(x, y) = a$, $f_y(x, y) = b$, so by (13.40),

$A(S) = \iint_D \sqrt{a^2 + b^2 + 1}\,dA = \sqrt{a^2 + b^2 + 1}\iint_D dA = \sqrt{a^2 + b^2 + 1}\,A(D)$

EXERCISES 13.7

1. $\int_0^1 \int_{-1}^2 \int_0^3 xyz^2\,dz\,dx\,dy = \int_0^1 \int_{-1}^2 xy(9)\,dx\,dy = 9\int_0^1 x(2 - 1/2)\,dx = (3/2)(9)(1/2) = 27/4$.

3. $\int_0^1 \int_0^z \int_0^y xyz\,dx\,dy\,dz = \int_0^1 \int_0^z (y^3 z/2)\,dy\,dz = \int_0^1 z^5/8\,dz = z^6/48\big|_0^1 = 1/48$.

5. $\int_0^\pi \int_0^2 \int_0^{\sqrt{4-z^2}} z\sin y\,dx\,dz\,dy = \int_0^\pi \int_0^2 z\sqrt{4 - z^2}\sin y\,dz\,dy$

$= \int_0^\pi ((-1/3)(4 - z^2)^{3/2}\sin y\big|_0^2)\,dy = \int_0^\pi (8/3)\sin y\,dy = -(8/3)\cos y\big|_0^\pi = 16/3$.

7. $\int_0^1 \int_0^{2z} \int_0^{z+2} yz\,dx\,dy\,dz = \int_0^1 \int_0^{2z} yz(z + 2)\,dy\,dz = \int_0^1 (2z^4 + 4z^3)\,dz = 7/5$.

9. $\int_0^1 \int_0^{x^2} \int_0^{x+2y} y\,dz\,dy\,dx = \int_0^1 \int_0^{x^2} (yx + 2y^2)\,dy\,dx = \int_0^1 (xy^2/2 + 2y^3/3\big|_0^{x^2})\,dx$

149

$$= \int_0^1 (x^5/2 + 2x^6/3)dx = (x^6/12 + 2x^7/21)\Big|_0^1 = 5/28.$$

11. Here E is the region that lies below the plane with x-, y-, and z-intercepts 1, 2, and 3 respectively, i.e. below the plane $2z + 6x + 3y = 6$ and above the region in the xy-plane bounded by the lines $x = 0$, $y = 0$ and $6x + 3y = 6$.

$$\text{So } \iiint_E xy\,dV = \int_0^1 \int_0^{2-2x} \int_0^{3-3x-3y/2} xy\,dz\,dy\,dx = \int_0^1 \int_0^{2-2x} (3xy - 3x^2y - 3xy^2/2)dy\,dx$$

$$= \int_0^1 (3xy^2/2 - 3x^2y^2/2 - xy^3/2)\Big|_0^{2-2x} dx = \int_0^1 (2x - 6x^2 + 6x^3 - 2x^4)\,dx$$

$$= x^2 - 2x^3 + 3x^4/2 - 2x^5/5\Big|_0^1 = 1/10.$$

13.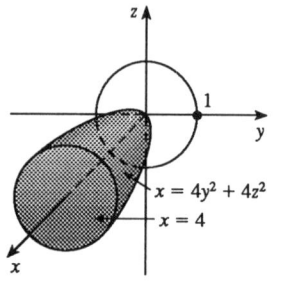

By symmetry $\iiint_E z\,dV = 2\iiint_{E'} z\,dV$ where E' is the part of E to the left [as viewed from $(10, 10, 0)$] of the plane $x = y$.

Hence $\iiint_E z\,dV = \int_0^1 \int_y^1 \int_0^{1-x} 2z\,dz\,dx\,dy$

$$= \int_0^1 \int_y^1 (1-x)^2 dx\,dy = \int_0^1 -(1-x)^3/3\Big|_y^1 dy$$

$$= \int_0^1 (1-y)^3/3\,dy = (1-y)^4/12\Big|_0^1 = 1/12.$$

15. The projection E on the yz-plane is the disk $y^2 + z^2 \le 1$. Using polar coordinates $y = r\cos\theta$ and $z = r\sin\theta$, we get

$$\iiint_E x\,dV = \iint_D \left[\int_{4y^2+4z^2}^4 x\,dx\right] dA$$

$$= \tfrac{1}{2}\iint_D [4^2 - (4y^2+4z^2)^2]\,dA = 8\int_0^{2\pi}\int_0^1 (1-r^4)r\,dr\,d\theta$$

$$= 8\int_0^{2\pi} d\theta \int_0^1 (r - r^5)\,d\theta = 8(2\pi)\left[\tfrac{r^2}{2} - \tfrac{r^6}{6}\right]_0^1 = 16\pi/3.$$

17. The plane $2x + 3y + 6z = 12$ intersects the xy-plane when $2x + 3y + 6(0) = 12 \Rightarrow$ $y = 4 - \tfrac{2}{3}x$. So $E = \{(x, y, z)\,|\,0 \le x \le 6,\ 0 \le y \le 4 - \tfrac{2}{3}x,\ 0 \le z \le (12 - 2x - 3y)/6\}$ and

$$V = \int_0^6 \int_0^{4-(2/3)x} \int_0^{(12-2x-3y)/6} dz\,dy\,dx = \tfrac{1}{6}\int_0^6 \int_0^{4-(2/3)x} (12 - 2x - 3y)\,dy\,dx$$

$$= \tfrac{1}{6}\int_0^6 \left(\tfrac{(12-2x)^2}{3} - \tfrac{3}{2}\tfrac{12-2x}{9}\right)dx = \tfrac{1}{36}\int_0^6 (12-2x)^2 dx = \tfrac{1}{36}\left(\tfrac{-1}{6}\right)(12-2x)^3\Big|_0^6 = 8$$

Section 13.7

19. $V = \int_0^1 \int_{-\sqrt{x}}^{\sqrt{x}} \int_0^{1-x} dz\,dy\,dx = \int_0^1 \int_{-\sqrt{x}}^{\sqrt{x}} (1-x)\,dy\,dx$

$= \int_0^1 2\sqrt{x}(1-x)\,dx = \int_0^1 2(\sqrt{x} - x^{3/2})\,dx$

$= 2\left(\frac{2}{3} - \frac{2}{5}\right) = \frac{8}{15}$

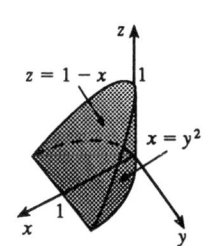

21.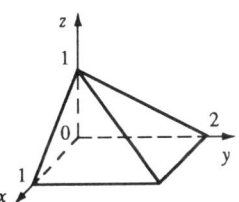

$V = \int_0^1 \int_0^x \int_0^{\sqrt{1-y^2}} dz\,dy\,dx = \int_0^1 \int_0^x \sqrt{1-y^2}\,dy\,dx$

$= \int_0^1 [y\sqrt{1-y^2} + (\sin^{-1} y)]/2 \Big|_0^x\,dx$

$= \int_0^1 [x\sqrt{1-x^2} + \sin^{-1} x]/2\,dx$

$= [-(1-x^2)^{3/2}/3 + x\sin^{-1} x + \sqrt{1-x^2}]/2 \Big|_0^1$

$= [(\pi/2) - (2/3)]/2 = (\pi/4) - (1/3).$

23. $E = \{(x, y, z) \mid 0 \le x \le 1,\ 0 \le z \le 1-x,\ 0 \le y \le 2-2z\}$

25.

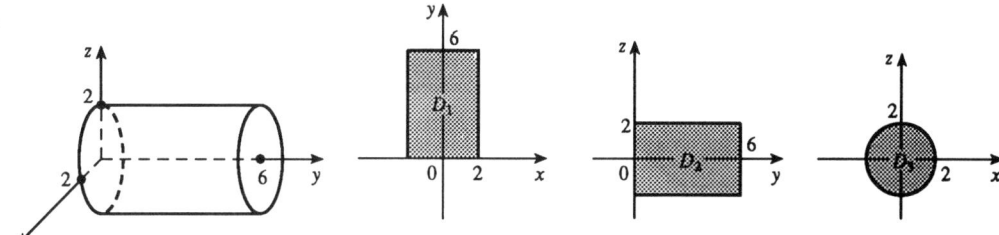

If D_1, D_2, D_3 are the projections of E on the xy-, yz-, and xz-planes, then

$D_1 = \{(x, y) \mid -2 \le x \le 2,\ 0 \le y \le 6\}$, $D_2 = \{(y, z) \mid -2 \le z \le 2,\ 0 \le y \le 6\}$,

$D_3 = \{(x, z) \mid x^2 + z^2 \le 4\}$. Therefore $E = \{(x, y, z) \mid -\sqrt{4-x^2} \le z \le \sqrt{4-x^2},$

$-2 \le x \le 2,\ 0 \le y \le 6\} = \{(x, y, z) \mid -\sqrt{4-z^2} \le x \le \sqrt{4-z^2},\ -2 \le z \le 2,\ 0 \le y \le 6\}$.

Section 13.7

$$\iiint_E f(x, y, z)\, dV = \int_{-2}^{2}\int_{0}^{6}\int_{-\sqrt{4-x^2}}^{\sqrt{4-x^2}} f(x,y,z)\, dz\, dy\, dx = \int_{0}^{6}\int_{-2}^{2}\int_{-\sqrt{4-x^2}}^{\sqrt{4-x^2}} f(x,y,z)\, dz\, dx\, dy$$

$$= \int_{0}^{6}\int_{-2}^{2}\int_{-\sqrt{4-z^2}}^{\sqrt{4-z^2}} f(x,y,z)\, dx\, dz\, dy = \int_{-2}^{2}\int_{0}^{6}\int_{-\sqrt{4-z^2}}^{\sqrt{4-z^2}} f(x,y,z)\, dx\, dy\, dz$$

$$= \int_{-2}^{2}\int_{-\sqrt{4-x^2}}^{\sqrt{4-x^2}}\int_{0}^{6} f(x,y,z)\, dy\, dz\, dx = \int_{-2}^{2}\int_{-\sqrt{4-z^2}}^{\sqrt{4-z^2}}\int_{0}^{6} f(x,y,z)\, dy\, dx\, dz.$$

27.

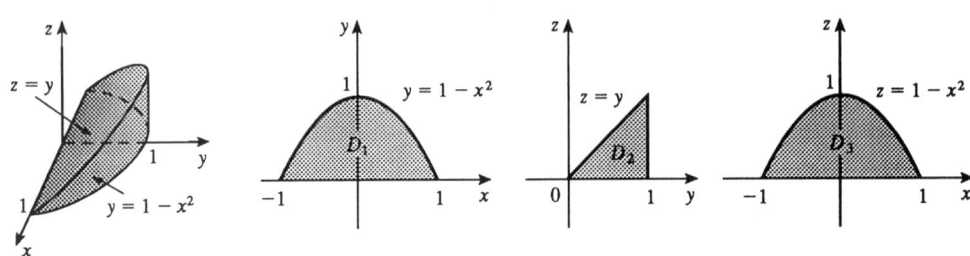

If D_1, D_2, and D_3 are the projections of E on the xy-, yz-, and xz-planes, then
$D_1 = \{(x, y)\,|\,-1 \leq x \leq 1,\ 0 \leq y \leq 1-x^2\} = \{(x,y)\,|\,0 \leq y \leq 1,\ -\sqrt{1-y} \leq x \leq \sqrt{1-y}\}$,
$D_2 = \{(y, z)\,|\,0 \leq y \leq 1,\ 0 \leq z \leq y\} = \{(y,z)\,|\,0 \leq z \leq 1,\ z \leq y \leq 1\}$,
$D_3 = \{(x, z)\,|\,-1 \leq x \leq 1,\ 0 \leq z \leq 1-x^2\} = \{(x,z)\,|\,0 \leq z \leq 1,\ -\sqrt{1-z} \leq x \leq \sqrt{1-z}\}$.
Therefore $E = \{(x, y, z)\,|\,-1 \leq x \leq 1,\ 0 \leq y \leq 1-x^2,\ 0 \leq z \leq y\}$
$= \{(x, y, z)\,|\,0 \leq y \leq 1,\ -\sqrt{1-y} \leq x \leq \sqrt{1-y},\ 0 \leq z \leq y\}$
$= \{(x, y, z)\,|\,0 \leq y \leq 1,\ 0 \leq z \leq y,\ -\sqrt{1-y} \leq x \leq \sqrt{1-y}\}$
$= \{(x, y, z)\,|\,0 \leq z \leq 1,\ z \leq y \leq 1,\ -\sqrt{1-y} \leq x \leq \sqrt{1-y}\}$
$= \{(x, y, z)\,|\,-1 \leq x \leq 1,\ 0 \leq z \leq 1-x^2,\ z \leq y \leq 1-x^2\}$
$= \{(x, y, z)\,|\,0 \leq z \leq 1,\ -\sqrt{1-z} \leq x \leq \sqrt{1-z},\ z \leq y \leq 1-x^2\}$.

Then $\iiint_E f(x, y, z)\, dV = \int_{-1}^{1}\int_{0}^{1-x^2}\int_{0}^{y} f(x,y,z)\, dz\, dy\, dx = \int_{0}^{1}\int_{-\sqrt{1-y}}^{\sqrt{1-y}}\int_{0}^{y} f(x,y,z)\, dz\, dx\, dy$

$= \int_{0}^{1}\int_{0}^{y}\int_{-\sqrt{1-y}}^{\sqrt{1-y}} f(x,y,z)\, dx\, dz\, dy = \int_{0}^{1}\int_{z}^{1}\int_{-\sqrt{1-y}}^{\sqrt{1-y}} f(x,y,z)\, dx\, dy\, dz$

$= \int_{-1}^{1}\int_{0}^{1-x^2}\int_{z}^{1-x^2} f(x,y,z)\, dy\, dz\, dx = \int_{0}^{1}\int_{-\sqrt{1-z}}^{\sqrt{1-z}}\int_{z}^{1-x^2} f(x,y,z)\, dy\, dx\, dz.$

29.

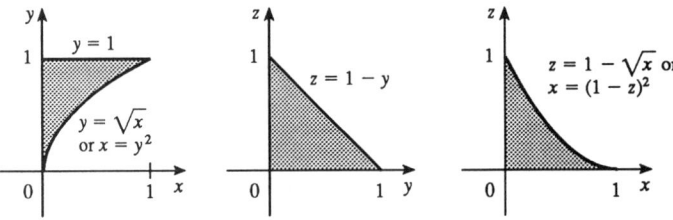

The diagrams show the projections of E on the xy-, yz-, and xz-planes. Therefore

$$\int_0^1 \int_{\sqrt{x}}^1 \int_0^{1-y} f(x,y,z)\,dz\,dy\,dx = \int_0^1 \int_0^{y^2} \int_0^{1-y} f(x,y,z)\,dz\,dx\,dy$$

$$= \int_0^1 \int_0^{1-z} \int_0^{y^2} f(x,y,z)\,dx\,dy\,dz = \int_0^1 \int_0^{1-y} \int_0^{y^2} f(x,y,z)\,dx\,dz\,dy$$

$$= \int_0^1 \int_0^{1-\sqrt{x}} \int_{\sqrt{x}}^{1-z} f(x,y,z)\,dy\,dz\,dx = \int_0^1 \int_0^{(1-z)^2} \int_{\sqrt{x}}^{1-z} f(x,y,z)\,dy\,dx\,dz.$$

31.

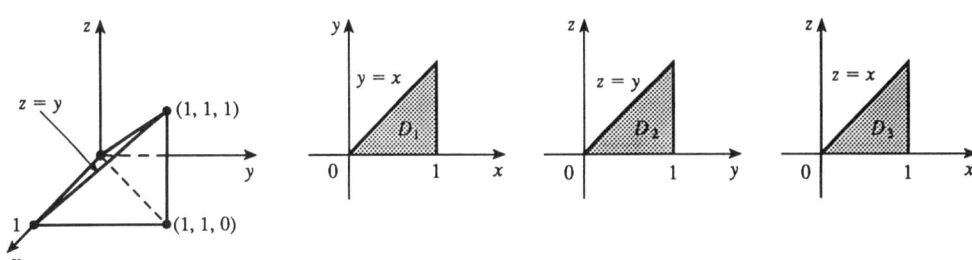

$$\int_0^1 \int_y^1 \int_0^y f(x,y,z)\,dz\,dx\,dy = \iiint_E f(x,y,z)\,dV \text{ where}$$

$E = \{(x,y,z)\,|\,0 \le z \le y,\ y \le x \le 1,\ 0 \le y \le 1\}$. If D_1, D_2, and D_3 are the projections of E on the xy-, yz- and xz-planes then $D_1 = \{(x,y)\,|\,0 \le y \le 1,\ y \le x \le 1\} = \{(x,y)\,|\,0 \le x \le 1,\ 0 \le y \le x\}$, $D_2 = \{(y,z)\,|\,0 \le y \le 1,\ 0 \le z \le y\} = \{(y,z)\,|\,0 \le z \le 1,\ z \le y \le 1\}$, and $D_3 = \{(x,z)\,|\,0 \le x \le 1,\ 0 \le z \le x\} = \{(x,z)\,|\,0 \le z \le 1,\ z \le x \le 1\}$. Therefore we also have $E = \{(x,y,z)\,|\,0 \le x \le 1,\ 0 \le y \le x,\ 0 \le z \le y\} = \{(x,y,z)\,|\,0 \le y \le 1,\ 0 \le z \le y,\ y \le x \le 1\} = \{(x,y,z)\,|\,0 \le z \le 1,\ z \le y \le 1,\ y \le x \le 1\} = \{(x,y,z)\,|\,0 \le x \le 1,\ 0 \le z \le x,\ z \le y \le x\} = \{(x,y,z)\,|\,0 \le z \le 1,\ z \le x \le 1,\ z \le y \le x\}$.

Then $\int_0^1 \int_y^1 \int_0^y f(x,y,z)\,dz\,dx\,dy = \int_0^1 \int_0^x \int_0^y f(x,y,z)\,dz\,dy\,dx = \int_0^1 \int_0^y \int_y^1 f(x,y,z)\,dx\,dz\,dy$

$= \int_0^1 \int_z^1 \int_y^1 f(x,y,z)\,dx\,dy\,dz = \int_0^1 \int_0^x \int_z^x f(x,y,z)\,dy\,dz\,dx = \int_0^1 \int_z^1 \int_z^x f(x,y,z)\,dy\,dx\,dz.$

33. $m = \int_0^1 \int_0^{x^2} \int_0^{x+2y} 2\,dz\,dy\,dx = 2\int_0^1 \int_0^{x^2} (x+2y)\,dy\,dx = 2\int_0^1 (x^3 + x^4)\,dx = 9/10,$

$M_{yz} = \int_0^1 \int_0^{x^2} \int_0^{x+2y} 2x\,dz\,dy\,dx = \int_0^1 \int_0^{x^2} 2(x^2 + 2xy)\,dy\,dx = \int_0^1 2(x^4 + x^5)\,dx = 11/15,$

$M_{xz} = 5/14$, and $M_{xy} = \int_0^1 \int_0^{x^2} \int_0^{x+2y} 2z\,dz\,dy\,dx = \int_0^1 \int_0^{x^2} 2(x+2y)^2/2\,dy\,dx$

$= \int_0^1 2(x^2 y/2 + xy^2 + 2y^3/3 \big|_0^{x^2})\,dx = \int_0^1 2(x^4/2 + x^5 + 2x^6/3)\,dx = 76/105.$ Hence $(\bar{x}, \bar{y}, \bar{z}) = (22/27, 25/63, 152/189).$

35. $m = \int_0^a \int_0^a \int_0^a (x^2+y^2+z^2)dx\,dy\,dz = \int_0^a \int_0^a [a^3/3 + a(y^2+z^2)]dy\,dz$

$= \int_0^a [(2a^4/3) + a^2 z^2]dz = 2a^5/3 + a^5/3 = a^5$, $M_{yz} = \int_0^a \int_0^a \int_0^a [x^3 + x(y^2+z^2)]dx\,dy\,dz$

$= \int_0^a \int_0^a [(a^4/4) + a^2(y^2+z^2)/2]dy\,dz = \int_0^a [(a^5/4) + (a^5/6) + a^3 z^2/2]dz$

$= a^6/4 + 2a^6/6 = 7a^6/12 = M_{xz} = M_{xy}$ by symmetry of E and $\rho(x,y,z)$. Hence $(\bar{x}, \bar{y}, \bar{z}) = (7a/12, 7a/12, 7a/12)$.

37. (a) $m = \int_0^1 \int_0^{\sqrt{1-x^2}} \int_0^y (1+x+y+z)dz\,dy\,dx$.

(b) $(\bar{x}, \bar{y}, \bar{z}) = ((1/m)\int_0^1 \int_0^{\sqrt{1-x^2}} \int_0^y x(1+x+y+z)dz\,dy\,dx,$

$(1/m)\int_0^1 \int_0^{\sqrt{1-x^2}} \int_0^y y(1+x+y+z)dz\,dy\,dx,$ $(1/m)\int_0^1 \int_0^{\sqrt{1-x^2}} \int_0^y z(1+x+y+z)dz\,dy\,dx)$.

(c) $I_z = \int_0^1 \int_0^{\sqrt{1-x^2}} \int_0^y (x^2+y^2)(1+x+y+z)dz\,dy\,dx$.

39. (a) $m = \int_{-1}^1 \int_{-\sqrt{1-y^2}}^{\sqrt{1-y^2}} \int_{4y^2+4z^2}^4 (x^2+y^2+z^2)dx\,dz\,dy$.

(b) $(\bar{x}, \bar{y}, \bar{z})$ where $\bar{x} = (1/m)\int_{-1}^1 \int_{-\sqrt{1-y^2}}^{\sqrt{1-y^2}} \int_{4y^2+4z^2}^4 x(x^2+y^2+z^2)dx\,dz\,dy$,

$\bar{y} = (1/m)\int_{-1}^1 \int_{-\sqrt{1-y^2}}^{\sqrt{1-y^2}} \int_{4y^2+4z^2}^4 y(x^2+y^2+z^2)dx\,dz\,dy$, and

$\bar{z} = (1/m)\int_{-1}^1 \int_{-\sqrt{1-y^2}}^{\sqrt{1-y^2}} \int_{4y^2+4z^2}^4 z(x^2+y^2+z^2)dx\,dz\,dy$.

(c) $I_z = \int_{-1}^1 \int_{-\sqrt{1-y^2}}^{\sqrt{1-y^2}} \int_{4y^2+4z^2}^4 (x^2+y^2)(x^2+y^2+z^2)dx\,dz\,dy$.

41. $I_x = \int_0^L \int_0^L \int_0^L k(y^2+z^2)dz\,dy\,dx = kL\int_0^L \left[y^2 L + \frac{L^3}{3}\right]dy = kL\left[\frac{L^4}{3} + \frac{L^4}{3}\right] = \frac{2kL^5}{3}$.

By symmetry, $I_x = I_y = I_z = \frac{2kL^5}{3}$.

43. $V(E) = L^3$. $f_{ave} = \frac{1}{L^3}\int_0^L \int_0^L \int_0^L xyz\,dx\,dy\,dz = \frac{1}{L^3}\int_0^L x^2 dx \int_0^L y^2 dy \int_0^L z^2 dz$

$= \frac{1}{L^3}\frac{L^2}{2}\frac{L^2}{2}\frac{L^2}{2} = \frac{L^3}{8}$.

Section 13.8

EXERCISES 13.8

1.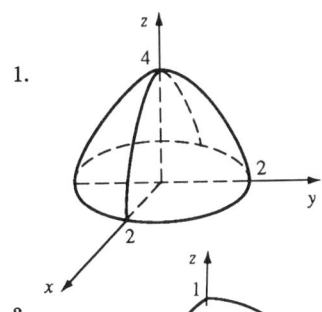

$$\int_0^{2\pi}\int_0^2\int_0^{4-r^2} r\,dz\,dr\,d\theta = \int_0^{2\pi}\int_0^2 (4r - r^3)\,dr\,d\theta$$

$$= \int_0^{2\pi}\left[2r^2 - \frac{r^4}{4}\right]_0^2 d\theta = \int_0^{2\pi}(8-4)\,d\theta = 8\pi.$$

3.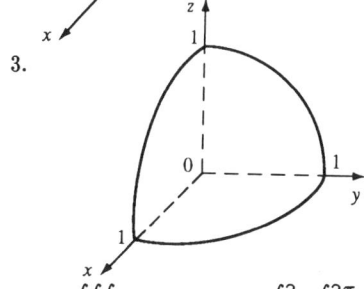

$$\int_0^{\pi/2}\int_0^{\pi/2}\int_0^1 \rho^2 \sin\phi\,d\rho\,d\theta\,d\phi = \int_0^{\pi/2}\int_0^{\pi/2} \tfrac{1}{3}\sin\phi\,d\theta\,d\phi$$

$$= \tfrac{1}{3}\int_0^{\pi/2} \tfrac{\pi}{2}\sin\phi\,d\phi = \tfrac{\pi}{6}\left[-\cos\phi\right]_0^{\pi/2} = \tfrac{\pi}{6}.$$

5. $\iiint_E (x^2+y^2)\,dV = \int_{-1}^{2}\int_0^{2\pi}\int_0^2 (r^2)r\,dr\,d\theta\,dz = (3)(2\pi)(r^4/4)\big|_0^2 = 24\pi.$

7. $\iiint_E y\,dV = \int_0^{2\pi}\int_1^2\int_0^{2+r\cos\theta} r^2\sin\theta\,dz\,dr\,d\theta = \int_0^{2\pi}\int_1^2 [2r^2\sin\theta + r^3\cos\theta]\,dr\,d\theta$

$= \int_0^{2\pi}[(2r^3\sin\theta)/3 + (r^4\cos\theta)/4]\big|_1^2 d\theta = \int_0^{2\pi}[(14/3)\sin\theta + (15/4)\cos\theta]\,d\theta = 0.$

9. $\iiint_E x^2\,dV = \int_0^{2\pi}\int_0^1\int_0^{2r} r^2\cos^2\theta\,r\,dz\,dr\,d\theta = \int_0^{2\pi}\int_0^1 r^3\cos^2\theta\,z\big|_0^{2r}\,dr\,d\theta$

$= \int_0^{2\pi}\int_0^1 2r^4\cos^2\theta\,dr\,d\theta = \int_0^{2\pi}\frac{2r^5}{5}\cos^2\theta\Big|_0^1 d\theta = \tfrac{2}{5}\int_0^{2\pi}\cos^2\theta\,d\theta = \tfrac{2}{5}\int_0^{2\pi}\frac{1+\cos 2\theta}{2}\,d\theta$

$= \tfrac{1}{5}\left[\theta + \tfrac{1}{2}\sin 2\theta\right]_0^{2\pi} = \tfrac{2\pi}{5}.$

11. The paraboloids intersect when $x^2 + y^2 = 36 - 3x^2 - 3y^2 \Rightarrow D = \{(x,y)\,|\,x^2 + y^2 \le 9\}$.
So, using cylindrical coordinates, $E = \{(r,\theta,z)\,|\,r^2 \le z \le 36 - r^2,\ 0 \le r \le 3,\ 0 \le \theta \le 2\pi\}$
and

$V = \int_0^{2\pi}\int_0^3\int_{r^2}^{36-3r^2} r\,dz\,dr\,d\theta = 2\pi\int_0^3 (36r - 4r^3)\,dr = 2\pi(18r^2 - r^4)\big|_0^3 = 162\pi.$

13. The paraboloid $z = 4x^2 + 4y^2$ intersects the plane $z = a$ when $a = 4x^2 + 4y^2$ or $x^2 + y^2 = \tfrac{a}{4}$.
So using cylindrical coordinates, $E = \{(r,\theta,z)\,|\,0 \le r \le \tfrac{\sqrt{a}}{2},\ 0 \le \theta \le 2\pi,\ 4r^2 \le z \le a\}$.
Thus

$m = \int_0^{2\pi}\int_0^{\sqrt{a}/2}\int_{4r^2}^{a} Kr\,dz\,dr\,d\theta = 2\pi K\int_0^{\sqrt{a}/2}(ar - 4r^3)\,dr = 2\pi K(ar^2/2 - r^4)\big|_0^{\sqrt{a}/2}$

Section 13.8

$= a^2 \pi K/8$. Since the region is homogeneous and symmetric, $M_{yz} = M_{xz} = 0$ and

$$M_{xy} = \int_0^{2\pi} \int_0^{\sqrt{a}/2} \int_{4r^2}^a Krz\,dz\,dr\,d\theta = 2\pi K \int_0^{\sqrt{a}/2} (a^2 r/2 - 8r^5)dr$$

$= 2\pi K(a^2 r^2/4 - 4r^6/3)\Big|_0^{\sqrt{a}/2} = a^3 \pi K/12$. Hence $(\bar{x},\bar{y},\bar{z}) = (0,0,2a/3)$.

15. $\iiint_B (x^2 + y^2 + z^2)dV = \int_0^\pi \int_0^{2\pi} \int_0^1 \rho^4 \sin\phi\,d\rho\,d\theta\,d\phi = 2\pi \int_0^\pi (\sin\phi)/5\,d\phi$

   $= (2\pi/5)(-\cos\phi)\Big|_0^\pi = 4\pi/5$.

17. $\iiint_E y^2\,dV = \int_0^{\pi/2} \int_0^{\pi/2} \int_0^1 (\rho^2 \sin^2\phi \sin^2\theta)(\rho^2 \sin\phi)d\rho\,d\phi\,d\theta$

   $= \int_0^{\pi/2} \int_0^{\pi/2} \int_0^1 \rho^4 \sin^3\phi \sin^2\theta\,d\rho\,d\phi\,d\theta = \int_0^{\pi/2} \int_0^{\pi/2} (\sin^3\phi \sin^2\theta)/5\,d\phi\,d\theta$

   $= \int_0^{\pi/2} (2/15)\sin^2\theta\,d\theta = (2/15)[\theta/2 - (\sin 2\theta)/4]\Big|_0^{\pi/2} = \pi/30$.

19. $\iiint_E \sqrt{x^2 + y^2 + z^2}\,dV = \int_0^{2\pi} \int_0^{\pi/6} \int_0^2 \rho^3 \sin\phi\,d\rho\,d\phi\,d\theta = 8\pi \int_0^{\pi/6} \sin\phi\,d\phi$

   $= 8\pi(-\cos\phi)\Big|_0^{\pi/6} = 8\pi(1 - (\sqrt{3}/2)) = 4\pi(2 - \sqrt{3})$.

21. Since $\rho = 4\cos\phi$ implies $\rho^2 = 4\rho\cos\phi$, the equation is that of a sphere of radius 2 with center at $(0,0,2)$. Thus $V = \int_0^{2\pi} \int_0^{\pi/3} \int_0^{4\cos\phi} \rho^2 \sin\phi\,d\rho\,d\phi\,d\theta$

   $= 2\pi \int_0^{\pi/3} (\sin\phi)(64\cos^3\phi)/3\,d\phi = 32\pi(-\cos^4\phi)/3\Big|_0^{\pi/3} = 10\pi$.

23. Placing the center of the base at $(0,0,0)$, $\rho(x,y,z) = K\sqrt{x^2 + y^2 + z^2}$ is the density function. Then $m = \int_0^{2\pi} \int_0^{\pi/2} \int_0^a K\rho^3 \sin\phi\,d\rho\,d\phi\,d\theta$

   $= 2\pi K \int_0^{\pi/2} (a^4/4)\sin\phi\,d\phi = (\pi K a^4/2)(-\cos\phi)\Big|_0^{\pi/2} = \pi K a^4/2$.

25. $I_z = \int_0^{2\pi} \int_0^{\pi/2} \int_0^a (K\rho^3 \sin\phi)(\rho^2 \sin^2\phi)d\rho\,d\phi\,d\theta = 2\pi K \int_0^{\pi/2} (a^6/6)\sin^3\phi\,d\phi$

   $= (\pi a^6 K/3)[-\cos\phi + (\cos^3\phi)/3]\Big|_0^{\pi/2} = 2\pi K a^6/9$.

27. Place the center of the base at $(0,0,0)$; the density function is $\rho(x,y,z) = K$. By symmetry, the moments of inertia about any two such diameters will be equal, so we just need to find I_x. $I_x = \int_0^{2\pi} \int_0^{\pi/2} \int_0^a (K\rho^2 \sin\phi)\rho^2(\sin^2\phi \sin^2\theta + \cos^2\phi)d\rho\,d\phi\,d\theta$

   $= K \int_0^{2\pi} \int_0^{\pi/2} (\sin^3\phi \sin^2\theta + \sin\phi \cos^2\phi)(a^5/5)d\phi\,d\theta$

   $= (Ka^5/5) \int_0^{2\pi} (\sin^2\theta[-\cos\phi + (\cos^3\phi)/3] + (-\cos^3\phi)/3)\Big|_0^{\pi/2} d\theta$

156

Section 13.8

$$= (Ka^5/5)\int_0^{2\pi}[(2\sin^2\theta)/3 + (1/3)]d\theta = (Ka^5/15)(4\pi) = 4Ka^5\pi/15.$$

29. In spherical coordinates $z = \sqrt{x^2+y^2}$ becomes $\cos\phi = \sin\phi$ or $\phi = \pi/4$. Then

$$V = \int_0^{2\pi}\int_0^{\pi/4}\int_0^1 \rho^2\sin\phi\,d\rho\,d\phi\,d\theta = 2\pi(\int_0^{\pi/4}\sin\phi\,d\phi)(\int_0^1 \rho^2 d\rho) = \pi(2-\sqrt{2})/3.$$

$$M_{xy} = \int_0^{2\pi}\int_0^{\pi/4}\int_0^1 \rho^3\sin\phi\cos\phi\,d\rho\,d\phi\,d\theta = 2\pi[(-\cos 2\phi)/4]_0^{\pi/4}(1/4) = \pi/8 \text{ and by}$$

symmetry $M_{yz} = M_{xz} = 0$. Hence $(\bar{x},\bar{y},\bar{z}) = (0,0,3/[8(2-\sqrt{2})])$.

31. In cylindrical coordinates the paraboloid is given by $z = r^2$ and the plane by $z = 2r\sin\theta$ and they intersect in the circle $r = 2\sin\theta$. Then $\iiint_E z\,dV = \int_0^\pi\int_0^{2\sin\theta}\int_{r^2}^{2r\sin\theta} rz\,dz\,dr\,d\theta$

$$= \int_0^\pi\int_0^{2\sin\theta} r(4r^2\sin^2\theta - r^4)/2\,dr\,d\theta = \int_0^\pi ((r^4\sin^2\theta - r^6/6)/2|_0^{2\sin\theta})d\theta$$

$$= \int_0^\pi [8\sin^6\theta - (16\sin^6\theta)/3]d\theta = (8/3)\int_0^\pi (\sin^6\theta)d\theta$$

$$= \int_0^\pi [(5/6) - (10\cos 2\theta)/3 + (2\cos 4\theta)/3 + (\cos 8\theta)/6 + (8\cos 2\theta\sin^2 2\theta)/3]d\theta$$

$$= (5/6)\theta - (5\sin 2\theta)/3 + (\sin 4\theta)/6 + (\sin 8\theta)/48 + (4\sin^3 2\theta)/9|_0^\pi = 5\pi/6.$$

33. $\int_{-1}^1\int_{-\sqrt{1-x^2}}^{\sqrt{1-x^2}}\int_{x^2+y^2}^{2-x^2-y^2}(x^2+y^2)^{3/2}\,dz\,dy\,dx = \int_0^{2\pi}\int_0^1\int_{r^2}^{2-r^2}(r^2)^{3/2}r\,dz\,dr\,d\theta$

$$= 2\pi\int_0^1 r^4 z|_{r^2}^{2-r^2}\,dr = 2\pi\int_0^1 (2r^4 - r^6 - r^6)\,dr = 4\pi\left[\frac{r^5}{5} - \frac{r^7}{7}\right]_0^1 = \frac{8\pi}{35}.$$

35. The region of integration E is the top half of the sphere $x^2+y^2+z^2 = 9$. So

$$\int_{-3}^3\int_{-\sqrt{9-x^2}}^{\sqrt{9-x^2}}\int_0^{\sqrt{9-x^2-y^2}} z\sqrt{x^2+y^2+z^2}\,dz\,dy\,dx = \iiint_E z\sqrt{x^2+y^2+z^2}\,dV$$

$$= \int_0^{2\pi}\int_0^{\pi/2}\int_0^3 (\rho^2\cos\phi)(\rho^2\sin\phi)d\rho\,d\phi\,d\theta = 2\pi\int_0^{\pi/2}(243/5)\cos\phi\sin\phi\,d\phi$$

$$= (486/5)\pi(-\cos 2\phi)/4|_0^{\pi/2} = 243\pi/5.$$

37. (a)

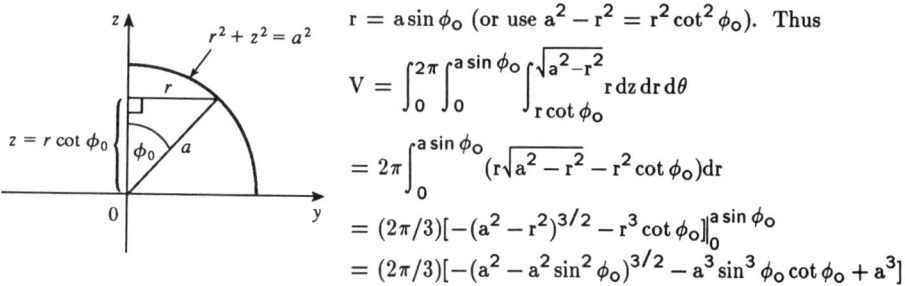

From the picture, $z = r\cot\phi_0$ to $z = \sqrt{a^2-r^2}$, $r = 0$ to $r = a\sin\phi_0$ (or use $a^2 - r^2 = r^2\cot^2\phi_0$). Thus

$$V = \int_0^{2\pi}\int_0^{a\sin\phi_0}\int_{r\cot\phi_0}^{\sqrt{a^2-r^2}} r\,dz\,dr\,d\theta$$

$$= 2\pi\int_0^{a\sin\phi_0}(r\sqrt{a^2-r^2} - r^2\cot\phi_0)dr$$

$$= (2\pi/3)[-(a^2-r^2)^{3/2} - r^3\cot\phi_0]_0^{a\sin\phi_0}$$

$$= (2\pi/3)[-(a^2 - a^2\sin^2\phi_0)^{3/2} - a^3\sin^3\phi_0\cot\phi_0 + a^3]$$

Section 13.9

$= (2\pi a^3/3)[1 - (\cos^3 \phi_0 + \sin^2 \phi_0 \cos \phi_0)] = (2\pi a^3/3)(1 - \cos \phi_0).$

(b)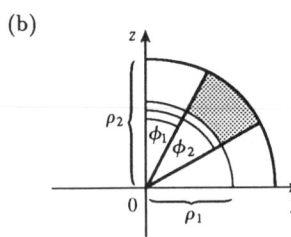
The wedge in question is the shaded area rotated from $\theta = \theta_1$ to $\theta = \theta_2$. Letting V_{ij} = volume bounded by the region bounded by the sphere of radius ρ_i and the cone with angle ϕ_j ($\theta = \theta_1$ to θ_2) and V be the volume of the wedge, we have

$V = (V_{22} - V_{21}) - (V_{12} - V_{11})$
$= [(\theta_2 - \theta_1)/3][\rho_2^3(1 - \cos \phi_2) - \rho_2^3(1 - \cos \phi_1) - \rho_1^3(1 - \cos \phi_2) + \rho_1^3(1 - \cos \phi_1)$
$= [(\theta_2 - \theta_1)/3][(\rho_2^3 - \rho_1^3)(1 - \cos \phi_2) - (\rho_2^3 - \rho_1^3)(1 - \cos \phi_1)]$
$= [(\theta_2 - \theta_1)/3][(\rho_2^3 - \rho_1^3)(\cos \phi_1 - \cos \phi_2)].$

(Or show $V = \int_{\theta_1}^{\theta_2} \int_{\rho_1 \sin \phi_1}^{\rho_2 \sin \phi_2} \int_{r \cot \phi_2}^{r \cot \phi_1} r \, dz \, dr \, d\theta$).

(c) By the Mean Value Theorem with $f(\rho) = \rho^3$ there exists $\tilde{\rho}$, $\rho_1 \leq \tilde{\rho} \leq \rho_2$, such that
$f(\rho_2) - f(\rho_1) = f'(\rho)(\rho_2 - \rho_1)$ or $\rho_1^3 - \rho_2^3 = 3\tilde{\rho}^2 \Delta \rho$. Similarly
$\cos \phi_2 - \cos \phi_1 = (-\sin \tilde{\phi})\Delta \phi$ with $\phi_1 \leq \tilde{\phi} \leq \phi_2$. Substituting into (b) gives
$V = (\tilde{\rho}^2 \Delta \rho)(\theta_2 - \theta_1)(\sin \tilde{\phi})\Delta \phi = \tilde{\rho}^2 \sin \tilde{\phi} \, \Delta \rho \, \Delta \phi \, \Delta \theta.$

EXERCISES 13.9

1. $\dfrac{\partial(x,y)}{\partial(u,v)} = \begin{vmatrix} \dfrac{\partial x}{\partial u} & \dfrac{\partial x}{\partial v} \\ \dfrac{\partial y}{\partial u} & \dfrac{\partial y}{\partial v} \end{vmatrix} = \begin{vmatrix} 1 & -2 \\ 2 & -1 \end{vmatrix} = 1(-1) - 2(-2) = 3.$

3. $\dfrac{\partial(x,y)}{\partial(u,v)} = \begin{vmatrix} \dfrac{\partial x}{\partial u} & \dfrac{\partial x}{\partial v} \\ \dfrac{\partial y}{\partial u} & \dfrac{\partial y}{\partial v} \end{vmatrix} = \begin{vmatrix} 2e^{2u} \cos v & -e^{2u} \sin v \\ 2e^{2u} \sin v & e^{2u} \cos v \end{vmatrix}$
$= 2e^{4u}(\cos^2 v + \sin^2 v) = 2e^{4u}.$

5. $\dfrac{\partial(x,y,z)}{\partial(u,v,w)} = \begin{vmatrix} 1 & 1 & 1 \\ 1 & 1 & -1 \\ 1 & -1 & 1 \end{vmatrix} = 1(1-1) - 1(1+1) + 1(-1-1) = -4.$

7. S_1: $v = 0$, $0 \leq u \leq 2$, so $x = u$, $y = 2u$ and $y = 2x$. S_2: $u = 2$, $0 \leq v \leq 1$, so $x = 2 - 2v$, $y = 4 - v$ and $x = 2y - 6$. S_3: $v = 1$, $0 \leq u \leq 2$, so $x = u - 2$, $y = 2u - 1$ and $y = 2x + 3$. S_4: $u = 0$, $0 \leq v \leq 1$, so $x = -2v$, $y = -v$ and $2y = x$.

Section 13.9

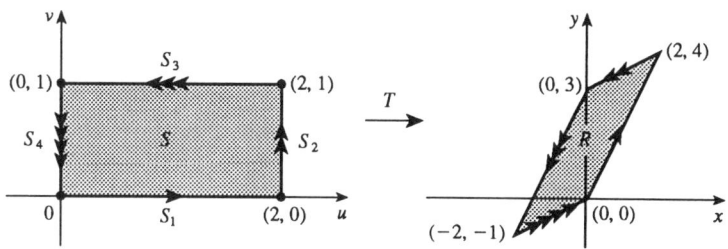

9. S_1: $0 \leq u \leq 1$, $v = 0$, $x = 4u$, $y = 0$ so $0 \leq x \leq 4$, $y = 0$ is the image of the 1st side.
S_2: $u + v = 1$, $x = 4(1-v) + 3v = 4 - v \Leftrightarrow v = 4 - x$, $y = 4v = 4(4-x) = 16 - 4x$,
$3 \leq x \leq 4$ so $y = 16 - 4x$, $3 \leq x \leq 4$ is the image of the 2nd side. S_3: $u = 0$, $0 \leq v \leq 1$,
$x = 3v$, $y = 4v = 4(x/3) = 4x/3$, $0 \leq x \leq 3$, so $y = 4x/3$, $0 \leq x \leq 3$ is the image of the 3rd side.

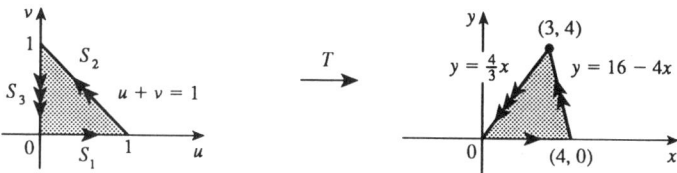

11. $\partial(x,y)/\partial(u,v) = \begin{vmatrix} 1/3 & 1/3 \\ -2/3 & 1/3 \end{vmatrix} = 1/3$ and $3x + 4y = (u+v) + (4/3)(v - 2u)$

$= (1/3)(7v - 5u)$. Then S is the region bounded by the lines $u = 0$, $(1/3)(v-2)$
$= (1/3)(u+v) - 2$ or $u = 2$, $(1/3)(v - 2u) = (-2/3)(u+v)$ or $v = 0$, and

$(1/3)(v - 2u) = 3 - (2/3)(u+v)$ or $v = 3$. Thus $\iint_R (3x+4y) dA$

$= \int_0^3 \int_0^2 (1/3)(7v - 5u)(1/3) du\, dv = (1/9)\int_0^3 (14v - 10) dv = (1/9)(33) = 11/3$.

13. $\partial(x,y)/\partial(u,v) = \begin{vmatrix} 2 & 0 \\ 0 & 3 \end{vmatrix} = 6$, $x^2 = 4u^2$ and the planar ellipse $9x^2 + 4y^2 \leq 36$ is the

image of the disc $u^2 + v^2 \leq 1$. Thus $\iint_R x^2\, dA = \iint_{u^2+v^2 \leq 1} (4u^2)(6) du\, dv$

$= \int_0^{2\pi} \int_0^1 (24r^2 \cos^2 \theta) r\, dr\, d\theta = 6\pi$.

Section 13.9

15. $\partial(x,y)/\partial(u,v) = \begin{vmatrix} 1/v & -u/v^2 \\ 0 & 1 \end{vmatrix} = 1/v$, $xy = u$, $y = x$ is the image of the parabola $v^2 = u$, $y = 3x$ of the parabola $v^2 = 3u$ and the hyperbolas $xy = 1$, $xy = 3$ are the images of the lines $u = 1$ and $u = 3$ respectively. Thus $\iint_R xy \, dA = \int_1^3 \int_{\sqrt{u}}^{\sqrt{3u}} u(1/v) dv \, du$

$= \int_1^3 u(\ln \sqrt{3u} - \ln \sqrt{u}) du = \int_1^3 u \ln \sqrt{3} \, du = 4 \ln \sqrt{3} = 2 \ln 3.$

17. $\partial(x,y,z)/\partial(u,v,w) = \begin{vmatrix} a & 0 & 0 \\ 0 & b & 0 \\ 0 & 0 & c \end{vmatrix} = abc$ and the solid enclosed by the ellipsoid is the

image of the ball $u^2 + v^2 + w^2 \leq 1$. Thus $\iiint_E dV = \iiint_{u^2+v^2+w^2 \leq 1} abc \, du \, dv \, dw$

$= (abc)(\text{volume of the ball}) = 4\pi abc/3.$

19. Letting $u = 2x - y$ and $v = 3x + y$, we have $x = (u+v)/5$, $y = (2v - 3u)/5$. Then

$\partial(x,y)/\partial(u,v) = \begin{vmatrix} 1/5 & 1/5 \\ -3/5 & 2/5 \end{vmatrix} = 1/5$ and $\iint_R xy \, dA$

$= \int_{-2}^{1} \int_{-3}^{1} [(u+v)(2v - 3u)/25](1/5) du \, dv = (1/125) \int_{-2}^{1} \int_{-3}^{1} (2v^2 - uv - 3u^2) du \, dv$

$= (1/125) \int_{-2}^{1} (8v^2 + 4v - 28) dv = -66/125.$

21. Letting $u = y - x$, $v = y + x$, we have $y = (u+v)/2$, $x = (v-u)/2$. Then

$\partial(x,y)/\partial(u,v) = \begin{vmatrix} -1/2 & 1/2 \\ 1/2 & 1/2 \end{vmatrix} = -1/2$ and R is the image of the trapezoidal region with

vertices $(-1, 1)$, $(-2, 2)$, $(2, 2)$, and $(1, 1)$. Thus $\iint_R \cos[(y-x)/(y+x)] dA$

$= \int_1^2 \int_{-v}^{v} |-1/2| \cos(u/v) du \, dv = (1/2) \int_1^2 v \sin(u/v) \Big|_{-v}^{v} dv = (1/2) \int_1^2 2v \sin(1) dv$

$= (\sin 1)(3/2).$

23. Let $u = x + y$ and $v = -x + y$ then $u + v = 2y \Rightarrow y = \frac{1}{2}(u+v)$ and $u - v = 2x \Rightarrow x = \frac{1}{2}(u-v)$.

$\dfrac{\partial(x,y)}{\partial(u,v)} = \begin{vmatrix} \frac{1}{2} & -\frac{1}{2} \\ \frac{1}{2} & \frac{1}{2} \end{vmatrix} = \frac{1}{2}.$ Now

$|u| = |x+y| \leq |x| + |y| \leq 1 \Rightarrow -1 \leq u \leq 1$, and

$|v| = |-x+y| \leq |x| + |y| \leq 1 \Rightarrow -1 \leq v \leq 1.$

R is the image of the square region with vertices $(1,1)$, $(1,-1)$, $(-1,-1)$, and $(-1,1)$.

So $\iint_R e^{x+y} dA = \frac{1}{2}\int_{-1}^{1}\int_{-1}^{1} e^u \, du \, dv = \frac{1}{2} 2 \cdot [e^u]_{-1}^{1} = e - e^{-1}$.

REVIEW EXERCISES FOR CHAPTER 13

1. This is true by Fubini's Theorem.

3. $\iint_D \sqrt{4 - x^2 - y^2} \, dA =$ the volume under the surface $x^2 + y^2 + z^2 = 4$ and above the xy-plane $= \frac{1}{2}$ (the volume of the sphere $x^2 + y^2 + z^2 = 4$) $= \frac{1}{2}(\frac{4}{3}\pi(2)^3) = \frac{16}{3}\pi$.

5. The volume enclosed by the cone $z = \sqrt{x^2 + y^2}$ and the plane $z = 2$ is, using cylindrical coordinates, $V = \int_0^{2\pi}\int_0^2\int_r^2 r \, dz \, dr \, d\theta \neq \int_0^{2\pi}\int_0^2\int_r^2 dz \, dr \, d\theta$ so the assertion is false.

7. $\int_{-2}^{2}\int_0^4 (4x^3 + 3xy^2) dx \, dy = \int_{-2}^{2}(256 + 24y^2) dy = 2[256y + 8y^3]_0^2 = 1152$.

9. $\int_1^2 \int_0^{x^2} [1/(x+y)] dy \, dx = \int_1^2 \ln|x+y|\Big|_0^{x^2} dx = \int_1^2 [\ln(1+x)] dx$
$= (1+x)\ln(1+x) - (1+x)\Big|_1^2 = \ln(27/4) - 1$.

11. $\int_0^1\int_0^{x^2}\int_0^y y^2 z \, dz \, dy \, dx = \int_0^1\int_0^{x^2}(y^4/2) dy \, dx = \int_0^1 (x^{10}/10) dx = 1/110$.

13. 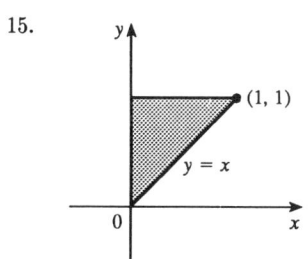
$r = 1 + \sin\theta$
$r = 1$

The region whose area is given by $\int_0^\pi \int_1^{1+\sin\theta} r \, dr \, d\theta$ is $\{(r,\theta) \mid 0 \leq \theta \leq \pi, 1 \leq r \leq 1 + \sin\theta\}$, which is the region outside the circle $r = 1$ and inside the cardioid $r = 1 + \sin\theta$.

15.
$\int_0^1 \int_x^1 e^{x/y} dy \, dx = \int_0^1 \int_0^y e^{x/y} dx \, dy$
$= \int_0^1 (ey - y) dy = (e-1)/2$.

(1, 1), $y = x$

Chapter 13 Review

17. $\int_2^4 \int_0^1 [1/(x-y)^2] dx\, dy = \int_2^4 [-(1/y) - (1/(1-y))] dy = -\ln y + \ln|1-y|\Big|_2^4$

$= -\ln 4 + \ln 3 + \ln 2 = \ln(3/2).$

19. The curves $y^2 = x^3$ and $y = x$ intersect when $x^3 = x$, that is when $x = 0$ and $x = 1$ ($x \neq -1$ since $x^3 = y^2 \Rightarrow x \geq 0$). So

$\int_0^1 \int_{x^{3/2}}^x xy\, dy\, dx = \int_0^1 [(x^3/2) - (x^4/2)] dx = x^4/8 - x^5/10\Big|_0^1 = 1/40.$

21. $\int_0^1 \int_0^{1-y^2} (xy + 2x + 3y) dx\, dy = \int_0^1 [(1-y^2)^2(y+2)/2 + 3y(1-y^2)] dy$

$= y^5/2 + y^4 - 4y^3 - 2y^2 + 7y/2 + 1 = (1/5) + (1/12) - (2/3) + (7/4) = 41/30.$

23. $\iint_D (x^2 + y^2)^{3/2} dA = \int_0^{\pi/3} \int_0^3 (r^2)^{3/2} r\, dr\, d\theta$

$= (\pi/3)(3^5/5) = 81\pi/5.$

25. $\iiint_E x^2 z\, dV = \int_0^2 \int_0^{2x} \int_0^x x^2 z\, dz\, dy\, dx = \int_0^2 \int_0^{2x} (x^4/2) dy\, dx = \int_0^2 x^5 dx = 2^6/6 = 32/3.$

27. $\iiint_E y^2 z^2\, dV = \int_{-1}^1 \int_{-\sqrt{1-y^2}}^{\sqrt{1-y^2}} \int_0^{1-y^2-z^2} y^2 z^2\, dx\, dz\, dy$

$= \int_{-1}^1 \int_{-\sqrt{1-y^2}}^{\sqrt{1-y^2}} y^2 z^2 (1 - y^2 - z^2) dz\, dy$

$= \int_0^{2\pi} \int_0^1 (r^2 \cos^2 \theta)(r^2 \sin^2 \theta)(1 - r^2) r\, dr\, d\theta$

$= \int_0^{2\pi} \int_0^1 (1/4) \sin^2 2\theta (r^5 - r^7) dr\, d\theta = \int_0^{2\pi} (1/8)(1 - \cos 4\theta)(r^6/6 - r^8/8)\Big|_0^1 d\theta$

$= (1/192)[\theta - (1/4) \sin 4\theta]\Big|_0^{2\pi} = 2\pi/192 = \pi/96.$

29. $\iiint_E yz\, dV = \int_{-2}^2 \int_0^{\sqrt{4-x^2}} \int_0^y yz\, dz\, dy\, dx = \int_{-2}^2 \int_0^{\sqrt{4-x^2}} (y^3/2) dy\, dx$

$= \int_0^\pi \int_0^2 [(r^3 \sin^3 \theta)/2] r\, dr\, d\theta = (16/5) \int_0^\pi \sin^3 \theta\, d\theta = (16/5)[-\cos \theta + (\cos^3 \theta)/3]\Big|_0^\pi = 64/15.$

31. $V = \int_0^2 \int_1^4 (x^2 + 4y^2) dy\, dx = \int_0^2 (3x^2 + 84) dx = 176.$

162

33.

$$V = \int_0^2 \int_0^y \int_0^{(2-y)/2} dz\,dx\,dy = \int_0^2 \int_0^y (1-y/2)dx\,dy$$

$$= \int_0^2 (y - y^2/2)dy = 2/3.$$

35. Using the wedge above the plane $z = 0$ and below the plane $z = mx$ and noting that we have the same volume for $m < 0$ as for $m > 0$ (so use $m > 0$), we have

$$V = 2\int_0^{a/3} \int_0^{\sqrt{a^2-9y^2}} mx\,dx\,dy = 2\int_0^{a/3} (m/2)(a^2 - 9y^2)dy = m(a^2 y - 3y^3)\Big|_0^{a/3}$$

$$= m(a^3/3 - a^3/9) = 2ma^3/9.$$

37. $m = \int_0^1 \int_0^{1-y^2} y\,dx\,dy = \int_0^1 (y - y^3)dy = (1/2) - (1/4) = 1/4$, $M_y = \int_0^1 \int_0^{1-y^2} yx\,dx\,dy$

$$= \int_0^1 y(1-y^2)^2/2\,dy = -(1-y^2)^3/12\Big|_0^1 = 1/12, \quad M_x = \int_0^1 \int_0^{1-y^2} y^2\,dx\,dy$$

$$= \int_0^1 (y^2 - y^4)dy = 2/15. \text{ Hence } (\bar{x}, \bar{y}) = (1/3, 8/15).$$

39. $m = \pi K a^2/4$ where K is constant, $M_y = \iint\limits_{x^2+y^2 \le a^2} Kx\,dA = K\int_0^{\pi/2} \int_0^a r^2 \cos\theta\,dr\,d\theta$

$$= (Ka^3/3)\int_0^{\pi/2} \cos\theta\,d\theta = a^3 K/3, \quad M_x = K\int_0^{\pi/2} \int_0^a r^2 \sin\theta\,dr\,d\theta = a^3 K/3$$

(by symmetry $M_x = M_y$). Hence the centroid is $(\bar{x}, \bar{y}) = (4a/(3\pi), 4a/(3\pi))$.

41. The equation of the cone with the suggested orientation is $(h - z) = (h/a)\sqrt{x^2 + y^2}$, $0 \le z \le h$. Then $V = \pi a^2 h/3$ is the volume of one frustum of a cone; by symmetry

$$M_{yz} = M_{xz} = 0; \text{ and } M_{xy} = \iint\limits_{x^2+y^2 \le a^2} \int_0^{h-(h/a)\sqrt{x^2+y^2}} z\,dz\,dA$$

$$= \int_0^{2\pi} \int_0^a \int_0^{(h/a)(a-r)} rz\,dz\,dr\,d\theta = \pi \int_0^a r(h^2/a^2)(a-r)^2 dr$$

$$= (\pi h^2/a^2)\int_0^a (a^2 r - 2ar^2 + r^3)dr = (\pi h^2/a^2)(a^4/2 - 2a^4/3 + a^4/4) = \pi h^2 a^2/12. \text{ Hence}$$

the centroid is $(\bar{x}, \bar{y}, \bar{z}) = (0, 0, h/4)$.

43. $z = 1 - \frac{1}{2}x - \frac{1}{3}y \Rightarrow \frac{\partial z}{\partial x} = -\frac{1}{2}$ and $\frac{\partial z}{\partial y} = -\frac{1}{3}$. The projection onto the xy-plane of the part of the given plane in the first octant is the triangular region D bounded by the x- and y-axes and the line $3x + 2y = 6$, which has intercepts 2 and 3. So $A(D) = \frac{1}{2} \cdot 2 \cdot 3 = 3$. Thus the surface area is $A(S) = \iint_D \sqrt{1 + (-\frac{1}{2})^2 + (-\frac{1}{3})^2}\, dA = \frac{7}{6}A(D) = \frac{7}{6} \cdot 3 = \frac{7}{2}$.

45.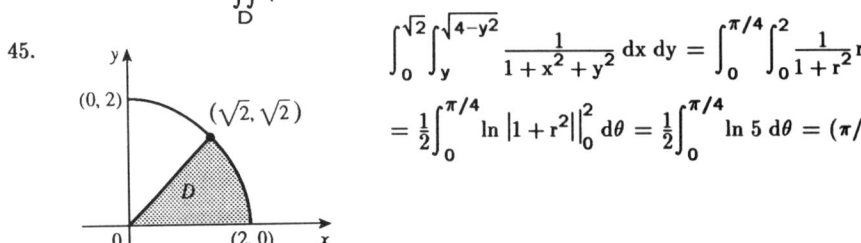

$$\int_0^{\sqrt{2}} \int_y^{\sqrt{4-y^2}} \frac{1}{1+x^2+y^2}\, dx\, dy = \int_0^{\pi/4} \int_0^2 \frac{1}{1+r^2} r\, dr\, d\theta$$

$$= \frac{1}{2} \int_0^{\pi/4} \ln|1+r^2|\Big|_0^2\, d\theta = \frac{1}{2} \int_0^{\pi/4} \ln 5\, d\theta = (\pi/8)\ln 5.$$

47.

$$\int_{-1}^1 \int_{x^2}^1 \int_0^{1-y} f(x,y,z)\, dz\, dy\, dx = \int_0^1 \int_0^{1-z} \int_{-\sqrt{y}}^{\sqrt{y}} f(x,y,z)\, dx\, dy\, dz$$

49. Since $u = x - y$, $v = x + y$, $x = (u+v)/2$ and $y = (v-u)/2$. Thus

$$\partial(x,y)/\partial(u,v) = \begin{vmatrix} 1/2 & 1/2 \\ -1/2 & 1/2 \end{vmatrix} = 1/2 \text{ and } \iint_R (x-y)/(x+y)\, dA$$

$$= \int_2^4 \int_{-2}^0 (u/v)(1/2)\, du\, dv = -\int_2^4 (1/v)\, dv = -\ln 2.$$

51. Let $u = y - x$ and $v = y + x$ so $x = y - u = (v-x) - u \Rightarrow x = (v-u)/2$ and $y = v - ((v-u)/2) = (v+u)/2$.

$\left|\frac{\partial(x,y)}{\partial(u,v)}\right| = \left|\frac{\partial x}{\partial u}\frac{\partial y}{\partial v} - \frac{\partial x}{\partial v}\frac{\partial y}{\partial u}\right| = \left|(-\frac{1}{2})(\frac{1}{2}) - (\frac{1}{2})(\frac{1}{2})\right| = \left|-\frac{1}{2}\right| = \frac{1}{2}$. R is the image under this transformation of the square with vertices $(u, v) = (0, 0), (-2, 0), (0, 2), (-2, 2)$. So

$$\iint_R xy\, dA = \int_0^2 \int_{-2}^0 \frac{v^2 - u^2}{4}\left(\frac{1}{2}\right) du\, dv$$

164

Chapter 13 Review

$$= \tfrac{1}{8}\int_0^2 \left[v^2 u - \tfrac{u^3}{3}\right]_{-2}^0 dv = \tfrac{1}{8}\int_0^2 \left(2v^2 - \tfrac{8}{3}\right) dv = \tfrac{1}{8}\left[\tfrac{2}{3}v^3 - \tfrac{8}{3}v\right]_0^2 = 0.$$

(This result could have been anticipated by symmetry since the integrand is an odd function of y and R is symmetric about the x–axis.)

53. For each r such that D_r lies within the domain D, $A(D_r) = \pi r^2$, and by the Mean-Value Theorem for double integrals there exists (x_r, y_r) in D_r such that

$$f(x_r, y_r) = \frac{1}{\pi r^2}\iint_{D_r} f(x, y)\, dA. \text{ But } \lim_{r \to 0^+}(x_r, y_r) = (a, b), \text{ so}$$

$$\lim_{r \to 0^+} \frac{1}{\pi r^2}\iint_{D_r} f(x, y)\, dA = \lim_{r \to 0^+} f(x_r, y_r) = f(a, b) \text{ by the continuity of f.}$$

APPLICATIONS PLUS (after Chapter 13)

1. (a) The area of a trapezoid is $\frac{1}{2}h(b_1 + b_2)$, where h is the height (the distance between the two parallel sides) and b_1, b_2 are the lengths of the bases (the parallel sides). From the figure in the text, we see that $h = x\sin\theta$, $b_1 = w - 2x$, and $b_2 = w - 2x + 2x\cos\theta$. Therefore the cross-sectional area of the rain gutter is
$$A(x,\theta) = \tfrac{1}{2}x\sin\theta[(w-2x)+(w-2x+2x\cos\theta)] = (x\sin\theta)(w-2x+x\cos\theta)$$
$$= wx\sin\theta - 2x^2\sin\theta + x^2\sin\theta\cos\theta, \quad 0 < x \le \tfrac{w}{2}, \ 0 < \theta \le \tfrac{\pi}{2}.$$ We look for the critical points of A: $\frac{\partial A}{\partial x} = w\sin\theta - 4x\sin\theta + 2x\sin\theta\cos\theta$ and
$\frac{\partial A}{\partial \theta} = wx\cos\theta - 2x^2\cos\theta + x^2(\cos^2\theta - \sin^2\theta)$, so
$\frac{\partial A}{\partial x} = 0 \Leftrightarrow \sin\theta(w - 4x + 2x\cos\theta) = 0 \Leftrightarrow \cos\theta = \frac{4x-w}{2x} = 2 - \frac{w}{2x}$
$(0 < \theta \le \tfrac{\pi}{2} \Rightarrow \sin\theta > 0)$. If, in addition, $\frac{\partial A}{\partial \theta} = 0$, then
$$0 = wx\cos\theta - 2x^2\cos\theta + x^2(2\cos^2\theta - 1)$$
$$= wx(2 - \tfrac{w}{2x}) - 2x^2(2 - \tfrac{w}{2x}) + x^2[2(2 - \tfrac{w}{2x})^2 - 1]$$
$$= 2wx - \tfrac{1}{2}w^2 - 4x^2 + wx + x^2[8 - \tfrac{4w}{x} + \tfrac{w^2}{2x^2} - 1] = -wx + 3x^2 = x(3x - w).$$
Since $x > 0$, we must have $x = \tfrac{w}{3}$, in which case $\cos\theta = \tfrac{1}{2}$, so $\theta = \tfrac{\pi}{3}$, $\sin\theta = \tfrac{\sqrt{3}}{2}$, $h = w\sqrt{3}/6$, $b_1 = \tfrac{w}{3}$, $b_2 = \tfrac{2w}{3}$, and $A = \tfrac{w^2\sqrt{3}}{12}$. As in Example 5 of Section 12.7, we can argue from the physical nature of this problem that we have found a relative maximum of A. Now checking the boundary of A, let
$g(\theta) = A(\tfrac{w}{2},\theta) = \tfrac{w^2}{2}\sin\theta - \tfrac{w^2}{2}\sin\theta + \tfrac{w^2}{4}\sin\theta\cos\theta = \tfrac{1}{8}w^2\sin 2\theta, \ 0 < \theta \le \tfrac{\pi}{2}$.
Clearly g is maximized when $\sin 2\theta = 1$ in which case $A = \tfrac{1}{8}w^2$. Also along the line $\theta = \tfrac{\pi}{2}$, let $h(x) = A(x,\tfrac{\pi}{2}) = wx - 2x^2$, $0 < x < \tfrac{w}{2} \Rightarrow h'(x) = w - 4x = 0$
$\Leftrightarrow x = \tfrac{w}{4}$, and $h(\tfrac{w}{4}) = w\tfrac{w}{4} - 2\tfrac{w^2}{16} = \tfrac{1}{8}w^2$. Since $\tfrac{1}{8}w^2 < \tfrac{w^2\sqrt{3}}{12}$, we conclude that the relative maximum found earlier was an absolute maximum.

(b) If the metal were bent into a semi-circular gutter of radius r, we would have $w = \pi r$ and $A = \tfrac{1}{2}\pi r^2 = \tfrac{1}{2}\pi(\tfrac{w}{\pi})^2 = \tfrac{w^2}{2\pi}$. Since $\tfrac{w^2}{2\pi} > \tfrac{w^2\sqrt{3}}{12}$, it *would* be better to bend the metal into a gutter with a semi-circular cross-section.

3. (a) The total amount of water supplied each hour to the region within R feet of the sprinkler is
$$V = \int_0^{2\pi}\int_0^R e^{-r}r\,dr\,d\theta = (2\pi)[-re^{-r} - e^{-r}]_0^R = 2\pi[-Re^{-R} - e^{-R} + 0 + 1]$$
$$= 2\pi(1 - Re^{-R} - e^{-R}) \text{ ft}^3/\text{h}.$$

(b) The average amount of water per hour per square foot supplied to the region within R feet of the sprinkler is V/(area of region) $= \dfrac{V}{\pi r^2} = \dfrac{2(1 - Re^{-R} - e^{-R})}{R^2}$ ft^3 per hour per square foot. [See the definition before Exercise 13.2.33.]

5. (a) If f(P, A) is the probability that an individual at A will be infected by an individual at P, and k dA is the number of infected individuals in an element of area dA, then f(P, A) k dA is the number of infections that should result from exposure of the individual at A to infected people in the element of area dA. Integration over D gives the number of infections of the person at A due to all the infected people in D. In rectangular coordinates (with origin at the city's center), the exposure of a person at A is

$$E = \iint_D k\, f(P, A)\, dA = k \iint_D \tfrac{1}{20}[20 - d(P, A)]\, dA = k \iint_D \left[1 - \tfrac{1}{20}\sqrt{(x - x_0)^2 + (y - y_0)^2}\right] dx\, dy.$$

(b) If $A = (0, 0)$, then $E = k \iint_D \left[1 - \tfrac{1}{20}\sqrt{x^2 + y^2}\right] dx\, dy = k \int_0^{2\pi} \int_0^{10} (1 - \tfrac{r}{20})\, r\, dr\, d\theta$

$= 2\pi k \left[\dfrac{r^2}{2} - \dfrac{r^3}{60}\right]_0^{10} = 2\pi k(50 - \tfrac{50}{3}) = \dfrac{200}{3}\pi k \approx 209\, k.$

For A at the edge of the city, it is convenient to use a polar coordinate system centered at A. Then the polar equation for the circular boundary of the city becomes $r = 20 \cos\theta$ instead of $r = 10$, and the distance from A to a point P in the city is again r (see the figure).

So $E = k \int_{-\pi/2}^{\pi/2} \int_0^{20\cos\theta} (1 - \tfrac{r}{20})\, r\, dr\, d\theta$

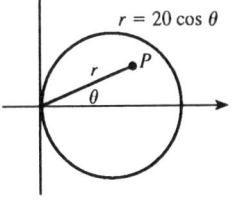

$r = 20 \cos\theta$

$= k \int_{-\pi/2}^{\pi/2} \left(\dfrac{r^2}{2} - \dfrac{r^3}{60}\right)_0^{20\cos\theta} d\theta$

$= k \int_{-\pi/2}^{\pi/2} \left(200 \cos^2\theta - \tfrac{400}{3} \cos^3\theta\right) d\theta$

$= 200 k \int_{-\pi/2}^{\pi/2} \left[\tfrac{1}{2} + \tfrac{1}{2}\cos 2\theta - \tfrac{2}{3}(1 - \sin^2\theta) \cos\theta\right] d\theta$

$= 200 k \left[\dfrac{\theta}{2} + \dfrac{\sin 2\theta}{4} - \tfrac{2}{3}\sin\theta + \tfrac{2}{3} \cdot \dfrac{\sin^3\theta}{3}\right]_{-\pi/2}^{\pi/2}$

$= 200 k \left[\tfrac{\pi}{4} + 0 - \tfrac{2}{3} + \tfrac{2}{9} + \tfrac{\pi}{4} + 0 - \tfrac{2}{3} + \tfrac{2}{9}\right] = 200 k(\tfrac{\pi}{2} - \tfrac{8}{9}) \approx 136\, k.$ Therefore the risk of infection is much lower at the edge of the city than in the middle, so it is better to live at the edge.

Applications Plus

7. (a) The mountain comprises a solid conical region C. The work done in lifting a small volume of material ΔV with density $g(P)$ to a height $h(P)$ above sea level is $h(P)g(P)\Delta V$. Summing over the whole mountain we get $W = \iiint_C h(P)g(P)\,dV$.

   (b) Here C is a solid right circular cone with radius $R = 62,000$ ft, height $H = 12,400$ ft, and density $g(P) = 200$ lb/ft^3 at all points P in C. We use cylindrical coordinates:

   $$W = \int_0^{2\pi}\int_0^H \int_0^{R(1-z/H)} z \cdot 200\, r\, dr\, dz\, d\theta$$

   $$= 2\pi \int_0^H 200z \left[\tfrac{1}{2}r^2\right]_0^{R(1-z/H)} dz$$

   $$= 400\pi \int_0^H z\,\frac{R^2}{2}(1-\tfrac{z}{H})^2\, dz$$

   $$= 200\pi R^2 \int_0^H (z - \tfrac{2z^2}{H} + \tfrac{z^3}{H^2})\, dz$$

   $$= 200\pi R^2 \left[\tfrac{z^2}{2} - \tfrac{2z^3}{3H} + \tfrac{z^4}{4H^2}\right]_0^H$$

   $$= 200\pi R^2 \left(\tfrac{H^2}{2} - \tfrac{2H^2}{3} + \tfrac{H^2}{4}\right) = \tfrac{50}{3}\pi R^2 H^2 = \tfrac{50}{3}\pi (62,000)^2 (12,400)^2$$

   $$\approx 3.1 \times 10^{19}\ \text{ft-lb.}$$

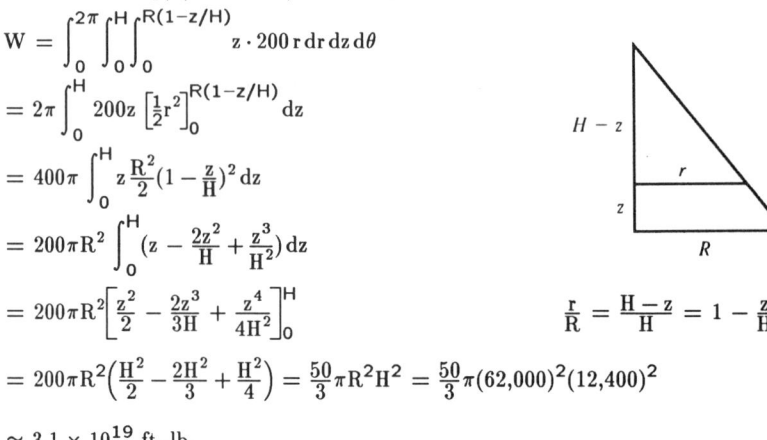

$\dfrac{r}{R} = \dfrac{H-z}{H} = 1 - \dfrac{z}{H}$

…

CHAPTER 14

EXERCISES 14.1

1. $\vec{F}(x,y) = x\vec{i} + y\vec{j}$.

   The length of the vector $x\vec{i} + y\vec{j}$ is the distance from $(0,0)$ to (x,y). Flow lines are rays emanating from the origin.

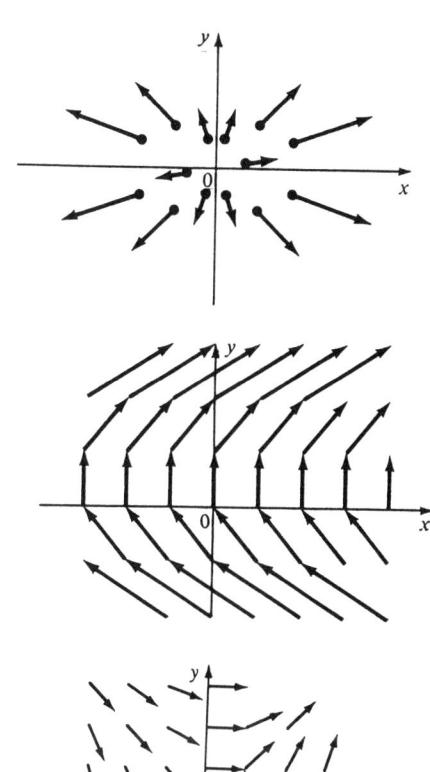

3. $\vec{F}(x,y) = y\vec{i} + \vec{j}$.

   The length of the vector $y\vec{i} + \vec{j}$ is $\sqrt{y^2 + 1}$. Flow lines are parabolas opening about the x-axis.

5. $\vec{F}(x,y) = \dfrac{y\vec{i} + x\vec{j}}{\sqrt{x^2 + y^2}}$

   The length of the vector $\dfrac{y\vec{i} + x\vec{j}}{\sqrt{x^2 + y^2}}$ is 1.

7. $\vec{F}(x,y,z) = \vec{j}$.

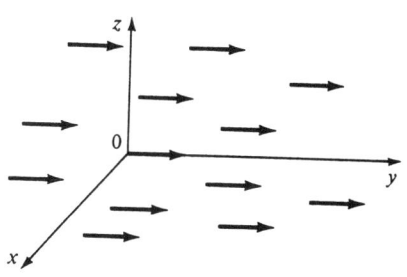

169

Section 14.1

9. $\vec{F}(x, y, z) = y\vec{j}$.

The length of $\vec{F}(x,y,z)$ is $|y|$. No vectors emanate from the xz-plane since $y = 0$ here. In each plane $y = b$, all the vectors are identical.

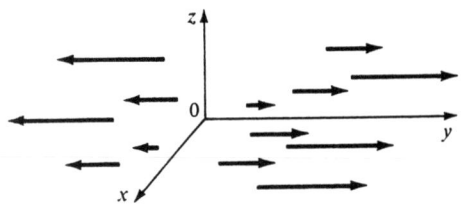

11. $\nabla f(x, y) = f_x(x,y)\vec{i} + f_y(x,y)\vec{j} = (5x^4 - 8xy^3)\vec{i} - (12x^2y^2)\vec{j}$.
13. $\nabla f(x, y) = <3e^{3x}\cos 4y, -4e^{3x}\sin 4y>$.
15. $\nabla f(x, y) = <y^2, 2xy - z^3, -3yz^2>$.
17. $f(x, y) = x^2 - \frac{1}{2}y^2$, $\nabla f(x, y) = 2x\vec{i} - y\vec{j}$. The length of $\nabla f(x,y)$ is $\sqrt{4x^2 + y^2}$ and $\nabla f(x,y)$ terminates on the x-axis at the point (3x, 0).

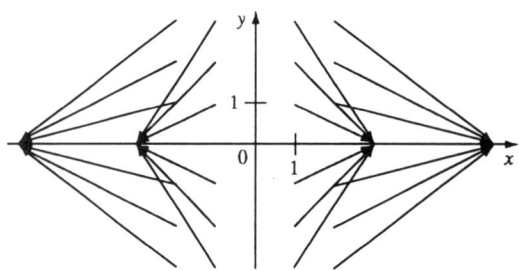

19. (a) 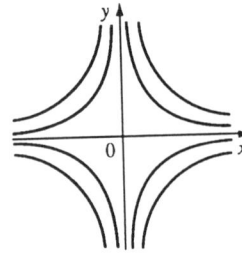 The flow lines appear to be hyperbolas with equations $y = \frac{C}{x}$.

(b) $\frac{dx}{dt} = x \Rightarrow \frac{dx}{x} = dt \Rightarrow \ln|x| = t + C \Rightarrow x = \pm e^{t+C} = Ae^t$ for some constant A.

$\frac{dy}{dt} = -y \Rightarrow \frac{dy}{y} = -dt \Rightarrow \ln|y| = -t + K \Rightarrow y = \pm e^{-t+K} = Be^{-t}$ for some constant B. Therefore $xy = Ae^t Be^{-t} = AB = $ constant. If the flow line passes through (1, 1) then $(1)(1) = $ constant $= 1 \Rightarrow xy = 1 \Rightarrow y = \frac{1}{x}$, $x > 0$.

170

EXERCISES 14.2

1. $\int_C x\,ds = \int_0^1 (t^3)\sqrt{9t^4+1}\,dt = (9t^4+1)^{3/2}/54\Big|_0^1 = (10^{3/2}-1)/54.$

3. $x = 4\cos t,\ y = 4\sin t,\ -\pi/2 \le t \le \pi/2.$

$\int_C xy^4\,ds = \int_{-\pi/2}^{\pi/2} [(4)^5 \cos t \sin^4 t](4)dt = (4)^6(\sin^5 t)/5\Big|_{-\pi/2}^{\pi/2}$

$= (2)4^6/5 = 1638.4.$

5. $x = x,\ y = x^2,\ -2 \le x \le 1.$ Then $\int_C (x-2y^2)dy = \int_{-2}^1 (x-2x^4)2x\,dx$

$= \int_{-2}^1 (2x^2 - 4x^5)dx = (2/3)(x^3 - x^6)\Big|_{-2}^1 = 48.$

7. $C = C_1 + C_2.$ On C_1: $x = x,\ y = 0,\ 0 \le x \le 2.$
On C_2: $x = x,\ y = 2x-4,\ 2 \le x \le 3.$

Then $\int_C xy\,dx + (x-y)dy = \int_{C_1} xy\,dx + (x-y)dy + \int_{C_2} xy\,dx + (x-y)dy$

$= \int_0^2 0\,dx + \int_2^3 [(2x^2 - 4x) + (-x+4)(2)]dx = \int_2^3 (2x^2 - 6x + 8)dx = 17/3.$

9. $\int_C xyz\,ds = \int_0^{\pi/2} (18t\sin t\cos t)\sqrt{4+9}\,dt = 18\sqrt{13}\int_0^{\pi/2} (t\sin t\cos t)dt$

$= 18\sqrt{13}\int_0^{\pi/2} (1/2)(t\sin 2t)dt = 9\sqrt{13}[(-t\cos 2t)/2 + (\sin 2t)/4]\Big|_0^{\pi/2} = 9\sqrt{13}\pi/4.$

11. $x = -t+1,\ y = 3t,\ z = 5t+1,\ 0 \le t \le 1.$ $\int_C xy^2 z\,ds =$

$\int_0^1 (1-t)(9t^2)(5t+1)\sqrt{1+9+25}\,dt = 9\sqrt{35}\int_0^1 (t^2 + 4t^3 - 5t^4)dt = 9\sqrt{35}/3 = 3\sqrt{35}.$

13. $\int_C x^3 y^2\,dz = \int_0^1 (8t^3)(t^4)(t^2)(2t)dt = \int_0^1 16t^{10}\,dt = 16/11.$

15. On C_1: $x = 0,\ y = t,\ z = t,\ 0 \le t \le 1,$
C_2: $x = t,\ y = t+1,\ z = 2t+1,\ 0 \le t \le 1,$
C_3: $x = 1,\ y = 2,\ z = t+3,\ 0 \le t \le 1.$

Then $\int_C z^2\,dx - z\,dy + 2y\,dz = \int_0^1 (0 - t + 2t)dt$

$+ \int_0^1 [(2t+1)^2 - (2t+1) + 2(t+1)(2)]dt$

Section 14.2

$$+ \int_0^1 (0 + 0 + 4)dt = (1/2) + [(4t^3/3 + 3t^2 + 4t)|_0^1 + 4 = 77/6.$$

17. $\vec{F}(\vec{r}(t)) = t^{10}\vec{i} - t^7\vec{j}, \vec{r}'(t) = 3t^2\vec{i} + 4t^3\vec{j}.$ $\int_C \vec{F} \cdot d\vec{r} = \int_0^1 (3t^{12} - 4t^{10})dt$
    $= 3/13 - 4/11 = -19/143.$

19. $\int_C \vec{F} \cdot d\vec{r} = \int_{-1}^1 (2t\vec{i} + t^2\vec{j} + 3t\vec{k}) \cdot (2\vec{i} + 3\vec{j} - 2t\vec{k})dt = \int_{-1}^1 (4t - 3t^2)dt = -2.$

21. $\int_C \vec{F} \cdot d\vec{r} = \int_0^1 <\sin t^3, \cos(-t^2), t^4> \cdot <3t^2, -2t, 1>dt$
    $= \int_0^1 (3t^2 \sin t^3 - 2t \cos t^2 + t^4)dt = -\cos t^3 - \sin t^2 + t^5/5|_0^1 = 6/5 - (\cos 1 + \sin 1).$

23. $x = 4\cos t, y = 4\sin t, -\pi/2 \le t \le \pi/2.$

    $m = \int_C k\,ds = k\int_{-\pi/2}^{\pi/2} (\sqrt{4\cos^2 t + 4\sin^2 t})dt = 2k(\pi),$

    $\bar{x} = \frac{1}{2\pi k}\int_C xk\,dS = \frac{1}{2\pi}\int_{-\pi/2}^{\pi/2} 4\cos t\,dt = \frac{1}{2\pi}(4\sin t)|_{-\pi/2}^{\pi/2} = 4/\pi,$

    $\bar{y} = \frac{1}{2\pi k}\int_C yk\,dS = \frac{1}{2\pi}\int_{-\pi/2}^{\pi/2} 4\sin t\,dt = 0.$ Hence $(\bar{x}, \bar{y}) = (4/\pi, 0).$

25. (a) $\bar{x} = (1/m)\int_C x\rho(x,y,z)\,ds, \bar{y} = (1/m)\int_C y\rho(x,y,z)\,ds,$
    $\bar{z} = (1/m)\int_C z\rho(x,y,z)\,ds$ where $m = \int_C \rho(x,y,z)\,ds.$

    (b) $m = \int_C k\,ds = k\int_0^{2\pi}\sqrt{4\sin^2 t + 4\cos^2 t + 9}\,dt = k\sqrt{13}\int_0^{2\pi} dt = 2\pi k\sqrt{13},$

    $\bar{x} = \frac{1}{2\pi k\sqrt{13}}\int_0^{2\pi} k2\sqrt{13}\sin t\,dt = 0, \bar{y} = \frac{1}{2\pi k\sqrt{13}}\int_0^{2\pi} k2\sqrt{13}\cos t\,dt = 0,$

    $\bar{z} = \frac{1}{2\pi k\sqrt{13}}\int_0^{2\pi}(k\sqrt{13})(3t)dt = \frac{3}{2\pi}(2\pi^2) = 3\pi.$ Hence $(\bar{x}, \bar{y}, \bar{z}) = (0, 0, 3\pi).$

27. From Example 3, $\rho(x,y) = k(1-y), x = \cos t, y = \sin t,$ and

    $ds = dt, 0 \le t \le \pi \Rightarrow I_x = \int_C y^2 \rho(x,y)\,ds = \int_0^\pi \sin^2 t[k(1-\sin t)]\,dt$
    $= k\int_0^\pi (\sin^2 t - \sin^3 t)\,dt = \frac{k}{2}\int_0^\pi (1 - \cos 2t)\,dt - k\int_0^\pi (1 - \cos^2 t)\sin t\,dt$

    Let $u = \cos t, du = -\sin t\,dt$ in the second integral to get

    $I_x = k\left[\frac{\pi}{2} + \int_1^{-1}(1-u^2)\,du\right] = k\left[\frac{\pi}{2} - \frac{4}{3}\right].$

    $I_y = \int_C x^2 \rho(x,y)\,ds = k\int_0^\pi \cos^2 t(1-\sin t)\,dt = \frac{k}{2}\int_0^\pi (1+\cos 2t)\,dt - k\int_0^\pi \cos^2 t \sin t\,dt.$

    Let $u = \cos t, du = -\sin t\,dt$ in the second integral to get $I_y = k\left[\frac{\pi}{2} - \frac{2}{3}\right].$

29. $W = \int_C \vec{F} \cdot d\vec{r} = \int_0^{2\pi} <t - \sin t, 3 - \cos t> \cdot <1 - \cos t, \sin t>dt$

Section 14.3

$$= \int_0^{2\pi}(t - t\cos t - \sin t + \sin t\cos t + 3\sin t - \sin t\cos t)dt = 2\pi^2.$$

31. $W = \int_0^1 <t^6, -t^5, -t^7> \cdot <2t, -3t^2, 4t^3>dt = \int_0^1 (5t^7 - 4t^{10})dt$
    $= (5/8) - (4/11) = 23/88.$

33. Let $\vec{F} = 185\vec{k}$. To parametrize the staircase, let $x = 20\cos t$, $y = 20\sin t$,
    $z = \frac{90}{6\pi}t = \frac{15}{\pi}t$, $0 \le t \le 6\pi \Rightarrow W = \int_C \vec{F} \cdot d\vec{r}$
    $= \int_0^{6\pi} <0, 0, 185> \cdot <-20\sin t, 20\cos t, \frac{15}{\pi}> dt$
    $= (185)\frac{15}{\pi}\int_0^{6\pi} dt = (185)(90) \approx 1.67 \times 10^4$ ft-lb.

35. Use the orientation pictured in the figure. Then since $\vec{B}$ is tangent to any circle that lies in the plane perpendicular to the wire, $\vec{B} = |\vec{B}|\vec{T}$ where $\vec{T}$ is the unit tangent to the circle, C: $x = r\cos\theta$, $y = r\sin\theta$. Thus $\vec{B} = |\vec{B}|<-\sin\theta, \cos\theta>$. Then
    $\int_C \vec{B} \cdot d\vec{r} = \int_0^{2\pi} |\vec{B}|<-\sin\theta, \cos\theta> \cdot <-r\sin\theta, r\cos\theta> d\theta = \int_0^{2\pi} |\vec{B}|r\, d\theta = 2\pi r|\vec{B}|.$ (Note $|\vec{B}|$ here is the magnitude of the field at a distance r from the wire center.) But by Ampere's Law $\int_C \vec{B} \cdot d\vec{r} = \mu_0 I$. Hence $|\vec{B}| = \mu_0 I/(2\pi r)$.

EXERCISES 14.3

1. $\partial(2x - 3y)/\partial y = -3 = \partial(2y - 3x)/\partial x$ and the domain of $\vec{F}$ is R^2 which is open and simply-connected, so $\vec{F}$ is conservative. Thus there exists f such that $\nabla f = \vec{F}$, i.e., $f_x(x,y) = 2x - 3y$ and $f_y(x,y) = 2y - 3x$. But $f_x(x,y) = 2x - 3y$ implies $f(x,y) = x^2 - 3yx + g(y)$ and differentiating both sides of this equation with respect to y gives $f_y(x,y) = -3x + g'(y)$. Thus $2y - 3x = -3x + g'(y)$ so $g'(y) = 2y$ and $g(y) = y^2 + K$ where K is a constant. Hence $f(x,y) = x^2 - 3xy + y^2 + K$ is a potential for $\vec{F}$.

3. $\partial(x^2 + y)/\partial y = 1$, $\partial(x^2)/\partial x = 2x$ and these are not equal, so $\vec{F}$ is not conservative.

5. $\partial(1 + 4x^3y^3)/\partial y = 12x^3y^2 = \partial(3x^4y^2)/\partial x$ and the domain of $\vec{F}$ is R^2 which is open and simply-connected. Thus $\vec{F}$ is conservative so there exists f such that $\nabla f = \vec{F}$. Then $f_x(x,y) = 1 + 4x^3y^3$ implies $f(x,y) = x + x^4y^3 + g(y)$ and $f_y(x,y) = 3x^4y^3 + g'(y)$. But $f_y(x,y) = 3x^4y^2$ implies $g(y) = K$. Hence a potential for $\vec{F}$ is $f(x,y) = x + x^4y^3 + K$.

7. $\partial(e^{2x} + x\sin y)/\partial y = x\cos y$, $\partial(x^2\cos y)/\partial x = 2x\cos y$, so $\vec{F}$ is not conservative.

Section 14.3

9. $\partial(ye^x + \sin y)/\partial y = e^x + \cos y = \partial(e^x + x\cos y)/\partial x$ and the domain of $\vec{F}$ is R^2. Hence $\vec{F}$ is conservative so there exists f such that $\nabla f = \vec{F}$. Then $f_x(x,y) = ye^x + \sin y$ implies $f(x,y) = ye^x + x\sin y + g(y)$ and $f_y(x,y) = e^x + x\cos y + g'(y)$. But $f_y(x,y) = e^x + x\cos y$ so $g(y) = K$ and $f(x,y) = ye^x + x\sin y + K$ is a potential for $\vec{F}$.

11. (a) $f_x(x,y) = x$ implies $f(x,y) = x^2/2 + g(y)$ and $f_y(x,y) = g'(y)$.
    But $f_y(x,y) = y$ so $g(y) = y^2/2 + K$ and $f(x,y) = x^2/2 + y^2/2 + K$ (*or set* $K = 0$).
    (b) $\int_C \vec{F}\cdot d\vec{r} = f(3,9) - f(-1,1) = 44$.

13. (a) $f_x(x,y) = 2xy^3$ implies $f(x,y) = x^2y^3 + g(y)$ and
    $f_y(x,y) = 3x^2y^2 + g'(y)$. But $f_y(x,y) = 3x^2y^2$ so $f(x,y) = x^2y^3$ (*setting* $K = 0$).
    (b) Since $\vec{r}(0) = <0,1>$ and $\vec{r}(\pi/2) = <1,(\pi^2+4)/4>$,
    $\int_C \vec{F}\cdot d\vec{r} = f(1,(\pi^2+4)/4) - f(0,1) = (\pi^2+4)^3/64$.

15. (a) $f_x(x,y,z) = y$ implies $f(x,y,z) = xy + g(y,z)$ and $f_y(x,y,z) = x + \partial g/\partial y$. But $f_y(x,y,z) = x + z$ so $\partial g/\partial y = z$ and $g(y,z) = yz + h(z)$. Thus $f(x,y,z) = xy + yz + h(z)$ and $f_z(x,y,z) = y + h'(z)$. But $f_z(x,y,z) = y$ so $h'(z) = 0$ or $h(z) = K$. Hence $f(x,y,z) = xy + yz$ (*letting* $K = 0$).
    (b) $\int_C \vec{F}\cdot d\vec{r} = f(8,3,-1) - f(2,1,4) = 21 - 6 = 15$.

17. (a) $f_x(x,y,z) = 2xz + \sin y$ implies $f(x,y,z) = x^2z + x\sin y + g(y,z)$ and
    $f_y(x,y,z) = x\cos y + g_y(y,z)$. But $f_y(x,y,z) = x\cos y$ so $g_y(y,z) = 0$ and
    $f(x,y,z) = x^2z + x\sin y + h(z)$.
    Thus $f_z(x,y,z) = x^2 + h'(z)$. But $f_z(x,y,z) = x^2$ so $h'(z) = 0$
    and $f(x,y,z) = x^2z + x\sin y$ (letting $K = 0$).
    (b) $\vec{r}(0) = <1,0,0>$, $\vec{r}(2\pi) = <1,0,2\pi>$. Thus $\int_C \vec{F}\cdot d\vec{r} = f(1,0,2\pi) - f(1,0,0) = 2\pi$.

19. Here $\vec{F}(x,y) = (2x\sin y)\vec{i} + (x^2\cos y - 3y^2)\vec{j}$. Then $f(x,y) = x^2\sin y - y^3$ is a potential for $\vec{F}$, i.e., $\nabla f = \vec{F}$ so $\vec{F}$ is conservative and thus its line integral is independent of path. Hence $\int_C 2x\sin y\, dx + (x^2\cos y - 3y^2)dy = \int_C \vec{F}\cdot d\vec{r} = f(5,1) - f(-1,0) = 25\sin 1 - 1$.

21. Here $\vec{F}(x,y) = x^2y^3\vec{i} + x^3y^2\vec{j}$. $W = \int_C \vec{F}\cdot d\vec{r}$. Since $\frac{\partial}{\partial y}(x^2y^3) = 3x^2y^2 = \frac{\partial}{\partial x}(x^3y^2)$, there exists f such that $\nabla f = \vec{F}$. In fact, $f_x = x^2y^3 \Rightarrow f(x,y) = \frac{1}{3}x^3y^3 + g(y) \Rightarrow f_y = x^3y^2 + g'(y)$ $\Rightarrow g'(y) = 0$, so we can take $f(x,y) = \frac{1}{3}x^3y^3$. Thus
$W = \int_C \vec{F}\cdot d\vec{r} = f(2,1) - f(0,0) = \frac{1}{3}(2^3)(1^3) - 0 = \frac{8}{3}$.

23. Since $\vec{F}$ is conservative, there exists a function f such that $\vec{F} = \nabla f$, that is, $P = f_x$, $Q = f_y$, and $R = f_z$. Since P, Q and R have continuous first order partial derivatives, Clairaut's Theorem says $\partial P/\partial y = f_{xy} = f_{yx} = \partial Q/\partial x$, $\partial P/\partial z = f_{xz} = f_{zx} = \partial R/\partial x$,

174

Section 14.4

and $\partial Q/\partial z = f_{yz} = f_{zy} = \partial R/\partial y$.

25. $D = \{(x,y) \mid x > 0, y > 0\}$ = the first quadrant (excluding the axes).
    (a) D is open because around every point in D we can put a disk that lies in D.
    (b) D is connected because the straight line segment joining any two points in D lies in D.
    (c) D is simply-connected because it's connected and has no holes.

27. $D = \{(x,y) \mid 1 < x^2 + y^2 < 4\}$ = the annular region between the circles with center $(0,0)$ and radii 1 and 2. (a) D is open. (b) D is connected. (c) D is not simply-connected. For example, $x^2 + y^2 = (1.5)^2$ is simple and closed and lies within D but encloses points that are not in D. [Or, D has a hole, so is not simply-connected.]

29. (a) $P = -y/(x^2+y^2)$, $\partial P/\partial y = (y^2-x^2)/(x^2+y^2)^2$ and $Q = x/(x^2+y^2)$, $\partial Q/\partial x = (y^2-x^2)/(x^2+y^2)$. Thus $\partial P/\partial y = \partial Q/\partial x$.

    (b) C_1: $x = \cos t$, $y = \sin t$, $0 \le t \le \pi$, C_2: $x = \cos t$, $y = \sin t$, $t = 2\pi$ to $t = \pi$. Then
    $$\int_{C_1} \vec{F} \cdot d\vec{r} = \int_0^\pi \{[(-\sin t)(-\sin t) + \cos t(\cos t)]/(\cos^2 t + \sin^2 t)\} dt$$
    $= \int_0^\pi dt = \pi$ and $\int_{C_2} \vec{F} \cdot d\vec{r} = \int_{2\pi}^\pi dt = -\pi$. Since these aren't equal, the line integral of $\vec{F}$ isn't independent of path. (Or $\int_{C_3} \vec{F} \cdot d\vec{r} = \int_0^{2\pi} dt = 2\pi$ where C_3 is the circle $x^2 + y^2 = 1$ and apply the contrapositive of Theorem 14.20.) This doesn't contradict Theorem 14.23, since the domain of $\vec{F}$, which is R^2 except the origin, isn't simply-connected.

EXERCISES 14.4

1. (a)

    $\oint_C x^2 y\, dx + xy^3\, dy$

    $= \oint_{C_1 + C_2 + C_3 + C_4} x^2 y\, dx + xy^3\, dy$

    $= \int_0^1 0\, dx + \int_0^1 y^3\, dy + \int_1^0 x^2\, dx + \int_1^0 0\, dy$

    $= (1/4) - (1/3) = -1/12$.

    (b) $\oint_C x^2 y\, dx + xy^3\, dy = \int_0^1 \int_0^1 (y^3 - x^2)\, dx\, dy = \int_0^1 [y^3 - (1/3)]\, dy$

    $= (1/4) - (1/3) = -1/12$.

Section 14.4

3. (a) [figure: y-axis, curves $x=x, y=x$ and $C_1: x=x, y=x^2$, $0 \leq x \leq 1$, point $(1,1)$, curve C_2]

$$\oint_C (x+2y)dx + (x-2y)dy$$
$$= \oint_{C_1+C_2}(x+2y)dx + (x-2y)dy$$
$$= \int_0^1 [x+2x^2 + (x-2x^2)(2x)]dx + \int_1^0 [3x + (-x)]dx$$
$$= \int_0^1 (x+4x^2-4x^3)dx + \int_1^0 2x\,dx$$
$$= (1/2 + 4/3 - 1) - 1 = -1/6.$$

(b) $\oint_C (x+2y)dx + (x-2y)dy = \int_0^1 \int_{x^2}^x (1-2)dy\,dx = \int_0^1 (x^2 - x)dx = 1/3 - 1/2$
$= -1/6.$

5. $\oint_C xy\,dx + y^5\,dy = \int_0^2 \int_0^{x/2} (0-x)dy\,dx = \int_0^2 (-x^2/2)dx = -4/3.$

7. $\int_0^1 \int_{y^2}^{\sqrt{y}} (2-1)dx\,dy = \int_0^1 (y^{1/2} - y^2)dy = 1/3.$

9. $\iint_D (0-0)dA = 0.$

11. $\iint_{\substack{0 \leq x^2+y^2 \leq 4 \\ y \geq 0}} (4x-x)dA = 3\int_0^\pi \int_0^2 r^2 \cos\theta\,dr\,d\theta = 0$ since $\int_0^\pi \cos\theta\,d\theta = 0$ or

$\iint_D 3x\,dA = 3M_y = 0.$

13. $\iint_D (2x-x)dA = \int_0^\pi \int_0^{\sin x} x\,dy\,dx = \int_0^\pi x \sin x\,dx$
$= -x\cos x + \sin x \big|_0^\pi = \pi.$

15. $\int_C \vec{F} \cdot d\vec{r} = \int_C (y^2 - x^2 y)dx + xy^2\,dy = \iint_{\substack{x^2+y^2 \leq 4 \\ 0 \leq y \leq x}} (y^2 - 2y + x^2)dA$

$= \int_0^{\pi/4} \int_0^2 (r^2 - 2r\sin\theta)r\,dr\,d\theta = \int_0^{\pi/4} [4 - (16/3)\sin\theta]d\theta$
$= 4\theta + (16/3)\cos\theta \big|_0^{\pi/4} = \pi + (8/3)(\sqrt{2} - 2).$

17. By Green's Theorem, $W = \int_C \vec{F} \cdot d\vec{r} = \int_C x(x+y)dx + xy^2\,dy = \iint_D (y^2 - x)dy\,dx$

where C is the path described in the question and D is the triangle bounded by C.

So $W = \int_0^1 \int_0^{1-x} (y^2 - x)dy\,dx = \int_0^1 \left[\frac{y^3}{3} - xy\right]_0^{1-x} dx = \int_0^1 \left(\frac{(1-x)^3}{3} - x(1-x)\right)dx$

$= \left[-\frac{(1-x)^4}{12} - \frac{x^2}{2} + \frac{x^3}{3}\right]_0^1 = (-\frac{1}{2} + \frac{1}{3}) - (-\frac{1}{12}) = -\frac{1}{12}.$

176

Section 14.4

19. $A = \oint_C x\,dy = \int_0^{2\pi}(\cos^3 t)(3\sin^2 t \cos t)dt = 3\int_0^{2\pi}(\cos^4 t \sin^2 t)dt$

    $= 3\{-(\sin t \cos^5 t)/6 + [(\sin t \cos^3 t)/4 + 3(\cos t \sin t)/8 + 3t/8]/6\}\Big|_0^{2\pi}$

    $= 3(6\pi/8)/6 = 3\pi/8.$

    (Or $\int_0^{2\pi}(\cos^4 t \sin^2 t)dt = \int_0^{2\pi}(1/8)[(1-\cos 4t)/2 + \sin^2 2t \cos 2t]dt = \pi/8.$)

21. (a) Using Equation 14.12, we write parametric equations of the line segment as

    $x = (1-t)x_1 + tx_2, \ y = (1-t)y_1 + ty_2, \ 0 \le t \le 1.$

    Then $dx = (x_2 - x_1)dt$ and $dy = (y_2 - y_1)dt$, so

    $\int_C x\,dy - y\,dx = \int_0^1 [(1-t)x_1 + tx_2](y_2 - y_1)dt + [(1-t)y_1 + ty_2](x_2 - x_1)dt$

    $= \int_0^1 \{x_1(y_2 - y_1) - y_1(x_2 - x_1) + t[(y_2 - y_1)(x_2 - x_1) - (x_2 - x_1)(y_2 - y_1)]\}dt$

    $= \int_0^1 (x_1 y_2 - x_2 y_1)dt = x_1 y_2 - x_2 y_1.$

    (b) We apply Green's Theorem to the path $C = C_1 \cup C_2 \cup \cdots \cup C_n$, where C_i is the line segment that joins (x_i, y_i) to (x_{i+1}, y_{i+1}) for $i = 1, 2, \ldots, n-1$, and C_n is the line segment that joins (x_n, y_n) to (x_1, y_1). From (14.31), $\frac{1}{2}\int_C x\,dy - y\,dx = \iint_D dA$, where D is the polygon bounded by C. Therefore area of polygon

    $= A(D) = \iint_D dA = \frac{1}{2}\int_C x\,dy - y\,dx$

    $= \frac{1}{2}\left(\int_{C_1} x\,dy - y\,dx + \int_{C_2} x\,dy - y\,dx + \cdots + \int_{C_{n-1}} x\,dy - y\,dx + \int_{C_n} x\,dy - y\,dx\right)$

    To evaluate these integrals we use the formula from (a) to get

    $A(D) = \frac{1}{2}[(x_1 y_2 - x_2 y_1) + (x_2 y_3 - x_3 y_2) + \cdots + (x_{n-1} y_n - x_n y_{n-1}) + (x_n y_1 - x_1 y_n)].$

    (c) $A = \frac{1}{2}[(0 \cdot 1 - 2 \cdot 0) + (2 \cdot 3 - 1 \cdot 1) + (1 \cdot 2 - 0 \cdot 3) + (0 \cdot 1 - (-1) \cdot 2) + (-1 \cdot 0 - 0 \cdot 1)]$

    $= \frac{1}{2}(0 + 5 + 2 + 2) = \frac{9}{2}.$

23. Here $A = \frac{1}{2}(1)(1) = \frac{1}{2}$ and $C = C_1 + C_2 + C_3$, where C_1: $x = x, y = 0, 0 \le x \le 1$; C_2: $x = x, y = 1 - x, x = 1$ to $x = 0$; C_3: $x = 0, y = 1$ to $y = 0$. Then

    $\bar{x} = \frac{1}{2A}\int_C x^2\,dy = \int_{C_1} x^2\,dy + \int_{C_2} x^2\,dy + \int_{C_3} x^2\,dy = 0 + \int_1^0 (x^2)(-dx) + 0 = \frac{1}{3}.$

    Similarly, $\bar{y} = -\frac{1}{2A}\int_C y^2\,dx = \int_{C_1} y^2\,dx + \int_{C_2} y^2\,dx + \int_{C_3} y^2\,dx$

    $= 0 + \int_1^0 (1-x)^2(-dx) + 0 = \frac{1}{3}.$ Therefore $(\bar{x}, \bar{y}) = (\frac{1}{3}, \frac{1}{3}).$

25. By Green's Theorem, $(-\rho/3)\oint_C y^3\,dx = (-\rho/3)\iint_D -3y^2\,dA = \iint_D y^2 \rho\,dA = I_x$ and

    $(\rho/3)\oint_C x^3\,dy = (\rho/3)\iint_D (3x^2)dA = \iint_D x^2 \rho\,dA = I_y.$

27. Since C is a simple closed path which doesn't pass through or enclose the origin, there exists an open region that doesn't contain the origin but does contain D. Thus $P = -y/(x^2 + y^2)$ and $Q = x/(x^2 + y^2)$ have continuous partials on this open region containing D and we can apply Green's Theorem. But by Exercise 29(a) of Section 14.3, $\partial P/\partial y = \partial Q/\partial x$ so $\oint_C \vec{F} \cdot d\vec{r} = \iint_D 0 \, dA = 0$.

29. Using the first part of (14.31) we have that $\iint_R dx\,dy = A(R) = \int_{\partial R} x\,dy$.
But $x = g(u,v)$, and $dy = \frac{\partial h}{\partial u} du + \frac{\partial h}{\partial v} dv$, and we orient ∂S by taking the positive direction to be that which corresponds, under the mapping, to the positive direction along $\partial R \Rightarrow \int_{\partial R} x\,dy = \int_{\partial S} g(u,v)\left(\frac{\partial h}{\partial u} du + \frac{\partial h}{\partial v} dv\right)$

$= \int_{\partial S} g(u,v) \frac{\partial h}{\partial u} du + g(u,v) \frac{\partial h}{\partial v} dv$

$= \pm \iint_S \left[\frac{\partial}{\partial u}\left(g(u,v) \frac{\partial h}{\partial v}\right) - \frac{\partial}{\partial v}\left(g(u,v) \frac{\partial h}{\partial u}\right)\right] dA$ (using Green's Theorem in the uv-plane)

$= \pm \iint_S \left(\frac{\partial g}{\partial u}\frac{\partial h}{\partial v} + g(u,v)\frac{\partial^2 h}{\partial u \partial v} - \frac{\partial g}{\partial v}\frac{\partial h}{\partial u} - g(u,v)\frac{\partial^2 h}{\partial v \partial u}\right) dA$ (using the Chain Rule)

$= \pm \iint_S \left(\frac{\partial x}{\partial u}\frac{\partial y}{\partial v} - \frac{\partial x}{\partial v}\frac{\partial y}{\partial u}\right) dA$ (by the equality of mixed partials)

$= \pm \iint_S \frac{\partial(x,y)}{\partial(u,v)} du\,dv$. The sign is chosen to be positive if the orientation that we gave to ∂S corresponds to the usual positive orientation, and it is negative otherwise. Either way, since $A(R)$ is positive, the sign chosen must be the same as the sign of $\frac{\partial(x,y)}{\partial(u,v)}$.

Therefore $A(R) = \iint_R dx\,dy = \iint_S \left|\frac{\partial(x,y)}{\partial(u,v)}\right| du\,dv$.

EXERCISES 14.5

1. (a) $\text{curl } \vec{F} = \nabla \times \vec{F} = \begin{vmatrix} \vec{i} & \vec{j} & \vec{k} \\ \partial/\partial x & \partial/\partial y & \partial/\partial z \\ x & y & z \end{vmatrix} = (0-0)\vec{i} + (0-0)\vec{j} + (0-0)\vec{k} = \vec{0}$.

   (b) $\text{div } \vec{F} = \nabla \cdot \vec{F} = \frac{\partial}{\partial x}(x) + \frac{\partial}{\partial y}(y) + \frac{\partial}{\partial z}(z) = 1 + 1 + 1 = 3$.

3. (a) $\text{curl } \vec{F} = \nabla \times \vec{F} = \begin{vmatrix} \vec{i} & \vec{j} & \vec{k} \\ \partial/\partial x & \partial/\partial y & \partial/\partial z \\ yz & xz & xy \end{vmatrix} = (x-x)\vec{i} + (y-y)\vec{j} + (z-z)\vec{k} = \vec{0}$.

Section 14.5

(b) $\text{div } \vec{F} = \nabla \cdot \vec{F} = \frac{\partial}{\partial x}(yz) + \frac{\partial}{\partial y}(xz) + \frac{\partial}{\partial z}(xy) = 0 + 0 + 0 = 0.$

5. (a) $\text{curl } \vec{F} = \nabla \times \vec{F} = \begin{vmatrix} \vec{i} & \vec{j} & \vec{k} \\ \partial/\partial x & \partial/\partial y & \partial/\partial z \\ 0 & xy & xyz \end{vmatrix} = xz\vec{i} - yz\vec{j} + y\vec{k}.$

(b) $\text{div } \vec{F} = \nabla \cdot \vec{F} = \frac{\partial}{\partial x}(0) + \frac{\partial}{\partial y}(xy) + \frac{\partial}{\partial z}(xyz) = 0 + x + xy = x(1+y).$

7. (a) $\nabla \times \vec{F} = \begin{vmatrix} \vec{i} & \vec{j} & \vec{k} \\ \partial/\partial x & \partial/\partial y & \partial/\partial z \\ e^{xz} & -2e^{yz} & 3xe^y \end{vmatrix} = (3xe^y + 2ye^{yz})\vec{i} + (xe^{xz} - 3e^y)\vec{j}$

(b) $\nabla \cdot \vec{F} = \frac{\partial}{\partial x}(e^{xz}) + \frac{\partial}{\partial y}(-2e^{yz}) + \frac{\partial}{\partial z}(3xe^y) = ze^{xz} - 2ze^{yz}.$

9. (a) $\text{curl } \vec{F} = \begin{vmatrix} \vec{i} & \vec{j} & \vec{k} \\ \partial/\partial x & \partial/\partial y & \partial/\partial z \\ xe^y & -ze^{-y} & y\ln z \end{vmatrix} = (e^{-y} + \ln z)\vec{i} - xe^y\vec{k}.$

(b) $\text{div } \vec{F} = \frac{\partial}{\partial x}(xe^y) + \frac{\partial}{\partial y}(-ze^{-y}) + \frac{\partial}{\partial z}(y\ln z) = e^y + ze^{-y} + (y/z).$

11. $\text{curl } \vec{F} = \begin{vmatrix} \vec{i} & \vec{j} & \vec{k} \\ \partial/\partial x & \partial/\partial y & \partial/\partial z \\ y & x & 1 \end{vmatrix} = \vec{0}$ and $\vec{F}$ is defined on all of R^3 with component

functions which have continuous partial derivatives, so by (14.35) $\vec{F}$ is conservtive. Thus there exists f such that $\vec{F} = \nabla f$. Then $f_x(x,y,z) = y$ implies $f(x,y,z) = xy + g(y,z)$ and $f_y(x,y,z) = x + g_y(y,z)$. But $f_y(x,y,z) = x$, so $g(y,z) = h(z)$ and $f(x,y,z) = xy + h(z)$. Thus $f_z(x,y,z) = h'(z)$ but $f_z(x,y,z) = 1$ so $h(z) = z + k$. Hence a potential for $\vec{F}$ is $f(x,y,z) = xy + z + k$.

13. $\text{curl } \vec{F} = \begin{vmatrix} \vec{i} & \vec{j} & \vec{k} \\ \partial/\partial x & \partial/\partial y & \partial/\partial z \\ yz & -z^2 & x^2 \end{vmatrix} = 2z\vec{i} + (y - 2x)\vec{j} - z\vec{k} \neq \vec{0}.$ Hence $\vec{F}$ isn't conservative.

15. $\text{curl } \vec{F} = \begin{vmatrix} \vec{i} & \vec{j} & \vec{k} \\ \partial/\partial x & \partial/\partial y & \partial/\partial z \\ \cos y & \sin x & \tan z \end{vmatrix} = (\cos x - \sin y)\vec{k} \neq \vec{0}.$ Hence $\vec{F}$ isn't conservative.

17. Since curl $\vec{F} = \begin{vmatrix} \vec{i} & \vec{j} & \vec{k} \\ \partial/\partial x & \partial/\partial y & \partial/\partial z \\ yz & y^2+xz & xy \end{vmatrix} = (x-x)\vec{i} + (y-y)\vec{j} + (z-z)\vec{k} = \vec{0}$, $\vec{F}$ is defined on R^3, and since the partial derivatives of the components of $\vec{F}$ are continuous, $\vec{F}$ is conservative. Thus there exists f such that $\nabla f = \vec{F}$. Then $f_x(x,y,z) = yz$ implies $f(x,y,z) = xyz + g(y,z)$ and $f_y(x,y,z) = xz + g_y(y,z)$. But $f_y(x,y,z) = xz + y^2$ so $g(y,z) = y^3/3 + h(z)$ and $f(x,y,z) = xyz + y^3/3 + h(z)$. Then $f_z(x,y,z) = xy + h'(z)$. But $f_z(x,y,z) = xy$ so $h(z) = k$. Hence $f(x,y,z) = xyz + y^3/3 + K$ is a potential for $\vec{F}$.

19. No. Assume there is such a $\vec{G}$, then $\text{div}(\text{curl }\vec{G}) = y^2 + z^2 + x^2 \neq 0$, which contradicts Theorem 14.38.

21. Curl $\vec{F} = \begin{vmatrix} \vec{i} & \vec{j} & \vec{k} \\ \partial/\partial x & \partial/\partial y & \partial/\partial z \\ f(x) & g(y) & h(z) \end{vmatrix} = (0-0)\vec{i} + (0-0)\vec{j} + (0-0)\vec{k} = \vec{0}$. Hence $\vec{F} = f(x)\vec{i} + g(y)\vec{j} + h(z)\vec{k}$ is irrotational.

For Exercises 23–29 let $\vec{F}(x,y,z) = P_1\vec{i} + Q_1\vec{j} + R_1\vec{k}$ and $\vec{G}(x,y,z) = P_2\vec{i} + Q_2\vec{j} + R_2\vec{k}$.

23. $\text{div}(F+G) = \partial(P_1+P_2)/\partial x + \partial(Q_1+Q_2)/\partial y + \partial(R_1+R_2)/\partial z$
$= (\partial P_1/\partial x + \partial Q_1/\partial y + \partial R_1/\partial z) + (\partial P_2/\partial x + \partial Q_2/\partial y + \partial R_3/\partial z) = \text{div }\vec{F} + \text{div }\vec{G}$.

25. $\text{div}(f\vec{F}) = \partial(fP_1)/\partial x + \partial(fQ_1)/\partial y + \partial(fR_1)/\partial z$
$= [f(\partial P_1/\partial x) + P_1(\partial f/\partial x)] + [f(\partial Q_1/\partial y) + Q_1(\partial f/\partial y)] + [f(\partial R_1/\partial z) + R_1(\partial f/\partial z)]$
$= f(\partial P_1/\partial x + \partial Q_1/\partial y + \partial R_1/\partial z) + <P_1, Q_1, R_1> \cdot <\partial f/\partial x, \partial f/\partial y, \partial f/\partial z>$
$= f\text{ div }\vec{F} + \vec{F}\cdot\nabla f$.

27. $\text{div}(\vec{F}\times\vec{G}) = \nabla\cdot(\vec{F}\times\vec{G}) = \begin{vmatrix} \partial/\partial x & \partial/\partial y & \partial/\partial z \\ P_1 & Q_1 & R_1 \\ P_2 & Q_2 & R_2 \end{vmatrix}$

$= (\partial/\partial x)\begin{vmatrix} Q_1 & R_1 \\ Q_2 & R_2 \end{vmatrix} - (\partial/\partial y)\begin{vmatrix} P_1 & R_1 \\ P_2 & R_2 \end{vmatrix} + (\partial/\partial z)\begin{vmatrix} P_1 & Q_1 \\ P_2 & Q_2 \end{vmatrix}$

$= [Q_1\partial R_2/\partial x + R_2\partial Q_1/\partial x - Q_2\partial R_1/\partial x - R_1\partial Q_2/\partial x]$

$- [P_1\partial R_2/\partial y + R_2\partial P_1/\partial y - P_2\partial R_1/\partial y - R_1\partial P_2/\partial y] + [P_1\partial Q_2/\partial z$
$+ Q_2\partial P_1/\partial z - P_2\partial Q_1/\partial z - Q_1\partial P_2/\partial z] = \{P_2[\partial R_1/\partial y - \partial Q_1/\partial z] + Q_2[\partial P_1/\partial z$
$- \partial R_1/\partial x] + R_2[\partial Q_1/\partial x - \partial P_1/\partial y]\} - \{P_1[\partial R_2/\partial y - \partial Q_2/\partial z] + Q_1[\partial P_2/\partial z$
$- \partial R_2/\partial x] + R_1[\partial Q_2/\partial x - \partial P_2/\partial y]\} = \vec{G}\cdot\text{curl }\vec{F} - \vec{F}\cdot\text{curl G}$.

Section 14.5

29. $\text{curl curl } \vec{F} = \nabla \times (\nabla \times \vec{F}) = \begin{vmatrix} \vec{i} & \vec{j} & \vec{k} \\ \partial/\partial x & \partial/\partial y & \partial/\partial z \\ \partial R_1/\partial y - \partial Q_1 \partial z & \partial P_1/\partial z - \partial R_1/\partial x & \partial Q_1/\partial x - \partial P_1/\partial y \end{vmatrix}$

$= (\partial^2 Q_1/\partial y \partial x - \partial^2 P_1/\partial y^2 - \partial^2 P_1/\partial z^2 + \partial^2 R_1/\partial z \partial x)\vec{i} + (\partial^2 R_1/\partial z \partial y - \partial^2 Q_1/\partial z^2$
$- \partial^2 Q_1/\partial x^2 + \partial^2 P_1/\partial x \partial y)\vec{j} + (\partial^2 P_1/\partial x \partial z - \partial^2 R_1/\partial x^2 - \partial^2 R_1/\partial y^2$
$+ \partial^2 Q_1/\partial y \partial z)\vec{k}$. Now let's consider grad div $\vec{F} - \nabla^2 \vec{F}$ and compare with
the above. grad div $\vec{F} - \nabla^2 \vec{F} = [(\partial^2 P_1/\partial x^2 + \partial^2 Q_1/\partial x \partial y + \partial^2 R_1/\partial x \partial z)\vec{i}$
$+ (\partial^2 P_1/\partial y \partial x + \partial^2 Q_1/\partial y^2 + \partial^2 R_1/\partial y \partial z)\vec{j} + (\partial^2 P_1/\partial z \partial x + \partial^2 Q_1/\partial z \partial y + \partial^2 R_1/\partial z^2)\vec{k}]$
$- [(\partial^2 P_1/\partial x^2 + \partial^2 P_1/\partial y^2 + \partial^2 P_1/\partial z^2)\vec{i} + (\partial^2 Q_1/\partial x^2 + \partial^2 Q_1/\partial y^2 + \partial^2 Q_1/\partial z^2)\vec{j}$
$+ (\partial^2 R_1/\partial x^2 + \partial^2 R_1/\partial y^2 + \partial^2 R_1/\partial z^2)\vec{k}] = (\partial^2 Q_1/\partial x \partial y + \partial^2 R_1/\partial x \partial z - \partial^2 P_1/\partial y^2$
$- \partial^2 P_1/\partial z^2)\vec{i} + (\partial^2 P_1/\partial y \partial x + \partial^2 R_1/\partial y \partial z - \partial^2 Q_1/\partial x^2 - \partial^2 Q_1/\partial z^2)\vec{j} + (\partial^2 P_1/\partial z \partial x$
$+ \partial^2 Q_1/\partial z \partial y - \partial^2 R_1/\partial x^2 - \partial^2 R_2/\partial y^2)\vec{k}$. Then applying Clairaut's Theorem to
reverse the order of differentiation in the second partial derivatives as needed and
comparing, we have curl curl $\vec{F}$ = grad div $\vec{F} - \nabla^2 F$ as desired.

31. (a) curl f = $\nabla \times$ f is meaningless since f is a scalar field
    (b) grad f is a vector field
    (c) div F is a scalar field
    (d) curl(grad f) is a vector field
    (e) grad F is meaningless
    (f) grad(div F) is a vector field
    (g) div(grad f) is a scalar field
    (h) grad(div f) is meaningless
    (i) curl(curl F) is a vector field
    (j) div(div F) is meaningless
    (k) (grad f) × (div F) is meaningless because div F is a scalar field
    (l) div(curl(grad f)) is a scalar field

33. $\nabla \cdot \vec{r} = \left(\frac{\partial}{\partial x}\vec{i} + \frac{\partial}{\partial y}\vec{j} + \frac{\partial}{\partial z}\vec{k}\right) \cdot (x\vec{i} + y\vec{j} + z\vec{k}) = 1 + 1 + 1 = 3$.

35. $\nabla(1/r) = \nabla(1/\sqrt{x^2 + y^2 + z^2}) = \frac{-[1/(2\sqrt{x^2 + y^2 + z^2})](2x)}{(x^2 + y^2 + z^2)}\vec{i}$

$- \frac{[1/(2\sqrt{x^2 + y^2 + z^2})](2y)}{(x^2 + y^2 + z^2)}\vec{j} - \frac{[1/(2\sqrt{x^2 + y^2 + z^2})](2z)}{(x^2 + y^2 + z^2)}\vec{k}$

$= -\frac{x\vec{i} + y\vec{j} + z\vec{k}}{(x^2 + y^2 + z^2)^{3/2}} = -\vec{r}/r^3$

181

Section 14.6

37. $\nabla \ln r = \nabla \ln(x^2+y^2+z^2)^{1/2} = \frac{1}{2}\nabla \ln(x^2+y^2+z^2)$

$= \frac{x}{x^2+y^2+z^2}\vec{i} + \frac{y}{x^2+y^2+z^2}\vec{j} + \frac{z}{x^2+y^2+z^2}\vec{k} = \frac{x\vec{i}+y\vec{j}+z\vec{k}}{x^2+y^2+z^2}$

$= \vec{r}/r^2$

39. By (14.40), $\oint_C f(\nabla g)\cdot \vec{n}\, ds = \iint_D \text{div}(f\nabla g)\, dA = \iint_D [f\,\text{div}(\nabla g) + \nabla g \cdot \nabla f]\, dA$

by Exercise 25. But $\text{div}(\nabla g) = \nabla^2 g$.

Hence $\iint_D f\nabla^2 g\, dA = \oint_C f(\nabla g)\cdot \vec{n}\, ds - \iint_D \nabla g \cdot \nabla f\, dA$.

41. (a) We know that $\omega = \frac{v}{d}$, and from the diagram

$\sin\theta = \frac{d}{r} \Rightarrow v = d\omega = (\sin\theta)r\omega = |\vec{\omega} \times \vec{r}|$. But $\vec{v}$ is perpendicular to both $\vec{w}$ and $\vec{r}$, so that $\vec{v} = \vec{w} \times \vec{r}$.

(b) From (a), $\vec{v} = \vec{w} \times \vec{r} = \begin{vmatrix} \vec{i} & \vec{j} & \vec{k} \\ 0 & 0 & \omega \\ x & y & z \end{vmatrix} = (0\cdot z - \omega y)\vec{i} + (\omega x - 0\cdot z)\vec{j} + (0\cdot y - x\cdot 0)\vec{k}$

$= -\omega y\vec{i} + \omega x\vec{j}$

(c) $\text{curl}\,\vec{v} = \vec{\nabla} \times \vec{v} = \begin{vmatrix} \vec{i} & \vec{j} & \vec{k} \\ \frac{\partial}{\partial x} & \frac{\partial}{\partial y} & \frac{\partial}{\partial z} \\ -\omega y & \omega x & 0 \end{vmatrix}$

$= [\frac{\partial}{\partial y}(0) - \frac{\partial}{\partial z}(\omega x)]\vec{i} + [\frac{\partial}{\partial z}(-\omega y) - \frac{\partial}{\partial x}(0)]\vec{j} + [\frac{\partial}{\partial x}(\omega x) - \frac{\partial}{\partial y}(-\omega y)]\vec{k}$

$= (\omega - (-\omega))\vec{k} = 2\omega\vec{k} = 2\vec{w}$.

EXERCISES 14.6

1. Letting x and y be the parameters, the parametric equations are

$x = x$, $y = y$, $z = \sqrt{1 - 3x^2 - 2y^2}$ where $-1/\sqrt{3} \le x \le 1/\sqrt{3}$ and $-1/\sqrt{2} \le y \le 1/\sqrt{2}$.
Then the vector equation of the surface is $\vec{r}(x,y) = x\vec{i} + y\vec{j} + \sqrt{1 - 3x^2 - 2y^2}\,\vec{k}$.

Alternate solution: Letting ϕ, θ be the parameters, the parametric equations are

$x = (1/\sqrt{3})\sin\phi\cos\theta$, $y = (1/\sqrt{2})\sin\phi\sin\theta$, $z = \cos\phi$ where $0 \le \phi \le \pi/2$ and $0 \le \theta \le 2\pi$. Note: there are many parametric representations of a given surface.

3. $x = x$, $y = 6 - 3x^2 - 2z^2$, $z = z$ where $3x^2 + 2z^2 \le 6$ since $y \ge 0$. Then the associated vector equation is $\vec{r}(x,y) = x\vec{i} + (6 - 3x^2 - 2z^2)\vec{j} + z\vec{k}$.

Section 14.6

5. Since the cone intersects the sphere in the circle $x^2 + y^2 = 2$, $z = 2$ and we want the portion of the sphere above this, we can parametrize the surface as $x = x$, $y = y$, $z = \sqrt{4 - x^2 - y^2}$ where $2 \leq x^2 + y^2 \leq 4$. Or using spherical coordinates $x = 2\sin\phi\cos\theta$, $y = 2\sin\phi\sin\theta$, $z = 2\cos\phi$ where $0 \leq \phi \leq \pi/4$ and $0 \leq \theta \leq 2\pi$.

7. The surface is a disc of radius 4 and center $(0,0,5)$. Thus $x = r\cos\theta$, $y = r\sin\theta$, $z = 5$ where $0 \leq r \leq 4$, $0 \leq \theta \leq 2\pi$ is a parametric representation of the surface. Or in rectangular coordinates we could represent the surface as $x = x$, $y = y$, $z = 5$ where $0 \leq x^2 + y^2 \leq 16$.

9. $\vec{r}_u = \vec{i} + 6u\vec{j} + \vec{k}$, $\vec{r}_v = \vec{i} - \vec{k}$, then $\vec{r}_u \times \vec{r}_v = -6u\vec{i} + 2\vec{j} - 6u\vec{k}$. Since the point $(2,3,0)$ corresponds to $u = 1$, $v = 1$, a normal vector to the surface at $(2,3,0)$ is $-6\vec{i} + 2\vec{j} - 6\vec{k}$ and the equation of the tangent plane is $-6x + 2y - 6z = -6$ or $3x - y + 3z = 3$.

11. $\vec{r}_u = <v, e^v, ve^u>$, $\vec{r}_v = <u, ue^v, e^u>$, and $\vec{r}_u \times \vec{r}_v = e^{u+v}(1 - uv)\vec{i} + e^u(uv - v)\vec{j} + e^v(uv - u)\vec{k}$. The point $(0,0,0)$ corresponds to $u = 0$, $v = 0$. Thus a normal vector to the surface at $(0,0,0)$ is $\vec{i}$ and the equation of the tangent plane is $x = 0$.

13. Here $z = f(x,y) = 4 - x - 2y$ with $0 \leq x^2 + y^2 \leq 4$. Thus by (14.48)
$$A(S) = \iint_D \sqrt{1 + (-1)^2 + (-2)^2}\, dA = \sqrt{6} \iint_{x^2+y^2 \leq 4} dA = 4\sqrt{6}\pi.$$

15. $z = f(x,y) = y^2 - x^2$ with $1 \leq x^2 + y^2 \leq 4$. Then $A(S) = \iint_D \sqrt{1 + 4x^2 + 4y^2}\, dA$
$= \int_0^{2\pi} \int_1^2 \sqrt{1 + 4r^2}\, r\, dr\, d\theta = 4\pi(1 + 4r^2)^{3/2}/24\big|_1^2 = (\pi/6)(17\sqrt{17} - 5\sqrt{5})$.

17. A parametric representation of the surface is $x = x$, $y = 4x + z^2$, $z = z$ with $0 \leq x \leq 1$, $0 \leq z \leq 1$. Hence $\vec{r}_x \times \vec{r}_z = 4\vec{i} - \vec{j} + 2z\vec{k}$.
(Note: in general if $y = f(x,z)$ then $\vec{r}_z \times \vec{r}_x = -(\partial f/\partial x)\vec{i} + \vec{j} - (\partial f/\partial x)\vec{k}$, and
$A(S) = \iint_D \sqrt{1 + (\partial f/\partial x)^2 + (\partial f/\partial z)^2}\, dA$.)
Then $A(S) = \int_0^1 \int_0^1 \sqrt{17 + 4z^2}\, dx\, dz = \int_0^1 \sqrt{17 + 4z^2}\, dz$
$= (z\sqrt{17 + 4z^2} + (17/2)\ln|2z + \sqrt{4z^2 + 17}|)/2\big|_0^1 = (\sqrt{21})/2 + (17/4)[\ln(2 + \sqrt{21}) - \ln\sqrt{17}]$.

19. Let $A(S_1)$ be the surface area of that portion of the surface which lies above the plane $z = 0$, then $A(S) = 2A(S_1)$. Following Example 5, a parametric representation of S_1 is $x = a\sin\phi\cos\theta$, $y = a\sin\phi\sin\theta$, $z = a\cos\phi$ and $|\vec{r}_\phi \times \vec{r}_\theta| = a^2\sin\phi$.
For D, $0 \leq \phi \leq \pi/2$ and for each fixed ϕ $(x - a/2)^2 + y^2 \leq (a/2)^2$ or $[(a\sin\phi\cos\theta) - (a/2)]^2 + a^2\sin^2\phi\sin^2\theta \leq (a/2)^2$ implies $a^2\sin^2\phi - a^2\sin\phi\cos\theta \leq 0$ or $\sin\phi(\sin\phi - \cos\theta) \leq 0$. But $0 \leq \phi \leq \pi/2$, so $\cos\theta \geq \sin\phi$ or $\sin[(\pi/2) + \theta] \geq \sin\phi$ or

183

Section 14.6

$\phi - (\pi/2) \leq \theta \leq (\pi/2) - \phi$.

Hence $D = \{(\phi, \theta) | 0 \leq \phi \leq \pi/2, \phi - (\pi/2) \leq \theta \leq (\pi/2) - \phi\}$.

Then $A(S_1) = \int_0^{\pi/2} \int_{\phi - \pi/2}^{(\pi/2)-\phi} a^2 \sin\phi \, d\theta \, d\phi = a^2 \int_0^{\pi/2} (\pi - 2\phi) \sin\phi \, d\phi$

$= a^2[(-\pi\cos\phi) - 2(-\phi\cos\phi + \sin\phi)]_0^{\pi/2} = a^2(\pi - 2)$. Thus $A(S) = 2a^2(\pi - 2)$.

<u>Alternate Solution</u>: Working on S_1 we could parametrize the portion of the sphere by

$x = x, y = y, z = \sqrt{a^2 - x^2 - y^2}$. Then

$|\vec{r}_x \times \vec{r}_y| = \sqrt{1 + (x^2/(a^2 - x^2 - y^2)) + (y^2/(a^2 - x^2 - y^2))} = a/\sqrt{a^2 - x^2 - y^2}$ and

$A(S_1) = \iint\limits_{0 \leq (x-a/2)^2 + y^2 \leq (a/2)^2} (a/\sqrt{a^2 - x^2 - y^2}) dA = \int_{-\pi/2}^{\pi/2} \int_0^{a\cos\theta} (a/\sqrt{a^2 - r^2}) r \, dr \, d\theta$

$= \int_{-\pi/2}^{\pi/2} -a(a^2 - r^2)^{1/2} \Big|_0^{a\cos\theta} d\theta = \int_{-\pi/2}^{\pi/2} a^2[1 - (1 - \cos^2\theta)^{1/2}] d\theta$

$= \int_{-\pi/2}^{\pi/2} a^2(1 - |\sin\theta|) = 2a^2 \int_0^{\pi/2} (1 - \sin\theta) d\theta = 2a^2(\pi/2 - 1)$.

Thus $A(S) = 4a^2(\pi/2 - 1) = 2a^2(\pi - 2)$.

Notes: 1) Perhaps working in spherical coordinates is the most obvious approach here. However you must be careful in setting up D.

2) In the alternate solution, you can avoid having to use $|\sin\theta|$ by working in the first octant and then multiplying by 8. However, if you set up S_1 as above and arrived at $A(S_1) = a^2\pi$ you now see your error.

21. $\vec{r}_u = <v, 1, 1>, \vec{r}_v = <u, 1, -1>$ and $\vec{r}_u \times \vec{r}_v = <-2, u + v, v - u>$. Then

$A(S) = \iint\limits_{u^2 + v^2 \leq 1} \sqrt{4 + 2u^2 + 2v^2} \, dA = \int_0^{2\pi} \int_0^1 r\sqrt{4 + 2r^2} \, dr \, d\theta = 2\pi(4 + 2r^2)^{3/2}/6 \Big|_0^1$

$= (\pi/3)(6\sqrt{6} - 8) = \pi(2\sqrt{6} - 8/3)$.

23. (a) $\vec{r}_u = \cos v \vec{i} + \sin v \vec{j} + c\vec{k}, \vec{r}_v = -u\sin v \vec{i} + u\cos v \vec{j} + 0\vec{k}$, and

$\vec{r}_u \times \vec{r}_v = -cu\cos v \vec{i} - cu\sin v \vec{j} + u\vec{k} \Rightarrow A(S) = \iint\limits_D |\vec{r}_u \times \vec{r}_v| dA$

$= \int_0^{2\pi} \int_0^h \sqrt{c^2 u^2 + u^2} \, du \, dv = \sqrt{c^2 + 1} \int_0^{2\pi} \left[\frac{u^2}{2}\right]_0^h dv = \pi h^2 \sqrt{c^2 + 1}$

(b) $x = u\cos v, y = u\sin v, z = cu \Rightarrow x^2 + y^2 = (\frac{z}{c})^2 \Rightarrow z = c\sqrt{x^2 + y^2}$, a cone.

To find D, notice that $0 \leq u \leq h \Rightarrow 0 \leq z \leq c \cdot h \Rightarrow 0 \leq c\sqrt{x^2 + y^2} \leq c \cdot h$

$\Rightarrow 0 \leq x^2 + y^2 \leq h^2$. So D is a disk of radius h centered at the origin.

Section 14.7

Therefore $A(S) = \iint_D \sqrt{1+\left(\frac{\partial z}{\partial x}\right)^2 + \left(\frac{\partial z}{\partial y}\right)^2}\,dA$

$= \iint_D \sqrt{1+\frac{c^2x^2}{x^2+y^2} + \frac{c^2y^2}{x^2+y^2}}\,dA = \iint_D \sqrt{1+c^2}\,dA = \sqrt{1+c^2}\,A(D) = \pi h^2 \sqrt{1+c^2}.$

25.

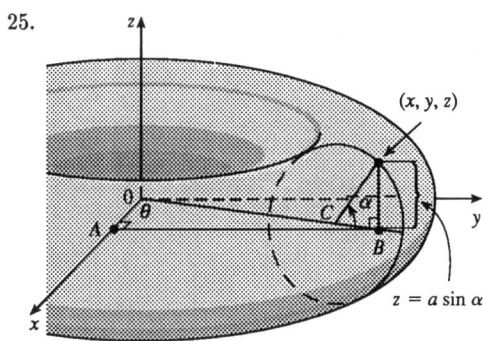

Here $z = a\sin\alpha$, $y = |AB|$, and $x = |OA|$. But $|OB| = |OC| + |CB| = b + a\cos\alpha$ and $\sin\theta = |AB|/|OB|$ so that
$y = |OB|\sin\theta = (b+a\cos\alpha)\sin\theta.$
Similarly $\cos\theta = |OA|/|OB|$ so
$x = (b+a\cos\alpha)\cos\theta.$
Hence a parametric representation for the torus is
$x = b\cos\theta + a\cos\alpha\cos\theta$, $y = b\sin\theta + a\cos\alpha\sin\theta$,
$z = a\sin\alpha$ where $0 \leq \alpha \leq 2\pi$, $0 \leq \theta \leq 2\pi$.

EXERCISES 14.7

1. $\vec{r}(x,y) = x\vec{i} + y\vec{j} + (6-3x-2y)\vec{k}$, $\vec{r}_x \times \vec{r}_y = 3\vec{i} + 2\vec{j} + \vec{k}$ (the normal to the plane) and $|\vec{r}_x \times \vec{r}_y| = \sqrt{14}$. The given plane meets the first octant in the line $3x + 2y = 6$, $z = 0$, $x \geq 0$, $y \geq 0$, so $D = \{(x,y) \mid 0 \leq x \leq (6-2y)/3, 0 \leq y \leq 3\}$.

   Then $\iint_S y\,dS = \int_0^3 \int_0^{(6-2y)/3} y\sqrt{14}\,dx\,dy = \sqrt{14}\int_0^3 (2y - 2y^2/3)\,dy = 3\sqrt{14}.$

3. $\vec{r}(x,z) = x\vec{i} + (x^2+4z)\vec{j} + z\vec{k}$, $0 \leq x \leq 2$, $0 \leq z \leq 2$, $|\vec{r}_x \times \vec{r}_z| = \sqrt{4x^2+17}$

   (see Exercise 14.6.17). Then $\iint_S x\,dS = \int_0^2 \int_0^2 x\sqrt{4x^2+17}\,dx\,dz = 2[(4x^2+17)^{3/2}/12]_0^2$
   $= (1/6)(33\sqrt{33} - 17\sqrt{17}).$

5. Since $z = y+3$, $|\vec{r}_x \times \vec{r}_y| = \sqrt{2}$ and $\iint_S yz\,dS = \iint_{x^2+y^2 \leq 1} \sqrt{2}y(y+3)\,dA$

   $= \sqrt{2}\int_0^{2\pi}\int_0^1 (r^2\sin^2\theta + 3r\sin\theta)r\,dr\,d\theta = \sqrt{2}\int_0^{2\pi} [(1/4)\sin^2\theta + \sin\theta]\,d\theta = \pi/(2\sqrt{2}).$

7. Using spherical coordinates and Example 14.6.5 we have
   $\vec{r}(\phi,\theta) = 2\sin\phi\cos\theta\,\vec{i} + 2\sin\phi\sin\theta\,\vec{j} + 2\cos\phi\,\vec{k}$ and $|\vec{r}_\phi \times \vec{r}_\theta| = 4\sin\phi.$

Section 14.7

Then $\iint_S (x^2z + y^2z)dS = \int_0^{2\pi}\int_0^{\pi/2}(4\sin^2\phi)(2\cos\phi)(4\sin\phi)d\phi\,d\theta = 16\pi\sin^4\phi\big|_0^{\pi/2} = 16\pi.$

9. Using cylindrical coordinates $\vec{r}(\theta, z) = 3\cos\theta\vec{i} + 3\sin\theta\vec{j} + z\vec{k}$, $0 \le \theta \le 2\pi$, $0 \le z \le 2$,

and $|\vec{r}_\theta \times \vec{r}_z| = 3$. $\iint_S (x^2y + z^2)dS = \int_0^{2\pi}\int_0^2 (27\cos^2\theta\sin\theta + z^2)3\,dz\,d\theta$

$= \int_0^{2\pi}(162\cos^2\theta\sin\theta + 8)d\theta = 16\pi.$

11. $\vec{r}(u, v) = uv\vec{i} + (u + v)\vec{j} + (u - v)\vec{k}$, $u^2 + v^2 \le 1$ and $|\vec{r}_u \times \vec{r}_v| = \sqrt{4 + 2u^2 + 2v^2}$

(see Exercise 14.6.21). Then $\iint_S yx\,dS = \iint_{u^2+v^2 \le 1}(u^2 - v^2)\sqrt{4 + 2u^2 + 2v^2}\,dA$

$= \int_0^{2\pi}\int_0^1 r^2(\cos^2\theta - \sin^2\theta)\sqrt{4 + 2r^2}\,r\,dr\,d\theta$

$= \int_0^{2\pi}(\cos^2\theta - \sin^2\theta)d\theta \int_0^1 r^3\sqrt{4 + 2r^2}\,dr = 0$ since the first integral is zero.

13. $\vec{F}(\vec{r}(x,y)) = e^y\vec{i} + ye^x\vec{j} + x^2y\vec{k}$ and $\vec{r}_x \times \vec{r}_y = -2x\vec{i} - 2y\vec{j} + \vec{k}$. Then
$\vec{F}(\vec{r}(x,y))\cdot(\vec{r}_x \times \vec{r}_y) = -2xe^y - 2y^2e^x + x^2y$ and

$\iint_S \vec{F}\cdot d\vec{S} = \int_0^1\int_0^1(-2xe^y - 2y^2e^x + x^2y)dx\,dy = \int_0^1(-e^y - 2ey^2 + y/3 + 2y^2)dy$
$= (11 - 10e)/6.$

15. As in Exercise 1, $D = \{(x, y) | 0 \le x \le 2, 0 \le y \le \frac{6-3x}{2}\}.$

$\iint_S \vec{F}\cdot d\vec{S} = \int_0^2\int_0^{(6-3x)/2}[x\vec{i} + xy\vec{j} + x(6 - 3x - 2y)\vec{k}]\cdot(3\vec{i} + 2\vec{j} + \vec{k})dy\,dx$

$= \int_0^2\int_0^{(6-3x)/2}(9x - 3x^2)dx\,dy = \int_0^2[27x - (45/2)x^2 + (9/2)x^3]dx = 12.$

17. $\vec{F}(\vec{r}(\phi, \theta)) = 3\sin\phi\cos\theta\vec{i} + 3\sin\phi\sin\theta\vec{j} + 3\cos\phi\vec{k}$ and

$\vec{r}_\phi \times \vec{r}_\theta = 9\sin^2\phi\cos\theta\vec{i} + 9\sin^2\phi\sin\theta\vec{j} + 9\sin\phi\cos\phi\vec{k}$. Then

$\vec{F}(\vec{r}(\phi,\theta))\cdot(\vec{r}_\phi \times \vec{r}_\theta) = 27\sin^3\phi\cos^2\theta + 27\sin^3\phi\sin^2\theta + 27\sin\phi\cos^2\phi = 27\sin\phi$

and $\iint_S \vec{F}\cdot d\vec{S} = \int_0^{2\pi}\int_0^\pi 27\sin\phi\,d\phi\,d\theta = (2\pi)(54) = 108\pi.$

19. Let S_1 be the paraboloid $y = x^2 + z^2$, $0 \le y \le 1$ and S_2 the disc $x^2 + z^2 \le 1$, $y = 1$.
Since S is a closed surface, we use the outward orientation.
On S_1: $\vec{F}(\vec{r}(x,z)) = (x^2 + z^2)\vec{j} - z\vec{k}$ and $\vec{r}_x \times \vec{r}_z = 2x\vec{i} - \vec{j} + 2z\vec{k}$ (since the
$\vec{j}$-component must be negative on S_1). Then

$\iint_{S_1}\vec{F}\cdot d\vec{S} = \iint_{x^2+z^2 \le 1}[-(x^2 + z^2) - 2z^2]dA = -\int_0^{2\pi}\int_0^1 (r^2 + 2r^2\cos^2\theta)r\,dr\,d\theta$

Section 14.7

$$= -\int_0^{2\pi} (1 + 2\cos^2\theta)/4 \, d\theta = -(\pi/2 + \pi/2) = -\pi.$$ On S_2: $\vec{F}(\vec{r}(x,z)) = \vec{j} - z\vec{k}$ and $\vec{r}_z \times \vec{r}_x = \vec{j}$. Then $\iint_{S_2} \vec{F} \cdot d\vec{S} = \iint_{x^2+z^2 \leq 1} (1) dA = \pi.$ Hence $\iint_S \vec{F} \cdot d\vec{S} = -\pi + \pi = 0.$

21.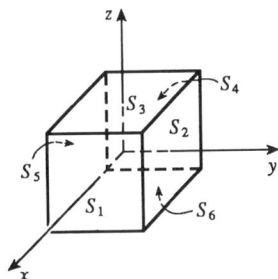

Here S consists of the 6 faces of the cube as labeled in the figure to the left. On
S_1: $\vec{F} = \vec{i} + 2y\vec{j} + 3z\vec{k}$, $\vec{r}_y \times \vec{r}_z = \vec{i}$ and
$$\iint_{S_1} \vec{F} \cdot d\vec{S} = \int_{-1}^1 \int_{-1}^1 dy \, dz = 4;$$
S_2: $\vec{F} = x\vec{i} + 2\vec{j} + 3z\vec{k}$, $\vec{r}_z \times \vec{r}_x = \vec{j}$ and
$$\iint_{S_2} \vec{F} \cdot d\vec{S} = \int_{-1}^1 \int_{-1}^1 2 \, dx \, dz = 8;$$

S_3: $\vec{F} = x\vec{i} + 2y\vec{j} + 3\vec{k}$, $\vec{r}_x \times \vec{r}_y = \vec{k}$ and $\iint_{S_3} \vec{F} \cdot d\vec{S} = \int_{-1}^1 \int_{-1}^1 3 \, dx \, dy = 12;$

S_4: $\vec{F} = -\vec{i} + 2y\vec{j} + 3z\vec{k}$, $\vec{r}_z \times \vec{r}_y = -\vec{i}$ and $\iint_{S_4} \vec{F} \cdot d\vec{S} = 4;$

S_5: $\vec{F} = x\vec{i} - 2\vec{j} + 3z\vec{k}$, $\vec{r}_x \times \vec{r}_z = -\vec{j}$ and $\iint_{S_5} \vec{F} \cdot d\vec{S} = 8;$

S_6: $\vec{F} = x\vec{i} + 2y\vec{j} - 3\vec{k}$, $\vec{r}_y \times \vec{r}_x = -\vec{k}$ and $\iint_{S_6} \vec{F} \cdot d\vec{S} = \int_{-1}^1 \int_{-1}^1 3 \, dx \, dy = 12;$

Hence $\iint_S \vec{F} \cdot d\vec{S} = \sum_{i=1}^6 \iint_{S_i} \vec{F} \cdot d\vec{S} = 48.$

23. If S is given by $y = h(x, z)$, then S is also the level surface $f(x, y, z) = y - h(x, z) = 0.$

$$\vec{n} = \frac{\nabla f(x,y,z)}{|\nabla f(x,y,z)|} = \frac{-h_x\vec{i} + \vec{j} - h_z\vec{k}}{\sqrt{h_x^2 + 1 + h_z^2}},$$ and we take the negative of this to get the unit

normal that points to the left. Now using (14.51) to evaluate the surface integral in

(14.56) as in the derivation of (14.57), we get $\iint_S \vec{F} \cdot d\vec{S} = \iint_S \vec{F} \cdot \vec{n} \, d\vec{S}$

$$= \iint_D (P\vec{i} + Q\vec{j} + R\vec{k}) \cdot \frac{\frac{\partial h}{\partial x}\vec{i} - \vec{j} + \frac{\partial h}{\partial z}\vec{k}}{\sqrt{\left(\frac{\partial h}{\partial x}\right)^2 + 1 + \left(\frac{\partial h}{\partial z}\right)^2}} \sqrt{\left(\frac{\partial h}{\partial x}\right)^2 + 1 + \left(\frac{\partial h}{\partial z}\right)^2} \, dA$$

where D is the projection of $f(x, y, z)$ onto the xz−plane,

$$\Rightarrow \iint_S \vec{F} \cdot d\vec{S} = \iint_D \left(P\frac{\partial h}{\partial x} - Q + R\frac{\partial h}{\partial z}\right) dA.$$

25. $m = \iint\limits_S K\,dS = K(4\pi a^2/2) = 2\pi a^2 K$, by symmetry $M_{xz} = M_{yz} = 0$, and

$$M_{xy} = \iint\limits_S zK\,dS = K\int_0^{2\pi}\int_0^{\pi/2}(a\cos\phi)(a^2\sin\phi)d\phi\,d\theta = 2\pi K a^3[-(\cos 2\phi)/4]_0^{\pi/2}$$

$= \pi K a^3$. Hence $(\bar{x},\bar{y},\bar{z}) = (0,0,a/2)$.

27. (a) $I_z = \iint\limits_S (x^2+y^2)\rho(x,y,z)dS$.

(b) $I_z = \iint\limits_S (x^2+y^2)(10-\sqrt{x^2+y^2})dS = \iint\limits_{1\leq x^2+y^2\leq 16}(x^2+y^2)(10-\sqrt{x^2+y^2})\sqrt{2}\,dA$

$= \int_0^{2\pi}\int_1^4 \sqrt{2}(10r^3-r^4)dr\,d\theta = 2\sqrt{2}\pi(4329/10) = 4329\sqrt{2}\pi/5$.

29. $\rho(x,y,z) = 1200$, $\vec{V} = y\vec{i}+\vec{j}+z\vec{k}$, $\vec{F} = \rho\vec{V} = (1200)(y\vec{i}+\vec{j}+z\vec{k})$. S is given by $\vec{r}(x,y) = x\vec{i}+y\vec{j}+[9-(x^2+y^2)/4]\vec{k}$, $0 \leq x^2+y^2 \leq 36$ and $\vec{r}_x \times \vec{r}_y = (x/2)\vec{i}+(y/2)\vec{j}+\vec{k}$. Thus the rate of flow is given by

$$\iint\limits_S \vec{F}\cdot d\vec{S} = \iint\limits_{0\leq x^2+y^2\leq 36}(1200)((xy/2)+(y/2)+[9-(x^2+y^2)/4])dA$$

$= 1200\int_0^6\int_0^{2\pi}[(r^2\sin\theta\cos\theta)/2+(r\sin\theta)/2+9-(r^2/4)]r\,d\theta\,dr$

$= 1200\int_0^6 2\pi(9r-r^3/4)dr = (1200)(2\pi)(81) = 194400\pi$.

31. S consists of the hemisphere S_1 given by $z = \sqrt{a^2-x^2-y^2}$ and the disk S_2 given by $0 \leq x^2+y^2 \leq a^2$, $z = 0$. On S_1: $\vec{E} = a\sin\phi\cos\theta\vec{i}+a\sin\phi\sin\theta\vec{j}+2a\cos\phi\vec{k}$, $T_\phi \times T_\theta = a^2\sin^2\phi\cos\theta\vec{i}+a^2\sin^2\phi\sin\theta\vec{j}+a^2\sin\phi\cos\phi\vec{k}$.
Thus $\iint\limits_{S_1}\vec{E}\cdot d\vec{S} = \int_0^{2\pi}\int_0^{\pi/2}(a^3\sin^3\phi+2a^3\sin\phi\cos^2\phi)d\phi\,d\theta$

$= \int_0^{2\pi}\int_0^{\pi/2}(a^3\sin\phi+a^3\sin\phi\cos^2\phi)d\phi\,d\theta = (2\pi)a^3(1+(1/3)) = 8\pi a^3/3$.

On S_2: $\vec{E} = x\vec{i}+y\vec{j}$ and $\vec{r}_y \times \vec{r}_x = -\vec{k}$ so $\iint\limits_{S_2}\vec{E}\cdot d\vec{S} = 0$. Hence the total charge is $q = \epsilon_o\iint\limits_S \vec{E}\cdot d\vec{S} = 8\pi a^3\epsilon_o/3$.

33. $K\nabla u = 6.5(4y\vec{j}+4z\vec{k})$. S is given by $\vec{r}(x,\theta) = x\vec{i}+\sqrt{6}\cos\theta\vec{j}+\sqrt{6}\sin\theta\vec{k}$ and since we want the inward heat flow use $\vec{r}_x \times \vec{r}_\theta = -\sqrt{6}\cos\theta\vec{j}-\sqrt{6}\sin\theta\vec{k}$. Then the rate of heat flow inward is given by

$$\iint\limits_S(-K\nabla u)\cdot d\vec{S} = \int_0^{2\pi}\int_0^4 -(6.5)(-24)dx\,d\theta = (2\pi)(156)(4) = 1248\pi.$$

Section 14.8

EXERCISES 14.8

1. The boundary curve is C: $x^2 + y^2 = 1$, $z = 0$ oriented in the counter–clockwise direction. The vector equation of C is $\vec{r}(t) = \cos t\vec{i} + \sin t\vec{j}$, $0 \le t \le 2\pi$. Then $\vec{F}(\vec{r}(t)) = \cos t\vec{j} + e^{\cos t \sin t}\vec{k}$ and $\vec{F}(\vec{r}(t)) \cdot \vec{r}'(t) = \cos^2 t$. Hence $\iint_S \text{curl } \vec{F} \cdot d\vec{S}$
$= \oint_C \vec{F} \cdot d\vec{r} = \int_0^{2\pi} \cos^2 t \, dt = \int_0^{2\pi} \tfrac{1}{2}(1 + \cos 2t) \, dt = \pi.$

3. C is the circle $x^2 + z^2 = 1$, $y = 0$ and the vector equation is $\vec{r}(t) = \cos t\vec{i} + \sin t\vec{k}$, $0 \le t \le 2\pi$ since the surface is oriented toward the xy-plane. Then $\vec{F}(\vec{r}(t)) = \cos^3 t\vec{k}$ and $\vec{F}(\vec{r}(t)) \cdot \vec{r}'(t) = \cos^4 t$. Hence
$\iint_S \text{curl } \vec{F} \cdot d\vec{S} = \oint_C \vec{F} \cdot d\vec{r} = \int_0^{2\pi} \cos^4 t \, dt = \int_0^{2\pi} [3/8 + (\cos 2t)/2 + (\cos 4t)/8] dt = 3\pi/4.$

5. C is the square in the plane $z = -1$. By (14.63) $\iint_{S_1} \text{curl } \vec{F} \cdot d\vec{S} = \oint_C \vec{F} \cdot d\vec{r} = \iint_{S_2} \text{curl } \vec{F} \cdot d\vec{S}$
where S_1 is the original cube without the bottom and S_2 is the bottom face of the cube.
$\text{curl } \vec{F} = x^2 z\vec{i} + (xy - 2xyz)\vec{j} + (y - xz)\vec{k}$. For S_2, $\vec{n} = -\vec{k}$ and
$\text{curl } \vec{F} \cdot \vec{n} = xz - y \equiv -x - y$ on S_2 ($z = -1$).
Then $\iint_{S_2} \text{curl } \vec{F} \cdot d\vec{S} = -\int_{-1}^1 \int_{-1}^1 (x+y) dx \, dy = 0$ so $\iint_{S_1} \text{curl } \vec{F} \cdot d\vec{S} = 0.$

7. $\text{curl } \vec{F} = 3x\vec{i} + (x-3y)\vec{j} + 2y\vec{k}$, $\vec{n} = (3\vec{i} + \vec{j} + \vec{k})/\sqrt{11}$ and $\oint_C \vec{F} \cdot d\vec{r} = \iint_S \text{curl } \vec{F} \cdot \vec{n} \, dS$
$= \int_0^1 \int_0^{3-3x} [[9x + (x-3y) + 2y]/\sqrt{11}](\sqrt{11}) dy \, dx = \int_0^1 \int_0^{3-3x} (10x - y) dy \, dx$
$= \int_0^1 [10(3x - 3x^2) - (3-3x)^2/2] dx = 15x^2 - 10x^3 + 3(1-x^3)/2 \big|_0^1 = 7/2.$

9. The curve of intersection is an ellipse in the plane $z = x + 4$ with unit normal
$\vec{n} = (-\vec{i} + \vec{k})/\sqrt{2}$ and $\text{curl } \vec{F} = 5\vec{i} + 2\vec{j} + 4\vec{k}$ so $\text{curl } \vec{F} \cdot \vec{n} = -1/\sqrt{2}.$
Then $\oint_C \vec{F} \cdot d\vec{r} = \iint_S (-1/\sqrt{2}) dS = (-1/\sqrt{2})$ (surface area of planar ellipse)
$= (-1/\sqrt{2})(\pi(2)(2\sqrt{2})) = -4\pi.$

11. S is the part of the surface $z = y^2 - x^2$ that lies above the unit disk D.
$\text{curl } \vec{F} = x\vec{i} - y\vec{j} + (x^2 - x^2)\vec{k} = x\vec{i} - y\vec{j}$. Using (14.57) with $g(x,y) = y^2 - x^2$, $P = x$, $Q = -y$,
we have $\int_C \vec{F} \cdot d\vec{r} = \iint_S \text{curl } \vec{F} \cdot d\vec{S} = \iint_D [-x(-2x) - (-y)(2y)] dA$
$= 2 \iint_D (x^2 + y^2) dA = 2 \int_0^{2\pi} \int_0^1 r^2 r \, dr \, d\theta = 2(2\pi)\left[\tfrac{1}{4} r^4\right]_0^1 = \pi.$

Section 14.8

13. The boundary curve C is the circle $x^2 + y^2 = 9$, $z = 0$ oriented in the counterclockwise direction as viewed from $(0,0,1)$. Then $\vec{r}(t) = 3\cos t\,\vec{i} + 3\sin t\,\vec{j}$, $0 \le t \le 2\pi$, $\vec{F}(\vec{r}(t)) = 9\sin t\,\vec{i} - 18\cos t\,\vec{k}$ and $\vec{F}\cdot\vec{r}'(t) = -27\sin^2 t$.

Thus $\oint_C \vec{F}\cdot d\vec{r} = \int_0^{2\pi} -27\sin^2 t\,dt = -27\pi$. Now curl $\vec{F} = -4\vec{i} + 6\vec{j} - 3\vec{k}$,

$\vec{r}_x \times \vec{r}_y = 2x\vec{i} + 2y\vec{j} + \vec{k}$ so $\iint_S \text{curl } \vec{F}\cdot d\vec{S} = \iint_{x^2+y^2\le 9}(-8x + 12y - 3)dA$

$= \int_0^{2\pi}\int_0^3 (-8r\cos\theta + 12r\sin\theta - 3)r\,dr\,d\theta = \int_0^3 (-3r)(2\pi)dr = -27\pi$.

15. The x-, y-, z-intercepts of the plane are all 1, so C consists of the three line segments
C_1: $\vec{r}_1(t) = (1-t)\vec{i} + t\vec{j}$, $0 \le t \le 1$, C_2: $\vec{r}_2(t) = (1-t)\vec{j} + t\vec{k}$, $0 \le t \le 1$,
and C_3: $\vec{r}_3(t) = t\vec{i} + (1-t)\vec{k}$, $0 \le t \le 1$.

Then $\oint_C \vec{F}\cdot d\vec{r} = \int_0^1 [t\vec{i} + (1-t)\vec{k}]\cdot(-\vec{i}+\vec{j})dt + \int_0^1 [(1-t)\vec{i} + t\vec{j}]\cdot(-\vec{j}+\vec{k})dt$

$+ \int_0^1 [(1-t)\vec{j} + t\vec{k}]\cdot(\vec{i}-\vec{k})dt = \int_0^1 -3t\,dt = -3/2$. Now curl $\vec{F} = -\vec{i} - \vec{j} - \vec{k}$ and

$\vec{r}_x \times \vec{r}_y = \vec{i} + \vec{j} + \vec{k}$. Hence $\iint_S \text{curl } \vec{F}\cdot d\vec{S} = \int_0^1\int_0^{1-x} -3\,dy\,dx = -3/2$.

17. $\text{curl } \vec{F} = \begin{vmatrix} \vec{i} & \vec{j} & \vec{k} \\ \partial/\partial x & \partial/\partial y & \partial/\partial z \\ x^x + z^2 & y^y + x^2 & z^z + y^2 \end{vmatrix} = 2y\vec{i} + 2z\vec{j} + 2x\vec{k}$

$W = \int_C \vec{F}\cdot d\vec{r} = \iint_S \text{curl } \vec{F}\cdot d\vec{S}$.

To parametrize the surface, let $x = 2\cos\theta\sin\phi$, $y = 2\sin\theta\sin\phi$, $z = 2\cos\phi$,

so that $\vec{r}(\phi,\theta) = 2\sin\phi\cos\theta\,\vec{i} + 2\sin\phi\sin\theta\,\vec{j} + 2\cos\phi\,\vec{k}$, $0 \le \phi \le \frac{\pi}{2}$, $0 \le \theta \le \frac{\pi}{2}$, and

$\vec{r}_\phi \times \vec{r}_\theta = 4\sin^2\phi\cos\theta\,\vec{i} + 4\sin^2\phi\sin\theta\,\vec{j} + 4\sin\phi\cos\phi\,\vec{k}$.

Then curl $\vec{F}(\vec{r}(\phi,\theta)) = 4\sin\phi\sin\theta\,\vec{i} + 4\cos\phi\,\vec{j} + 4\sin\phi\cos\theta\,\vec{k}$, and

curl $\vec{F}\cdot(\vec{r}_\phi \times \vec{r}_\theta) = 16\sin^3\phi\sin\theta\cos\theta + 16\cos\phi\sin^2\phi\sin\theta + 16\sin^2\phi\cos\phi\cos\theta$

Therefore $\iint_S \text{curl } \vec{F}\cdot d\vec{S} = \iint_D \text{curl } \vec{F}\cdot(\vec{r}_\phi \times \vec{r}_\theta)dA = 16\int_0^{\pi/2}\sin\theta\cos\theta\,d\theta\int_0^{\pi/2}\sin^3\phi\,d\phi$

$+ 16\int_0^{\pi/2}\sin\theta\,d\theta\int_0^{\pi/2}\sin^2\phi\cos\phi\,d\phi + 16\int_0^{\pi/2}\cos\theta\,d\theta\int_0^{\pi/2}\sin^2\phi\cos\phi\,d\phi$

$= 8\left[-\cos\phi + \frac{\cos^3\phi}{3}\right]_0^{\pi/2} + 16(1)\left[\frac{\sin^3\phi}{3}\right]_0^{\pi/2} + 16(1)\left[\frac{\sin^3\phi}{3}\right]_0^{\pi/2}$

$= 8[0 + 1 + 0 - \frac{1}{3}] + 16(\frac{1}{3}) + 16(\frac{1}{3}) = \frac{16}{3} + \frac{16}{3} + \frac{16}{3} = 16$.

Section 14.9

19. Assume S is centered at the origin with radius a and let H_1 and H_2 be the upper and lower hemispheres, respectively, of S. Then $\iint_S \text{curl } \vec{F} \cdot d\vec{S} = \iint_{H_1} \text{curl } \vec{F} \cdot d\vec{S} + \iint_{H_2} \text{curl } \vec{F} \cdot d\vec{S}$
$= \oint_{C_1} \vec{F} \cdot d\vec{r} + \oint_{C_2} \vec{F} \cdot d\vec{r}$ by Stokes' Theorem. But C_1 is the circle $x^2 + y^2 = a^2$ oriented in the counterclockwise direction while C_2 is the same circle but oriented in the clockwise direction. Hence $\oint_{C_2} \vec{F} \cdot d\vec{r} = -\oint_{C_1} \vec{F} \cdot d\vec{r}$ so $\iint_S \text{curl } \vec{F} \cdot d\vec{S} = 0$ as desired.

EXERCISES 14.9

1. div $\vec{F} = 3 + x + 2x = 3 + 3x$, so $\iiint_E \text{div } \vec{F} \, dV$
$= \int_0^1 \int_0^1 \int_0^1 (3x + 3) dx \, dy \, dz = 9/2$ (notice the triple integral is three times the volume of the cube plus three times $\bar{x}$). To compute $\iint_S \vec{F} \cdot d\vec{S}$: on S_1: $\vec{n} = \vec{i}$,
$\vec{F} = 3\vec{i} + y\vec{j} + 2z\vec{k}$, and $\iint_{S_1} \vec{F} \cdot d\vec{S} = \iint_{S_1} 3 \, dS = 3$;

S_2: $\vec{F} = 3x\vec{i} + x\vec{j} + 2xz\vec{k}$, $\vec{n} = \vec{j}$ and $\iint_{S_2} \vec{F} \cdot d\vec{S} = \iint_{S_2} x \, dS = 1/2$;

S_3: $\vec{F} = 3x\vec{i} + xy\vec{j} + 2x\vec{k}$, $\vec{n} = \vec{k}$ and $\iint_{S_3} \vec{F} \cdot d\vec{S} = \iint_{S_3} 2x \, dS = 1$;

S_4: $\vec{F} = \vec{0}$, $\iint_{S_4} \vec{F} \cdot d\vec{S} = 0$;

S_5: $\vec{F} = 3x\vec{i} + 2x\vec{k}$, $\vec{n} = -\vec{j}$ and $\iint_{S_5} \vec{F} \cdot d\vec{S} = \iint_{S_5} 0 \, dS = 0$;

S_6: $\vec{F} = 3x\vec{i} + xy\vec{j}$, $\vec{n} = -\vec{k}$ and $\iint_{S_6} \vec{F} \cdot d\vec{S} = \iint_{S_6} 0 \, dS = 0$.

Thus $\iint_S \vec{F} \cdot d\vec{S} = 9/2$.

3. div $\vec{F} = \frac{\partial}{\partial x}(3y^2 z^3) + \frac{\partial}{\partial y}(9x^2 yz^2) + \frac{\partial}{\partial z}(4xy^2) = 9x^2 z^2$, so by the Divergence Theorem,
$\iint_S \vec{F} \cdot d\vec{S} = \iiint_E 9x^2 z^2 \, dV = \int_{-1}^1 \int_{-1}^1 \int_{-1}^1 9x^2 z^2 \, dx \, dy \, dz = 8$.

Section 14.9

5. $\iint_S \vec{F}\cdot d\vec{S} = \iiint_E (-z - z + 2z)dV = 0.$

7. $\iint_S \vec{F}\cdot d\vec{S} = \iiint_E x\,dV = \int_0^1 \int_0^{2-2x} \int_0^{2-2x-y} x\,dz\,dy\,dx = \int_0^1 \int_0^{2-2x} [x(2-2x) - xy]dy\,dx$

$= \int_0^1 [x(2-2x)^2 - x(2-2x)^2/2]dx = 1/6.$

9. $\iint_S \vec{F}\cdot d\vec{S} = \iiint_E 3(x^2+y^2+z^2)dV = \int_0^{2\pi}\int_0^\pi \int_0^1 3\rho^4 \sin\phi\,d\rho\,d\phi\,d\theta$

$= 2\pi \int_0^\pi (3/5)\sin\phi\,d\phi = 12\pi/5.$

11. $\iint_S \vec{F}\cdot d\vec{S} = \iiint_E 2y\,dV = \iint_{x^2+y^2\le 9} \int_{y-3}^0 2y\,dz\,dA = \int_0^{2\pi}\int_0^3 \int_{-3+r\sin\theta}^0 (2r^2\sin\theta)dz\,dr\,d\theta$

$= \int_0^{2\pi}\int_0^3 (6r^2\sin\theta - 2r^3\sin^2\theta)dr\,d\theta = \int_0^{2\pi} [54\sin\theta - 81(\sin^2\theta)/2]d\theta = -81\pi/2.$

13. $\iint_S \vec{F}\cdot d\vec{S} = \iiint_E (x^2+y^2+z)dV = \int_0^{2\pi}\int_1^2 \int_1^3 (r^2+z)r\,dz\,dr\,d\theta = 2\pi\int_1^2 (2r^3 + 4r)dr = 27\pi.$

15. For S_1 we have $\vec{n} = -\vec{k}$, so $\vec{F}\cdot\vec{n} = \vec{F}\cdot(-\vec{k}) = -x^2 z - y^2 = -y^2$ (since $z = 0$ on S_1).

So if D is the unit disk, we get $\iint_{S_1} \vec{F}\cdot d\vec{S} = \iint_{S_1} \vec{F}\cdot\vec{n}\,dS = \iint_D (-y^2)dA$

$= -\int_0^{2\pi}\int_0^1 r^2\sin^2\theta\,r\,dr\,d\theta = -\tfrac{1}{4}\pi.$ Now since S_2 is closed, we can use the Divergence Theorem. Since $\text{div}\,\vec{F} = \frac{\partial}{\partial x}(z^2 x) + \frac{\partial}{\partial y}(\tfrac{1}{3}y^3 + \tan z) + \frac{\partial}{\partial z}(x^2 z + y^2) = z^2 + y^2 + x^2$, we use spherical coordinates to get $\iint_{S_2} \vec{F}\cdot d\vec{S} = \iiint_E \text{div}\,\vec{F}\,dV$

$= \int_0^{2\pi}\int_0^{\pi/2}\int_0^1 \rho^2\cdot\rho^2\sin\phi\,d\rho\,d\phi\,d\theta = \tfrac{2}{5}\pi.$ Finally

$\iint_S \vec{F}\cdot d\vec{S} = \iint_{S_2} \vec{F}\cdot d\vec{S} - \iint_{S_1} \vec{F}\cdot d\vec{S} = \tfrac{2}{5}\pi - (-\tfrac{1}{4}\pi) = \tfrac{13}{20}\pi.$

17. Since $\vec{x}/|\vec{x}|^3 = (x\vec{i} + y\vec{j} + z\vec{k})/(x^2+y^2+z^2)^{3/2}$ and
$(\partial/\partial x)(x/(x^2+y^2+z^2)^{3/2}) = [(x^2+y^2+z^2) - 3x^2]/(x^2+y^2+z^2)^{5/2}$ with similar expressions for $(\partial/\partial y)(y/(x^2+y^2+z^2)^{3/2})$ and $(\partial/\partial z)(z/(x^2+y^2+z^2)^{3/2})$, we have
$\text{div}[\vec{x}/|\vec{x}|^3] = [3(x^2+y^2+z^2) - 3(x^2+y^2+z^2)]/(x^2+y^2+z^2)^{5/2} = 0$, except at $(0,0,0)$ where it is not defined.

19. $\iint_S \vec{a}\cdot\vec{n}\,dS = \iiint_E \text{div}\,\vec{a}\,dV = 0$ since $\text{div}\,\vec{a} = 0.$

Chapter 14 Review

21. $\iint_S \text{curl } \vec{F} \cdot d\vec{S} = \iiint_E \text{div}(\text{curl } \vec{F})dV = 0$ by 14.38.

23. $\iint_S (f\nabla g) \cdot \vec{n}\, dS = \iiint_E \text{div}(f\nabla g)dV = \iiint_E (f\nabla^2 g + \nabla g \cdot \nabla f)dV$ by Exercise 14.5.25.

REVIEW EXERCISES FOR CHAPTER 14

1. False, div $\vec{F}$ is a scalar field.

3. True, by (14.34) and the fact that div $\vec{0} = 0$.

5. False. See Exercise 14.3.29. (But the assertion is true if D is simply-connected. See Theorem 14.23.)

7. True. Apply the Divergence Theorem and use the fact that div $\vec{F} = 0$.

9. $x = y^2/2,\ y = y,\ \int_C y\, dS = \int_0^2 y\sqrt{y^2 + 1}\, dy = (5\sqrt{5} - 1)/3$.

11. $\int_C x^3 z\, ds = \int_0^{\pi/2} (16\sin^3 t \cos t)\sqrt{5}\, dt = 4\sqrt{5}\, \sin^4 t\Big|_0^{\pi/2} = 4\sqrt{5}$.

13. $x = \cos t,\ y = \sin t,\ 0 \le t \le 2\pi$ and $\int_C x^3 y\, dx - x\, dy$
$= \int_0^{2\pi} (-\cos^3 t \sin^2 t - \cos^2 t)dt = -\pi$. (Or since C is a simple closed curve, apply Green's Theorem giving $\iint_{x^2+y^2 \le 1} (-1 - x^3)dA = \int_0^1 \int_0^{2\pi} (-r - r^4 \cos^3 \theta)d\theta = -\pi$.)

15. C_1: $x = t,\ y = t,\ z = 2t,\ 0 \le t \le 1$;

C_2: $x = 1 + 2t,\ y = 1,\ z = 2 + 2t,\ 0 \le t \le 1$.

Then $\int_C y\, dx + z\, dy + x\, dz = \int_0^1 5t\, dt$
$+ \int_0^1 (4 + 4t)dt = 17/2$.

17. $\vec{F}(\vec{r}(t)) = (2t + t^2)\vec{i} + t^4\vec{j} + 4t^4\vec{k},\ \vec{F} \cdot \vec{r}'(t) = 4t + 2t^2 + 2t^5 + 16t^7$
and $\int_C \vec{F} \cdot d\vec{r} = \int_0^1 (4t + 2t^2 + 2t^5 + 16t^7)dt = 5$.

19. $\partial(\sin y)/\partial y = \cos y$ and $\partial(x\cos y + \sin y)/\partial x = \cos y$ and the domain of $\vec{F}$ is R^2 so $\vec{F}$ is conservative. Hence there exists f such that $\nabla f = \vec{F}$. Then $f_x(x, y) = \sin y$ implies

193

$f(x, y) = x \sin y + g(y)$ and $f_y(x, y) = x \cos y + g'(y)$. But $f_y(x, y) = x \cos y + \sin y$ so $g'(y) = \sin y$ and $f(x, y) = x \sin y - \cos y + K$ is a potential for $\vec{F}$.

21. Since $\partial(2x + y^2 + 3x^2y)/\partial y = 2y + 3x^2 = \partial(2xy + x^3 + 3y^2)/\partial x$ and the domain of $\vec{F}$ is R^2, $\vec{F}$ is conservative. Furthermore $f(x, y) = x^2 + xy^2 + x^3y + y^3 + K$ is a potential for $\vec{F}$. Then $\int_C \vec{F} \cdot d\vec{r} = f(\pi, 0) - f(0, 0) = \pi^2$.

23.

C_1: $0 \leq x \leq 1$, $y = 0$; C_2: $x = 1$, $0 \leq y \leq 2$;
C_3: $x = x$, $y = 2x$, $x = 1$ to $x = 0$. Then

$$\oint_C xy\,dx + x^2\,dy = \int_0^1 0\,dx + \int_0^2 (0+1)\,dy$$
$$+ \int_1^0 (2x^2 + 2x^2)\,dx = 2/3. \text{ And } \iint_D (2x - x)\,dA$$
$$= \int_0^1 \int_0^{2x} x\,dy\,dx = 2/3.$$

25. $\int_C x^2 y\,dx - xy^2\,dy = \iint_{x^2+y^2 \leq 4} (-y^2 - x^2)\,dA = -\int_0^{2\pi} \int_0^2 r^3\,dr\,d\theta = -8\pi.$

27. Assume there is such a vector field $\vec{G}$, then div(curl $\vec{G}$) = $2 + 3z - 2xz$ but div(curl $\vec{F}$) = 0 for all vector fields $\vec{F}$. Thus such a $\vec{G}$ can't exist.

29. For any piecewise–smooth simple closed plane curve C, bounding a region D, we can apply Green's Theorem to $\vec{F}(x, y) = f(x)\vec{i} + g(y)\vec{j}$ to get $\int_C f(x)\,dx + g(y)\,dy$
$= \iint_D \left(\frac{\partial}{\partial x}[g(y)] - \frac{\partial}{\partial y}[f(x)] \right) dA = \iint_D 0\,dA = 0.$

31. $\nabla^2 f = 0$ means that $\frac{\partial^2 f}{\partial x^2} + \frac{\partial^2 f}{\partial y^2} = 0$. Now if $\vec{F} = f_y \vec{i} - f_x \vec{j}$ and C is any closed path in D, then applying Green's Theorem, we get $\int_C \vec{F} \cdot d\vec{r} = \int_C f_y\,dx - f_x\,dy = \iint_D \left(\frac{\partial}{\partial x}(-f_x) - \frac{\partial}{\partial y}f_y \right) dA$
$= -\iint_D (f_{xx} + f_{yy})\,dA = -\iint_D 0\,dA = 0.$ Therefore the line integral is independent of path by Theorem 14.20.

33. $\vec{r}_u = -v\vec{j} + 2u\vec{k}$, $\vec{r}_v = 2v\vec{i} - u\vec{j}$ and $\vec{r}_u \times \vec{r}_v = 2u^2\vec{i} + 4uv\vec{j} + 2v^2\vec{k}$. Since the point $(4, -2, 1)$ corresponds to $u = 1$, $v = 2$ (or $u = -1$, $v = -2$ but $\vec{r}_u \times \vec{r}_v$ is the same for both), a normal vector to the surface at $(4, -2, 1)$ is $2\vec{i} + 8\vec{j} + 8\vec{k}$ and the equation of the tangent plane is $2x + 8y + 8z = 0$ or $x + 4y + 4z = 0$.

35. $z = f(x,y) = x^2 + y^2$ with $0 \le x^2 + y^2 \le 4$ so $\vec{r}_x \times \vec{r}_y = -2x\vec{i} - 2y\vec{j} + \vec{k}$
(using upward orientation). Then $\iint\limits_S z\,dS = \iint\limits_{x^2+y^2 \le 4} (x^2+y^2)\sqrt{4x^2+4y^2+1}\,dA$

$= \int_0^{2\pi} \int_0^2 r^3\sqrt{1+4r^2}\,dr\,d\theta$

$= \pi(391\sqrt{17}+1)/60$ (Substitute $2r = \tan\theta$ or use tables.)

37. Since the sphere bounds a simple solid region, the divergence theorem applies and

$\iint\limits_S \vec{F}\cdot d\vec{S} = \iiint\limits_E (z-2)\,dV = \iiint\limits_E z\,dV - 2\iiint\limits_E dV = m\bar{z} - 2(4\pi 2^3/3) = -64\pi/3$.

Alternate Solution: $\vec{F}(\vec{r}(\phi,\theta)) = 4\sin\phi\cos\theta\cos\phi\vec{i} - 4\sin\phi\sin\theta\vec{j} + 6\sin\phi\cos\theta\vec{k}$,

$\vec{r}_\phi \times \vec{r}_\theta = 4\sin^2\phi\cos\theta\vec{i} + 4\sin^2\phi\sin\theta\vec{j} + 4\sin\phi\cos\phi\vec{k}$, and

$\vec{F}\cdot(\vec{r}_\phi \times \vec{r}_\theta) = 16\sin^3\phi\cos^2\theta\cos\phi - 16\sin^3\phi\sin^2\theta + 24\sin^2\phi\cos\phi\cos\theta$. Then

$\iint\limits_S \vec{F}\cdot d\vec{S} = \int_0^{2\pi}\int_0^\pi (16\sin^3\phi\cos\phi\cos^2\theta - 16\sin^3\phi\sin^2\theta + 24\sin^2\phi\cos\phi\cos\theta)d\phi\,d\theta$

$= \int_0^{2\pi}(4/3)(-16\sin^2\theta)d\theta = (-64/3)(\pi) = -64\pi/3$.

39. Since curl $\vec{F} = \vec{0}$, $\iint\limits_S (\text{curl } \vec{F})\cdot d\vec{S} = 0$. And $C: \vec{r}(t) = \cos t\vec{i} + \sin t\vec{j}$,

$0 \le t \le 2\pi$ and $\oint_C \vec{F}\cdot d\vec{r} = \int_0^{2\pi}(-\cos^2 t\sin t + \sin^2 t\cos t)dt$

$= (\cos^3 t)/3 + (\sin^3 t)/3 \big|_0^{2\pi} = 0$.

41. The surface is given by $x+y+z = 1$ or $z = 1-x-y$, $0 \le x \le 1$, $0 \le y \le 1-x$ and

$\vec{r}_x \times \vec{r}_y = \vec{i} + \vec{j} + \vec{k}$. Then $\oint_C \vec{F}\cdot d\vec{r} = \iint\limits_S \text{curl } \vec{F}\cdot d\vec{S}$

$= \iint\limits_D (-y\vec{i} - z\vec{j} - x\vec{k})\cdot(\vec{i}+\vec{j}+\vec{k})dA = \iint\limits_D (-1)dA = -\text{area of } D = -1/2$.

43. $\iiint\limits_E \text{div } \vec{F}\,dV = \iiint\limits_{x^2+y^2+z^2 \le 1} 3\,dV = 3(\text{volume of sphere}) = 4\pi$. Then

$\vec{F}(\vec{r}(\phi,\theta))\cdot(\vec{r}_\phi \times \vec{r}_\theta) = \sin^3\phi\cos^2\theta + \sin^3\phi\sin^2\theta + \sin\phi\cos^2\phi = \sin\phi$ and

$\iint\limits_S \vec{F}\cdot d\vec{S} = \int_0^{2\pi}\int_0^\pi \sin\phi\,d\phi\,d\theta = (2\pi)(2) = 4\pi$.

45. Since curl $\vec{F} = \vec{0}$, $\vec{F}$ is conservative and setting $f(x,y,z) = x^3yz - 3xy + z^2$, we have

$\nabla f = \vec{F}$. Hence $\int_C \vec{F}\cdot d\vec{r} = \int_C \nabla f\cdot d\vec{r} = f(0,3,0) - f(0,0,2) = 0 - 4 = -4$.

195

47. By the Divergence Theorem $\iint_S \vec{F} \cdot \vec{n}\, dS = \iiint_E \text{div } \vec{F}\, dV = 3(\text{volume of } E)$
$= 3(8-1) = 21$.

49. Let $\vec{H} = \langle h_1, h_2, h_3 \rangle$ and $\vec{E} = \langle E_1, E_2, E_3 \rangle$

(a) $\nabla \times (\nabla \times \vec{E}) = \nabla \times (\text{curl } \vec{E}) = \nabla \times \left(-\frac{1}{c}\frac{\partial \vec{H}}{\partial t}\right) = -\frac{1}{c}\begin{vmatrix} \vec{i} & \vec{j} & \vec{k} \\ \frac{\partial}{\partial x} & \frac{\partial}{\partial y} & \frac{\partial}{\partial z} \\ \frac{\partial h_1}{\partial t} & \frac{\partial h_2}{\partial t} & \frac{\partial h_3}{\partial t} \end{vmatrix}$

$= -\frac{1}{c}\left[\left(\frac{\partial^2 h_3}{\partial y \partial t} - \frac{\partial^2 h_2}{\partial z \partial t}\right)\vec{i} + \left(\frac{\partial^2 h_1}{\partial z \partial t} - \frac{\partial^2 h_3}{\partial x \partial t}\right)\vec{j} + \left(\frac{\partial^2 h_2}{\partial x \partial t} - \frac{\partial^2 h_1}{\partial y \partial t}\right)\vec{k}\right]$

$= -\frac{1}{c}\frac{\partial}{\partial t}\left[\left(\frac{\partial h_3}{\partial y} - \frac{\partial h_2}{\partial z}\right)\vec{i} + \left(\frac{\partial h_1}{\partial z} - \frac{\partial h_3}{\partial x}\right)\vec{j} + \left(\frac{\partial h_2}{\partial x} - \frac{\partial h_1}{\partial y}\right)\right]$ (*assuming that the partial derivatives are continuous so that the order of differentiation does not matter*)

$= -\frac{1}{c}\frac{\partial}{\partial t}\text{curl } \vec{H} = -\frac{1}{c}\frac{\partial}{\partial t}\left(\frac{1}{c}\frac{\partial \vec{E}}{\partial t}\right) = -\frac{1}{c^2}\frac{\partial^2 \vec{E}}{\partial t^2}$

(b) $\nabla \times (\nabla \times \vec{H}) = \nabla \times (\text{curl } \vec{H}) = \nabla \times \left(\frac{1}{c}\frac{\partial \vec{E}}{\partial t}\right) = \frac{1}{c}\begin{vmatrix} \vec{i} & \vec{j} & \vec{k} \\ \frac{\partial}{\partial x} & \frac{\partial}{\partial y} & \frac{\partial}{\partial z} \\ \frac{\partial E_1}{\partial t} & \frac{\partial E_2}{\partial t} & \frac{\partial E_3}{\partial t} \end{vmatrix}$

$= \frac{1}{c}\left[\left(\frac{\partial^2 E_3}{\partial y \partial t} - \frac{\partial^2 E_2}{\partial z \partial t}\right)\vec{i} + \left(\frac{\partial^2 E_1}{\partial z \partial t} - \frac{\partial^2 E_3}{\partial x \partial t}\right)\vec{j} + \left(\frac{\partial^2 E_2}{\partial x \partial t} - \frac{\partial^2 E_1}{\partial y \partial t}\right)\vec{k}\right]$

$= \frac{1}{c}\frac{\partial}{\partial t}\left[\left(\frac{\partial E_3}{\partial y} - \frac{\partial E_2}{\partial z}\right)\vec{i} + \left(\frac{\partial E_1}{\partial z} - \frac{\partial E_3}{\partial x}\right)\vec{j} + \left(\frac{\partial E_2}{\partial x} - \frac{\partial E_1}{\partial y}\right)\right]$ (*assuming that the partial derivatives are continuous so that the order of differentiation does not matter*)

$= \frac{1}{c}\frac{\partial}{\partial t}\text{curl } \vec{E} = \frac{1}{c}\frac{\partial}{\partial t}\left(-\frac{1}{c}\frac{\partial \vec{H}}{\partial t}\right) = -\frac{1}{c^2}\frac{\partial^2 \vec{H}}{\partial t^2}$

(c) Using Exercise 29 in Section 14.5 we have that

curl curl $\vec{E} = \text{grad div } \vec{E} - \nabla^2 \vec{E} \Rightarrow \nabla^2 \vec{E} = \text{grad div } \vec{E} - \text{curl curl } \vec{E}$

$= \text{grad } 0 + \frac{1}{c^2}\frac{\partial^2 \vec{E}}{\partial t^2}$ (*from part (a)*) $= \frac{1}{c^2}\frac{\partial^2 \vec{E}}{\partial t^2}$

(d) As in part (c) $\nabla^2 \vec{H} = \text{grad div } \vec{H} - \text{curl curl } \vec{H} = \text{grad } 0 + \frac{1}{c^2}\frac{\partial^2 \vec{H}}{\partial t^2}$ (*using part (b)*)

$= \frac{1}{c^2}\frac{\partial^2 \vec{H}}{\partial t^2}$

Problems Plus (after Chapter 14)

1. Since $|xy| < 1$, except at $(1,1)$, the formula for the sum of a geometric series gives

$$\frac{1}{1-xy} = \sum_{n=0}^{\infty} (xy)^n, \text{ so } \int_0^1 \int_0^1 \frac{1}{1-xy}\,dx\,dy = \int_0^1 \int_0^1 \sum_{n=0}^{\infty} (xy)^n\,dx\,dy$$

$$= \sum_{n=0}^{\infty} \int_0^1 \int_0^1 (xy)^n\,dx\,dy = \sum_{n=0}^{\infty} \int_0^1 x^n\,dx \int_0^1 y^n\,dy = \sum_{n=0}^{\infty} \frac{1}{n+1}\cdot\frac{1}{n+1}$$

$$= \sum_{n=0}^{\infty} \frac{1}{(n+1)^2} = \frac{1}{1^2} + \frac{1}{2^2} + \frac{1}{3^2} + \cdots = \sum_{n=1}^{\infty} \frac{1}{n^2}.$$

3. Since $|-xyz| < 1$, except at $(1,1,1)$, the formula for the sum of a geometric series gives

$$\frac{1}{1+xyz} = \sum_{n=0}^{\infty} (-xyz)^n, \text{ so } \int_0^1\int_0^1\int_0^1 \frac{1}{1+xyz}\,dx\,dy\,dz = \int_0^1\int_0^1\int_0^1 \sum_{n=0}^{\infty} (-xyz)^n\,dx\,dy\,dz$$

$$= \sum_{n=0}^{\infty} \int_0^1\int_0^1\int_0^1 (-xyz)^n\,dx\,dy\,dz = \sum_{n=0}^{\infty} (-1)^n \int_0^1 x^n\,dx \int_0^1 y^n\,dy \int_0^1 z^n\,dz$$

$$= \sum_{n=0}^{\infty} (-1)^n \frac{1}{n+1}\cdot\frac{1}{n+1}\cdot\frac{1}{n+1}$$

$$= \sum_{n=0}^{\infty} \frac{(-1)^n}{(n+1)^3} = \frac{1}{1^3} - \frac{1}{2^3} + \frac{1}{3^3} - \cdots = \sum_{n=1}^{\infty} \frac{(-1)^{n-1}}{n^3}.$$

To evaluate this sum, we first write out a few terms:

$$s = 1 - \frac{1}{2^3} + \frac{1}{3^3} - \frac{1}{4^3} + \frac{1}{5^3} - \frac{1}{6^3} \approx 0.8998.$$

Notice that $a_7 = \frac{1}{7^3} < 0.003$. By Theorem 10.28, we have $|s - s_6| \leq a_7 < 0.003$. This error of 0.003 will not affect the second decimal place, so we have $s \approx 0.90$.

5. Let $R = \bigcup_{i=1}^{5} R_i$, where $R_i = \{(x,y) \mid x+y \geq i+2,\ x+y < i+3,\ 1 \leq x \leq 3,\ 2 \leq y \leq 5\}$.

$$\iint_R [\![x+y]\!]\,dA = \sum_{i=1}^{5} \iint_{R_i} [\![x+y]\!]\,dA$$

$$= \sum_{i=1}^{5} [\![x+y]\!] \iint_{R_i} dA, \text{ since } [\![x+y]\!] = \text{constant} = i+2$$

for $(x,y) \in R_i$. Therefore $\iint_R [\![x+y]\!]\,dA = \sum_{i=1}^{5} (i+2)[A(R_i)]$

$$= 3A(R_1) + 4A(R_2) + 5A(R_3) + 6A(R_4) + 7A(R_5)$$

$$= 3(\tfrac{1}{2}) + 4(\tfrac{3}{2}) + 5(2) + 6(\tfrac{3}{2}) + 7(\tfrac{1}{2}) = 30.$$

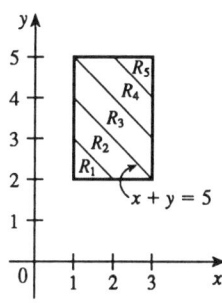

Problems Plus

7. $f_{ave} = \dfrac{1}{b-a}\int_a^b f(x)\,dx = \dfrac{1}{1-0}\int_0^1\left[\int_x^1 \cos(t^2)\,dt\right]dx = \int_0^1\int_x^1 \cos(t^2)\,dt\,dx$

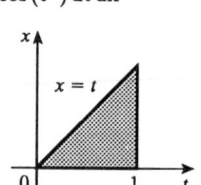

$= \int_0^1\int_0^t \cos(t^2)\,dx\,dt$ (*changing the order of integration*)

$= \int_0^1 t\cos(t^2)\,dt = \left[\tfrac{1}{2}\sin(t^2)\right]_0^1 = \tfrac{1}{2}\sin 1.$

9.

$\int_0^x\int_0^y\int_0^z f(t)\,dt\,dz\,dy = \iiint_E f(t)\,dV$, where

$E = \{(t,z,y)\,|\,0\le t\le z,\ 0\le z\le y,\ 0\le y\le x\}$. If we let D be the projection of E on the yt–plane then $D = \{(y,t)\,|\,0\le t\le x,\ t\le y\le x\}$. And we see from the diagram that $E = \{(t,z,y)\,|\,t\le z\le y,\ t\le y\le x,\ 0\le t\le x\}$.

So $\int_0^x\int_0^y\int_0^z f(t)\,dt\,dz\,dy = \int_0^x\int_t^x\int_t^y f(t)\,dz\,dy\,dt = \int_0^x\left[\int_t^x(y-t)f(t)\,dy\right]dt$

$= \int_0^x\left[(\tfrac{1}{2}y^2 - ty)f(t)\right]_t^x dt = \int_0^x [\tfrac{1}{2}x^2 - tx - \tfrac{1}{2}t^2 + t^2]f(t)\,dt = \int_0^x[\tfrac{1}{2}x^2 - tx + \tfrac{1}{2}t^2]f(t)\,dt$

$= \int_0^x \tfrac{1}{2}(x^2 - 2tx + t^2)f(t)\,dt = \tfrac{1}{2}\int_0^x (x-t)^2 f(t)\,dt.$

11. Let $u = \vec{a}\cdot\vec{r}$, $v = \vec{b}\cdot\vec{r}$, $w = \vec{c}\cdot\vec{r}$, where $\vec{a} = <a_1, a_2, a_3>$, $\vec{b} = <b_1, b_2, b_3>$, $\vec{c} = <c_1, c_2, c_3>$. Under this change of variables, E corresponds to the rectangular box $0\le u\le\alpha$, $0\le v\le\beta$, $0\le w\le\gamma$. So, by the Change of Variables Theorem,

$\int_0^\gamma\int_0^\beta\int_0^\alpha uvw\,du\,dv\,dw = \iiint_E (\vec{a}\cdot\vec{r})(\vec{b}\cdot\vec{r})(\vec{c}\cdot\vec{r})\left|\dfrac{\partial(u,v,w)}{\partial(x,y,z)}\right|dV$

But $\left|\dfrac{\partial(u,v,w)}{\partial(x,y,z)}\right| = \begin{Vmatrix} a_1 & a_2 & a_3 \\ b_1 & b_2 & b_3 \\ c_1 & c_2 & c_3 \end{Vmatrix} = |\vec{a}\cdot\vec{b}\times\vec{c}|$

$\Rightarrow \iiint_E (\vec{a}\cdot\vec{r})(\vec{b}\cdot\vec{r})(\vec{c}\cdot\vec{r})\,dV = \dfrac{1}{|\vec{a}\cdot\vec{b}\times\vec{c}|}\int_0^\gamma\int_0^\beta\int_0^\alpha uvw\,du\,dv\,dw$

$\Rightarrow \iiint_E (\vec{a}\cdot\vec{r})(\vec{b}\cdot\vec{r})(\vec{c}\cdot\vec{r})\,dV = \dfrac{1}{|\vec{a}\cdot\vec{b}\times\vec{c}|}\left(\dfrac{\alpha^2}{2}\right)\left(\dfrac{\beta^2}{2}\right)\left(\dfrac{\gamma^2}{2}\right) = \dfrac{(\alpha\beta\gamma)^2}{8|\vec{a}\cdot\vec{b}\times\vec{c}|}.$

CHAPTER 15

EXERCISES 15.1

1. $\frac{dy}{dx} = \frac{3x^2 + e^x}{4y^3} \Rightarrow 4y^3\, dy = (3x^2 + e^x)\, dx \Rightarrow \int 4y^3\, dy = \int (3x^2 + e^x)\, dx$

   $\Rightarrow y^4 = x^3 + e^x + C \Rightarrow y = \pm(x^3 + e^x + C)^{1/4}$.

3. $x^2 y' + y = 0 \Rightarrow \frac{dy}{dx} = \frac{-y}{x^2} \Rightarrow \int \frac{dy}{y} = \int \frac{-dx}{x^2}\ (y \neq 0) \Rightarrow \ln|y| = \frac{1}{x} + K$

   $\Rightarrow |y| = e^K e^{1/x} \Rightarrow y = Ce^{1/x}$, where now we allow C to be any constant.

5. $\frac{dy}{dx} = \frac{x\sqrt{x^2+1}}{ye^y} \Rightarrow \int ye^y\, dy = \int x\sqrt{x^2+1}\, dx \Rightarrow (y-1)e^y = \frac{1}{3}(x^2+1)^{3/2} + C$

7. $\frac{dy}{dx} = y^2 + 1$, $y(1) = 0$. $\int \frac{dy}{y^2 + 1} = \int dx \Rightarrow \tan^{-1} y = x + C$. $y = 0$ when $x = 1$,

   so $1 + C = \tan^{-1} 0 = 0$ and $C = -1$. Thus $\tan^{-1} y = x - 1$ and $y = \tan(x - 1)$.

9. $\frac{du}{dt} = \frac{2t+1}{2(u-1)}$, $u(0) = -1$. $\int 2(u-1)\, du = \int (2t+1)\, dt \Rightarrow u^2 - 2u = t^2 + t + C$.

   $u(0) = -1$ so $(-1)^2 - 2(-1) = 0^2 + 0 + C$ and $C = 3$. Thus $u^2 - 2u = t^2 + t + 3$;

   the quadratic formula gives $u = 1 - \sqrt{t^2 + t + 4}$.

11. $x^2 + 1 + 2xyy' = 0$ isn't homogeneous since $y' = -\frac{x}{2y} - \frac{1}{2xy}$ can't be written as a function of $\frac{y}{x}$.

13. $y' = \ln y - \ln x \Rightarrow y' = \ln(y/x)$ and as a result the equation is homogeneous.

15. Since $y' = 1 - y/x$, setting $v = y/x$ gives $v' = (1-2v)/x$ or $(1-2v)^{-1} dv = x^{-1} dx$ or

    $(-1/2)\ln|1-2v| = c_1 + \ln|x|\ (v \neq 1/2)$. Then $\ln|1-2v| = c_2 + \ln x^{-2}$ or

    $v = (1 - kx^{-2})/2$ and $y = xv = (x - kx^{-1})/2$.

17. $y' = (x/y) + (y/x) = (y/x)^{-1} + (y/x)$. Setting $v = y/x$ gives $v' = 1/(vx)$ or

    $v\, dv = (1/x)dx$. Hence $v^2 = (\ln x^2) + C$ or $v = \pm(C + \ln x^2)^{1/2}$. Thus

    $y = xv = \pm x\sqrt{C + \ln x^2}$ is the solution.

19. $y' = (y/x) + e^{y/x}$, so setting $v = y/x$ gives $v' = e^v/x$ or $e^{-v} dv = (1/x)dx$. Hence

    $-e^{-v} = \ln|x| + c_1$, or $-v = \ln[c - \ln|x|]$ or $v = -\ln[c - \ln|x|]$. Thus the solution is

    $y = -x\ln[c - \ln|x|]$.

21.

x	0	0	0	1	−1	1	1	1	1	−1	−1	...
y	0	1	−1	0	0	−1	1	2	−2	2	−2	...
$y' = y^2$	0	1	1	0	0	1	1	4	4	4	4	...

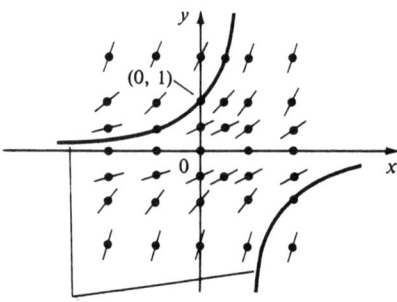

The solution curve through (0,1)

23.

x	0	0	1	1	−1	0	2	2	2	−2	1	...
y	0	1	0	1	1	2	0	2	1	−1	2	...
$y' = x^2 + y^2$	0	1	1	2	2	4	4	8	5	5	5	...

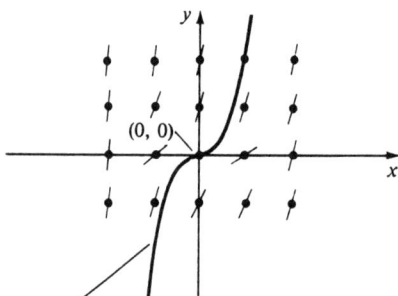

The solution curve through (0, 0)

Section 15.1

25. (a)(b)

x	0	0	0	0	0	0	0	0	0	0	...
y	.5	−.5	1	−1	2	−2	4	3	.25	$.\overline{3}$	...
$y' = 1/y$	2	−2	1	−1	.5	−.5	.25	$.\overline{3}$	4	3	...

(c) $y\,dy = dx$ so $y^2/2 = x + c$
or $y = \pm\sqrt{2(x+c)}$

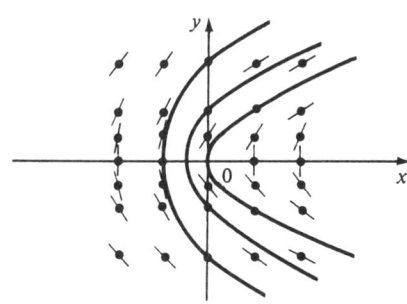

(d)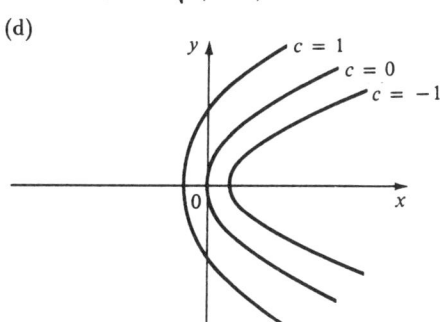

27. (a)(b)

x	.5	.5	−.5	−.5	1	1	−1	−1	2	1	−2	2	...
y	1	−1	1	−1	1	−1	1	−1	1	2	1	2	...
$y' = -y/x$	−2	2	2	−2	−1	1	1	−1	−.5	−2	.5	−1	...

(c) $(1/y)dy = (-1/x)dx$ so
$\ln|y| = -\ln|x| + c_1$, or
$y = c/x$.

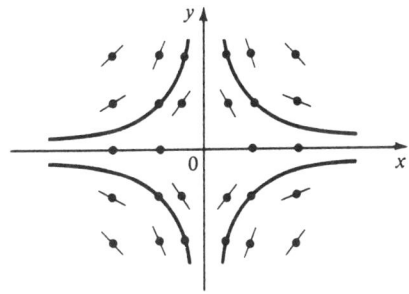

(d)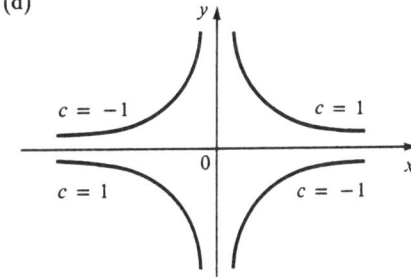

29. The curves $y = kx^2$ form a family of parabolas with axis the y-axis. Differentiating gives $y' = 2kx$, but $k = y/x^2$, so $y' = 2y/x$. Thus the slope of the tangent line at any point (x,y) on one of the parabolas is $y' = 2y/x$ so the orthogonal trajectories must satisfy $y' = -x/(2y)$ or $y^2 = -x^2/2 + c_1$. This is a family of ellipses: $x^2 + 2y^2 = c$.

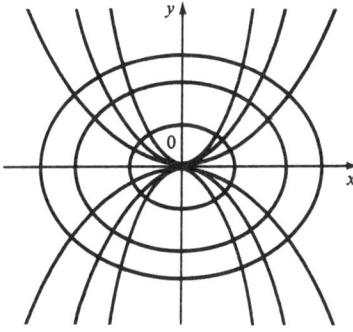

31. Differentiating gives $y' = -(x+k)^{-2}$ but $k = (1/y) - x$ so $y' = -(1/y)^{-2} = -y^2$. So the orthogonal trajectories must satisfy $y' = 1/y^2$ or $y^3/3 = x + c$ or $y = [3(x+c)]^{1/3}$.

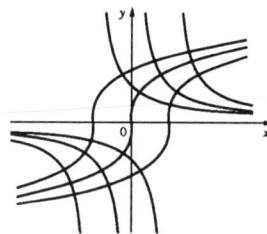

33. The curves $x^2 - 2y^2 = k$ form a family of hyperbolas. Differentiating gives $2x - 4y(dy/dx) = 0$ or $y' = x/(2y)$. Thus the orthogonal trajectories must satisfy $y' = -2y/x$ or $\ln|2y| = -\ln|x| + c_1$ or $y = c/(2x)$. This is a family of hyperbolas.

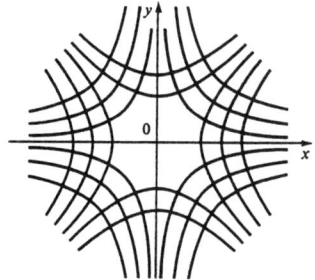

35. (a) We have $V(t) = \pi r^2 y(t) \Rightarrow \frac{dV}{dy} = \pi r^2 = 4\pi$ where $\frac{dV}{dt} = \frac{dV}{dy}\frac{dy}{dt}$ and $a = \pi(1/12)^2$. Thus, $\frac{dV}{dt} = -a\sqrt{2gy} \Rightarrow \frac{dV}{dy}\frac{dy}{dt} = -\pi(1/12)^2\sqrt{64y} \Rightarrow \frac{dy}{dt}(4\pi) = -\pi(8/144)\sqrt{y}$
$\Rightarrow \frac{dy}{dt} = -(1/72)\sqrt{y}$.

(b) $\frac{dy}{dt} = -(1/72)\sqrt{y} \Rightarrow (y^{-1/2})dy = -(1/72)dt \Rightarrow 2\sqrt{y} = -(1/72)t + C$ where $y(0) = 6$. Thus, $2\sqrt{6} = 0 + C \Rightarrow C = 2\sqrt{6} \Rightarrow y = (-\frac{1}{144}t + \sqrt{6})^2$.

(c) We want to find t when $y = 0$. Thus, $0 = (-\frac{1}{144}t + \sqrt{6})^2 \Rightarrow t = 144\sqrt{6}$
≈ 5 min 53 s.

EXERCISES 15.2

1. $y' + x^2y = y^2$ is not linear since it cannot be put into the standard linear form of (15.4).

3. $xy' = x - y \Rightarrow xy' + y = x \Rightarrow y' + (1/x)y = 1$, which is in the form of (15.4), and thus the differential equation is linear.

5. $I(x) = e^{\int -3\,dx} = e^{-3x}$ and thus multiplying the differential equation by $I(x)$:
$e^{-3x}y' - 3e^{-3x}y = e^{-2x} \Rightarrow (e^{-3x}y)' = e^{-2x} \Rightarrow y = e^{3x}[\int (e^{-2x})dx + C] = Ce^{3x} - e^x/2$.

7. $I(x) = e^{\int -2x\,dx} = e^{-x^2}$ and thus multiplying the differential equation by $I(x)$:
$e^{-x^2}y' - 2xe^{-x^2}y = xe^{-x^2} \Rightarrow (e^{-x^2}y)' = xe^{-x^2} \Rightarrow y = e^{x^2}[\int xe^{-x^2}dx + C] = Ce^{x^2} - (1/2)$.

9. Since $y' + (1/x)y = \cos x$ ($x \neq 0$), $I(x) = e^{\int (1/x)\,dx} = e^{\ln|x|} = x$ (for $x > 0$).
Multiplying the differential equation by $I(x)$: $xy' + (x/x)y = x\cos x$
$\Rightarrow (xy)' = x\cos x$. Thus,
$y = (1/x)[\int x\cos x\,dx + C] = (1/x)[x\sin x + \cos x + C] = \sin x + (\cos x)/x + C/x$.

11. $y' - (\tan x)y = (\sin 2x)/\cos x = 2\sin x$ so $I(x) = e^{\int -\tan x\,dx} = e^{\ln|\cos x|} = \cos x$
(when $\cos x > 0$). Multiplying the differential equation by $I(x)$:
$(\cos x)y' - (\cos x)(\tan x)y = (\cos x)(\sin 2x)/\cos x$
$\Rightarrow [(\cos x)y]' = (\cos x)(\sin 2x)/\cos x$. Thus, $y = (1/\cos x)[\int \sin 2x\,dx + C]$
$= (1/\cos x)[-(\cos 2x)/2 + C] = (1/\cos x)[(1/2) - \cos^2 x + C] = (\sec x)/2 - \cos x + C\sec x$.

13. $I(x) = e^{\int 2x\,dx} = e^{x^2}$ and multiplying the differential equation by $I(x)$:
$e^{x^2}y' + 2xe^{x^2} = x^2e^{x^2} \Rightarrow (e^{x^2}y)' = x^2e^{x^2}$. Thus, $y = e^{-x^2}[\int x^2e^{x^2}dx + C]$
$= e^{-x^2}[xe^{x^2}/2 - \int (e^{x^2}/2)dx + C] = (x/2) + Ce^{-x^2} - e^{-x^2}\int (e^{x^2}/2)dx$.

15. $I(\theta) = e^{-\int \tan\theta\,d\theta} = e^{-\ln\sec\theta} = \cos\theta$ and thus multiplying the differential equation by $I(\theta)$:
$\cos\theta \frac{dy}{d\theta} - y\sin\theta = \cos\theta \Rightarrow (y\cos\theta)' = \cos\theta \Rightarrow y\cos\theta = \int \cos\theta\,d\theta$
$\Rightarrow y\cos\theta = \sin\theta + C \Rightarrow y = \tan\theta + C\sec\theta$.

17. $I(x) = e^{\int dx} = e^x$ and thus multiplying the differential equation by $I(x)$:
$e^xy' + e^xy = e^x(x + e^x) \Rightarrow (e^xy)' = e^x(x + e^x)$. Thus,
$y = e^{-x}[\int e^x(x + e^x)dx + C] = e^{-x}[xe^x - e^x + (e^{2x}/2) + C]$
$= x - 1 + (e^x/2) + (C/e^x)$. But $0 = y(0) = -1 + (1/2) + C$ so $C = 1/2$
and the solution to the initial-value problem is
$y = x - 1 + e^x/2 + e^{-x}/2 = x - 1 + \cosh x$.

Section 15.2

19. $I(x) = e^{\int -2x\,dx} = e^{-x^2}$ and thus multiplying the differential equation by $I(x)$:

$e^{-x^2}y' - 2xe^{-x^2} = 2x \Rightarrow (e^{-x^2}y)' = 2x \Rightarrow y = e^{x^2}[\int 2x\,dx + C] = x^2 e^{x^2} + Ce^{x^2}$.

But $3 = y(0) = C$, so the solution to the initial-value problem is $y = (x^2 + 3)e^{x^2}$.

21. Since $y' + 2(y/x) = (\cos x)/x^2$ $(x \neq 0)$, $I(x) = e^{\int (2/x)\,dx} = x^2$ and thus multiplying the differential equation by $I(x)$: $x^2 y' + 2xy = \cos x \Rightarrow (x^2 y)' = \cos x$

$\Rightarrow y = x^{-2}[\int (\cos x)dx + C] = x^{-2}(\sin x + C)$ $(x \neq 0)$. But $0 = y(\pi) = C$, so the solution to the initial-value problem is $y = (\sin x)/x^2$.

23. Setting $u = y^{1-n}$, $du/dx = (1-n)y^{-n}(dy/dx)$ or $dy/dx = [y^n/(1-n)]du/dx$

$= [u^{n/(1-n)}/(1-n)](du/dx)$. Then the Bernoulli differential equation becomes

$[u^{n/(1-n)}/(1-n)](du/dx) + P(x)(u^{1/(1-n)}) = Q(x)(u^{n/(1-n)})$ or

$(du/dx) + (1-n)P(x)u = Q(x)(1-n)$.

25. Here $n = 3$, $P(x) = 2/x$, $Q(x) = 1/x^2$ and setting $u = y^{-2}$, u satisfies

$u' - (4u/x) = -2/x^2$. Then $I(x) = e^{\int -4/x\,dx} = x^{-4}$ and $u = x^4[\int(-2/x^6)dx + C]$

$= x^4[2/(5x^5) + C] = Cx^4 + 2/(5x)$. Thus $y = \pm[Cx^4 + 2/(5x)]^{-1/2}$.

27. (a) $2(dI/dt) + 10I = 40$ or $dI/dt + 5I = 20$. Then the integrating factor is $e^{\int 5\,dt} = e^{5t}$

and multiplying the differential equation by the integrating factor:

$e^{5t}(dI/dt) + 5Ie^{5t} = 20e^{5t} \Rightarrow (e^{5t}I)' = 20e^{5t} \Rightarrow I(t) = e^{-5t}[\int 20e^{5t}dt + C] = 4 + Ce^{-5t}$.

But $0 = I(0) = 4 + C$ so $I(t) = 4 - 4e^{-5t}$.

(b) $I(0.1) = 4 - 4e^{-0.5} \approx 1.57$ A.

29. $5(dQ/dt) + 20Q = 60$ with $Q(0) = 0$ C. Then the integrating factor is $e^{\int 4\,dt} = e^{4t}$ and multiplying the differential equation by the integrating factor:

$e^{4t}(dQ/dt) + 4e^{4t}Q = 12e^{4t} \Rightarrow (e^{4t}Q)' = 12e^{4t} \Rightarrow Q(t) = e^{-4t}[\int 12e^{4t}dt + C]$

$= 3 + Ce^{-4t}$. But $0 = Q(0) = 3 + C$ so $Q(t) = 3(1 - e^{-4t})$ is the charge at time t and

$I = dQ/dt = 12e^{-4t}$ is the current at time t.

31. $dP/dt + kP = kM$ so $I(t) = e^{\int k\,dt}$

$= e^{kt}$ and multiplying the DE by $I(x)$:

$e^{kt}dP/dt + kPe^{kt} = kMe^{kt} \Rightarrow (e^{kt}P)' = kMe^{kt}$

$\Rightarrow P(t) = e^{-kt}[\int kMe^{kt}dt + C]$

$= M + Ce^{-kt}$, $k > 0$. Furthermore

it is reasonable to assume

$0 \leq P(0) \leq M$, so $-M \leq C \leq 0$.

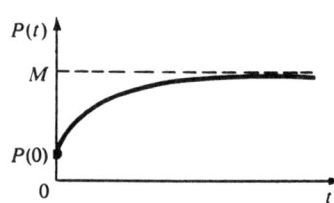

Section 15.3

33. (a) $dv/dt + (c/m)v = g$ and $I(t) = e^{\int (c/m)\,dt} = e^{(c/m)t}$ and multiplying the differential equation by $I(x)$: $e^{(c/m)t}(dv/dt) + v(c/m)e^{(c/m)t} = ge^{(c/m)t}$
$\Rightarrow (e^{(c/m)t}v)' = ge^{(c/m)t}$.
Hence, $v(t) = e^{-(c/m)t}[\int ge^{(c/m)t}dt + K] = (mg/c) + Ke^{-(c/m)t}$. But the object is dropped from rest so $v(0) = 0$ and $K = -mg/c$. Thus the velocity at time t is
$v(t) = (mg/c)(1 - e^{-(c/m)t})$.

(b) $\lim_{t \to \infty} v(t) = mg/c$.

(c) $s(t) = \int v(t)dt = (mg/c)[t + (m/c)e^{-(c/m)t}] + c_1$ where $c_1 = s(0) - m^2g/c^2$, $s(0)$ is the initial position so $s(0) = 0$ and $s(t) = (mg/c)[t + (m/c)e^{-(c/m)t}] - m^2g/c^2$.

EXERCISES 15.3

1. (a) $\partial(x \sin y)/\partial y = x \cos y$ and $\partial(y \cos x)/\partial x = -y \sin x$, so the equation is not exact.
(b) $\partial(2xy^3)/\partial y = 6xy^2$, $\partial(3x^2y^2)/\partial x = 6xy^2$, so the equation is exact.
(c) $\partial(x - y)/\partial y = -1$, $\partial(x + y) = 1$, so the equation is not exact.

3. $\partial(2x + y)/\partial y = 1 = \partial(x + 2y)\partial x$, so the equation is exact. Thus there exists f such that $f_x(x, y) = 2x + y$ implies $f(x, y) = x^2 + xy + g(y)$ and $f_y(x, y) = x + g'(y)$. But $f_y(x, y) = x + 2y$ so $g(y) = y^2$ and $f(x, y) = x^2 + xy + y^2$. Thus the solution to the differential equation is given implicitly by $x^2 + xy + y^2 = C$ or $y = (-x \pm \sqrt{4C - 3x^2})/2$.

5. $\partial(3xy - 2)/\partial y = 3x$, $\partial(3y^2 - x^2)/\partial x = -2x$, so the equation isn't exact.

7. $\partial(\sin y)/\partial y = \cos y = \partial(1 + x \cos y)/\partial x$, so the equation is exact and there exists f such that $f_x(x, y) = \sin y$ implies $f(x, y) = x \sin y + g(y)$ and $f_y(x, y) = x \cos y + g'(y)$. But $f_y(x, y) = 1 + x \cos y$ so $g'(y) = 1$ and $f(x, y) = x \sin y + y$. Hence the solution is $x \sin y + y = C$.

9. $\partial[(x+y)e^{x/y}]/\partial y = e^{x/y}[1 + (x+y)(-xy^{-2})] = e^{x/y}(1 - x^2y^{-2} - xy^{-1})$ and $\partial[(x - x^2y^{-1})e^{x/y}] = e^{x/y}[1 - 2xy^{-1} + (x - x^2y^{-1})y^{-1}] = e^{x/y}$, so the equation is exact and there exists f such that $f_x(x, y) = (x + y)e^{x/y}$ implies $f(x, y) = e^{x/y}[(x + y)y - y^2] + g(y) = xye^{x/y} + g(y)$ and $f_y(x, y) = e^{x/y}[x + (xy)(-xy^{-2})] + g'(y) = e^{x/y}(x + x^2y^{-1}) + g'(y)$. But $f_y(x, y) = (x - x^2y^{-1})e^{x/y}$ so $g'(y) = 0$ and $f(x, y) = xye^{x/y}$. Thus the solution is given by $xye^{x/y} = C$.

11. $\partial(x \ln y)/\partial y = x/y$, $\partial(-x - y \ln x)/\partial x = -1 - y/x$, so the equation isn't exact.

205

Section 15.3

13. $\partial(y^{-1} + 2yx^{-3})/\partial y = -y^{-2} + 2x^{-3} = \partial(-xy^{-2} - x^{-2})/\partial x$, so the equation is exact.
Hence there exists f such that $f_x(x,y) = y^{-1} + 2yx^{-3}$ implies $f(x,y) = xy^{-1} - yx^{-2} + g(y)$
and $f_y(x,y) = -xy^{-2} - x^{-2} + g'(y)$. But $f_y(x,y) = -xy^{-2} - x^{-2}$ so $g'(y) = 0$ and the
solution is given by $xy^{-1} - yx^{-2} = C$. For an explicit formula we could write $x^3 - y^2 = Cx^2y$
or $y^2 + Cx^2y - x^3 = 0$ and solve for y using the quadratic formula. This gives
$y = \frac{1}{2}[-Cx^2 \pm \sqrt{C^2x^4 + 4x^3}] = -Kx^2 \pm \sqrt{x^3 + K^2x^4}$, where $K = C/2$.

15. Here the equation is exact and f such that $f_x = P$ and $f_y = Q$ is given by
$f(x,y) = x^3 + x^2y + 3xy^2$. Thus the general solution is given by $3xy^2 + x^2y + x^3 = C$.
But when $x = 1$, $y = 2$, $C = 15$ so the solution is $3xy^2 + x^2y + x^3 = 15$. But this is a
quadratic in y so $y = [-x^2 \pm \sqrt{x^4 - 4(3x)(x^3 - 15)}]/(6x) = [-x^2 \pm \sqrt{180x - 11x^4}]/(6x)$.
However $2 = y(1)$ so we must take the positive sign, i.e., the solution explicitly is
$y = (-x^2 + \sqrt{180x - 11x^4})/(6x)$.

17. The equation is exact so there exists f such that $f_x(x,y) = 1 + y\cos xy$ implies
$f(x,y) = x + \sin xy + g(y)$ and $f_y(x,y) = x\cos xy + g'(y)$. But $f_y(x,y) = x\cos xy$ so the
general solution is given by $x + \sin xy = C$. But when $x = 1$, $y = 0$ so $C = 1$ and the solution
to the initial-value problem is $x + \sin xy = 1$ which we could rewrite as $y = [\sin^{-1}(1-x)]/x$.

19. $\partial(y^2)/\partial y = 2y$ while $\partial(1 + xy)/\partial x = y$; but $\partial(y^2 e^{xy})/\partial y = e^{xy}(2y + y^2 x)$ and
$\partial[(1 + xy)e^{xy}]/\partial x = e^{xy}(y + y(1 + xy)) = e^{xy}(2y + y^2 x)$. Hence the new equation
$y^2 e^{xy} + (1 + xy)e^{xy}(dy/dx) = 0$ is exact so there exists f such that $f_x(x,y) = y^2 e^{xy}$
implies $f(x,y) = ye^{xy} + g(y)$ and $f_y(x,y) = e^{xy}(1 + yx) + g'(y)$. But $f_y(x,y) = e^{xy}(1 + xy)$
so the solution is given by $ye^{xy} = C$.

21. $\partial(y + y^3)/\partial y = 1 + 3y^2$ while $\partial(x + x^3)/\partial x = 1 + 3x^2$; but
$\partial[(y + y^3)/(1 + x^2 + y^2)^{3/2}]/\partial y = (1 + x^2 + 3x^2y^2 + y^2)/(1 + x^2 + y^2)^{5/2}$
$= \partial[(x + x^3)/(1 + x^2 + y^2)^{3/2}]/\partial x$. So the new equation is exact and so there exists f
such that $f_x(x,y) = (y + y^3)/(1 + x^2 + y^2)^{3/2}$ implies
$f(x,y) = (y + y^3)[x/(1 + y^2)(1 + x^2 + y^2)^{1/2}] + g(y) = xy/(1 + x^2 + y^2)^{1/2} + g(y)$ and
$f_y(x,y) = (x + x^3)/(1 + x^2 + y^2)^{3/2} + g'(y)$. But
$f_y(x,y) = (x + x^3)/(1 + x^2 + y^2)^{3/2}$ so $g'(y) = 0$ and the solution is given by
$xy/(1 + x^2 + y^2)^{1/2} = C$.

23. $P_y = 3x + 4y$ and $Q_x = 2x + 2y$ so the equation isn't exact. But
$(P_y - Q_x)/Q = (x + 2y)/(x^2 + 2xy) = 1/x$. Thus there exists an integrating factor
which satisfies $dI/dx = I/x$ so $I(x) = x$ and the new equation
$3x^2y + 2xy^2 + (x^3 + 2x^2y)y' = 0$ is exact. So there exists f such that
$f_x(x,y) = 3x^2y + 2xy^2$ and $f_y(x,y) = x^3 + 2x^2y$. Hence

206

Section 15.4

$f(x, y) = x^3y + x^2y^2$ and the solution is $x^3y + x^2y^2 = C$.

25. $(P_y - Q_x)/Q = [(2x + 3x^2 + 6y) - (2x)]/(x^2 + 2y) = 3$. Thus there exists an integrating factor which satisfies $dI/dx = 3I$ or $I(x) = e^{3x}$ and the new equation
$e^{3x}(2xy + 3x^2y + 3y^2) + e^{3x}(x^2 + 2y)y' = 0$ is exact. So there exists f such that
$f_x(x, y) = e^{3x}(2xy + 3x^2y + 3y^2)$ and $f_y(x, y) = e^{3x}(x^2 + 2y)$. The second equation implies $f(x, y) = e^{3x}(x^2y + y^2) + h(x)$ and $f_x(x, y) = e^{3x}(3x^2y + 3y^2 + 2xy) + h'(x)$ so $f(x, y) = e^{3x}(x^2y + y^2)$ and the solution is given by $e^{3x}(x^2y + y^2) = C$.

27. The general form of a first order separable equation is $dy/dx = g(x)f(y)$ or $[1/f(y)](dy/dx) = g(x)$ or $-g(x) + [1/f(y)](dy/dx) = 0$. Hence $\partial(-g(x))/\partial y = 0 = \partial(1/f(y))/\partial x$, so the equation is exact.

EXERCISES 15.4

1. (a) $yy' + x^2 = 0 \Rightarrow y\, dy = -x^2\, dx$ and so the equation is separable.
   (b) $y' = (-x^2)/y$ which cannot be written as a function of y/x. Thus, it is not homogeneous.
   (c) This equation is not linear since it cannot be placed into standard linear form (15.14).
   (d) $\partial(x^2)/\partial y = 0$, $\partial(y)/\partial x = 0$ and so the equation is exact.

3. (a) $xy' + y = 0 \Rightarrow (1/y)dy = -(1/x)dx$ and so the equation is separable.
   (b) $y' = -(y/x)$ and so it is homogeneous.
   (c) $xy' + y = 0 \Rightarrow y' + (1/x)y = 0$ which is in standard linear form and so the equation is linear.
   (d) $\partial(y)/\partial y = 1$, $\partial(x)/\partial x = 1$ and so the equation is exact.

5. $y' - y = \sin x$ is linear with integrating factor $I(x) = e^{-\int dx} = e^{-x}$ and multiplying the differential equation by $I(x): e^{-x}y' - e^{-x}y = e^{-x}\sin x$
$\Rightarrow (e^{-x}y)' = e^{-x}\sin x \Rightarrow y = e^x[\int e^{-x}\sin x\, dx + C] = e^x[e^{-x}(-\sin x - \cos x)/2 + C]$
$= Ce^x - (\sin x + \cos x)/2$.

7. Since $y' = (e^x - x)/(4y^3)$ the equation is separable as well as exact. Then $4y^3 dy = (e^x - x)dx$ or $y^4 = e^x - (x^2/2) + C$ or $y = \pm[e^x - (x^2/2) + C]^{1/4}$.

9. Since $y' = (x^2 + 2xy - y^2)/(x^2 - 2xy - y^2)$
$= [1 + 2(y/x) - (y^2/x^2)]/[1 - 2(y/x) - (y^2/x^2)]$ the equation is homogeneous
(note: it isn't exact since $P_y = -Q_x$). Setting $v = y/x$ gives

207

Section 15.4

$v' = [(1 + 2v - v^2)/(1 - 2v - v^2) - v]/x = (1 + v + v^2 + v^3)/[(1 - 2v - v^2)x]$. Then
$[(1 - 2v - v^2)/(v^3 + v^2 + v + 1)]dv = (1/x)dx$ or
$[-2v/(v^2 + 1) + 1/(v + 1)]dv = (1/x)dx$ so $-\ln(v^2 + 1) + \ln|v + 1| = \ln|x| + C$. Hence
$(v + 1)/(v^2 + 1) = kx$ and the solution is given by $x(y + x)/(x^2 + y^2) = kx$ or
$x + y = k(x^2 + y^2)$.

11. Since $\partial(-3y\sec^2 x + y^2)/\partial y = -3\sec^2 x + 2y = \partial(2xy - 3\tan x)/\partial x$ the equation is exact so there exists f such that $f_x(x,y) = y^2 - 3y\sec^2 x$ and $f_y(x,y) = 2xy - 3\tan x$. Thus $f(x,y) = xy^2 - 3y\tan x$ and the solution is given by $xy^2 - 3y\tan x = C$.

13. Since $y' = x/(x^2y + y) = [x/(x^2 + 1)](1/y)$ the equation is separable and exact. Then $y\,dy = [x/(x^2 + 1)]dx$, so $y^2/2 = (1/2)\ln(x^2 + 1) + C$ or $y^2 = \ln(x^2 + 1) + K$.

15. $y' - (2/x)y = \sqrt{1 + x^2}/x$ so the equation is linear. $I(x) = e^{\int(-2/x)dx} = x^{-2}$ and multiplying the differential equation by $I(x)$: $x^{-2}y' - 2x^{-3} = x^{-3}\sqrt{1 + x^2}$
$\Rightarrow (x^{-2}y)' = x^{-3}\sqrt{1 + x^2} \Rightarrow y = x^2[\int(\sqrt{1 + x^2}/x^3)dx + C]$
$= x^2[-\sqrt{1 + x^2}/(2x^2) - (1/2)\ln|(\sqrt{1 + x^2} + 1)/x| + C]$ by trig substitution.
Hence the solution is $y = Cx^2 - (1/2)\sqrt{1 + x^2} - (x^2/2)\ln|(\sqrt{1 + x^2} + 1)/x|$.

17. $y' = -y^2\sqrt{1 + x^3}$ is separable and exact so $(1/y^2)dy = -\sqrt{1 + x^3}dx$ or
$-y^{-1} = -\int\sqrt{1 + x^3}dx + C$ or $y = 1/[K + \int\sqrt{1 + x^3}dx]$.

19. Rewriting the equation as $(2x + e^y) + (2y + xe^y)y' = 0$ we see the equation is exact. Thus there exists f such that $f_x(x,y) = 2x + e^y$ and $f_y(x,y) = 2y + xe^y$. So $f(x,y) = x^2 + y^2 + xe^y$ and the solution is given by $x^2 + y^2 + xe^y = C$.

21. $\partial[y^2\cos(xy) - 1]/\partial y = 2y\cos(xy) - y^2x\sin(xy) = \partial[\sin(xy) + xy\cos(xy)]/\partial x$ so the equation is exact and there exists f such that $f_x(x,y) = y^2\cos(xy) - 1$ and $f_y(x,y) = \sin(xy) + xy\cos(xy)$. Thus $f(x,y) = y\sin(xy) - x$ and the solution is given by $y\sin(xy) - x = C$.

23. $y' = (2\sqrt{xy} - y)/x = 2\sqrt{y/x} - (y/x)$ is homogeneous. Setting $v = y/x$ gives
$v' = (2\sqrt{v} - 2v)/x$ or $[1/2(\sqrt{v} - v)]dv = (1/x)dx$ or
$[(1/2)v^{-1/2}/(1 - v^{1/2})]dv = (1/x)dx$. Thus
$-\ln|1 - v^{1/2}| = \ln|x| + C_1$ implies $\ln|1 - v^{1/2}| = K - \ln|x|$ or $1 - \sqrt{v} = C/x$ or
$v = [1 - (C/x)]^2$. And the solution is $y = x[1 - (C/x)]^2$.

Section 15.5

25. (a) $y' = 3x(y + x^n)$ is separable only if $n = 0$ since only then can we separate x and y factors to get an equation of the form (15.19).

(b) $y' = 3x(y + x^n) \Rightarrow y' - 3xy = 3x^{n+1}$ which is linear [of the form (15.20)] for any integer n (and in fact for any real number n).

EXERCISES 15.5

1. The auxiliary equation is $r^2 - 3r + 2 = (r-2)(r-1) = 0$ so $y = c_1 e^x + c_2 e^{2x}$.

3. The auxiliary equation is $3r^2 - 8r - 3 = (3r+1)(r-3) = 0$ so $y = c_1 e^{-x/3} + c_2 e^{3x}$.

5. $r^2 + 2r + 10 = 0$ or $r = -1 \pm 3i$ and the solution is $y = e^{-x}(c_1 \cos 3x + c_2 \sin 3x)$.

7. $r^2 - 1 = (r-1)(r+1) = 0$ so $y = c_1 e^x + c_2 e^{-x}$.

9. $r^2 + 25 = 0$ or $r = \pm 5i$ and the solution is $y = c_1 \cos 5x + c_2 \sin 5x$.

11. $2r^2 + r = r(2r+1) = 0$ so $y = c_1 + c_2 e^{-x/2}$.

13. $r^2 - r + 2 = 0$ or $r = (1 \pm \sqrt{7}i)/2$ and the solution is
$y = e^{x/2}(c_1 \cos(\sqrt{7}x/2) + c_2 \sin(\sqrt{7}x/2))$.

15. $r^2 + 2r - 1 = (r - (-1 + \sqrt{2}))(r - (-1 - \sqrt{2}))$ so $y = c_1 e^{(-1+\sqrt{2})x} + c_2 e^{(-1-\sqrt{2})x}$

17. $r^2 - 8r + 16 = (r-4)^2 = 0$ so $y = c_1 e^{4x} + c_2 x e^{4x}$.

19. $r^2 - 2r + 5 = 0$ or $r = 1 \pm 2i$ and the solution is $y = e^x(c_1 \cos 2x + c_2 \sin 2x)$.

21. $r^2 + 3r - 4 = (r+4)(r-1) = 0$ so the general solution is $y = c_1 e^x + c_2 e^{-4x}$. Then $2 = y(0) = c_1 + c_2$ and $-3 = y'(0) = c_1 - 4c_2$ so $c_1 = 1$, $c_2 = 1$ and the solution to the initial-value problem is $y = e^x + e^{-4x}$.

23. $r^2 - 2r + 2 = 0$ or $r = 1 \pm i$ and the general solution is $y = e^x(c_1 \cos x + c_2 \sin x)$. But $1 = y(0) = c_1$ and $2 = y'(0) = c_1 + c_2$ so the solution to the initial-value problem is $y = e^x(\cos x + \sin x)$.

25. $r^2 - 2r - 3 = (r-3)(r+1) = 0$ so the general solution is $y = c_1 e^{-x} + c_2 e^{3x}$. However the conditions are given at $x = 1$ so rewrite the general solution as
$y = k_1 e^{-(x-1)} + k_2 e^{3(x-1)}$. Then $3 = y(1) = k_1 + k_2$ and $1 = y'(1) = -k_1 + 3k_2$ so $k_1 = 2$, $k_2 = 1$ and the solution to the initial-value problem is $y = 2e^{-(x-1)} + e^{3(x-1)}$.

27. $r^2 + 9 = 0$ or $r = \pm 3i$ and the general solution is $y = c_1 \cos 3x + c_2 \sin 3x$. But

209

Section 15.5

$0 = y(\pi/3) = -c_1$ and $1 = y'(\pi/3) = -3c_2$, so the solution to the initial-value problem is $y = (-1/3)\sin 3x$.

29. $r^2 + 4r + 4 = (r+2)^2 = 0$ so the general solution is $y = c_1 e^{-2x} + c_2 x e^{-2x}$. Then $0 = y(0) = c_1$ and $3 = y(1) = 0 + c_2 e^{-2}$ so $c_2 = 3e^2$ and the solution of the boundary-value problem is $y = 3xe^{-2x+2}$.

31. $r^2 + 1 = 0$ or $r = \pm i$ and the general solution is $y = c_1 \cos x + c_2 \sin x$. But $1 = y(0) = c_1$ and $0 = y(\pi) = -c_1$ so there is NO solution.

33. $r^2 - r - 2 = (r-2)(r+1) = 0$ so the general solution is $y = c_1 e^{-x} + c_2 e^{2x}$. Then $1 = y(-1) = c_1 e + c_2 e^{-2}$ and $0 = y(1) = c_1 e^{-1} + c_2 e^2$ so $c_1 = e^5/(e^6 - 1)$ and $c_2 = e^2/(1 - e^6)$. Hence the solution to the boundary-value problem is
$y = [e^5/(e^6-1)]e^{-x} + [e^2/(1-e^6)]e^{2x} = (1/(e^6-1))[e^{5-x} - e^{2(1+x)}]$.

35. $r^2 + 4r + 13 = 0$ or $r = -2 \pm 3i$ and the general solution is
$y = e^{-2x}(c_1 \cos 3x + c_2 \sin 3x)$.
But $2 = y(0) = c_1$ and $1 = y(\pi/2) = e^{-\pi}(-c_2)$, so the solution to the boundary-value problem is $y = e^{-2x}(2\cos 3x - e^{\pi} \sin 3x)$.

37. (a) CASE I : $\lambda = 0$. $y'' + \lambda y = 0 \Rightarrow y'' = 0$ which has an auxiliary equation $r^2 = 0$ $\Rightarrow r = 0, 0 \Rightarrow y = c_1 + c_2 x$ where $y(0) = 0$ and $y(L) = 0$. Thus, $0 = y(0) = c_1$ and $0 = y(L) = c_2 L \Rightarrow c_1 = c_2 = 0$. Thus, $y = 0$.
CASE II : $\lambda < 0$. $y'' + \lambda y = 0$ has auxiliary equation $r^2 = -\lambda \Rightarrow r = \pm\sqrt{-\lambda}$ (distinct and real since $\lambda < 0$) $\Rightarrow y = c_1 e^{\sqrt{-\lambda}x} + c_2 e^{-\sqrt{-\lambda}x}$ where $y(0) = 0$ and $y(L) = 0$.
Thus, $0 = y(0) = c_1 + c_2$ (1) and $0 = y(L) = c_1 e^{\sqrt{-\lambda}L} + c_2 e^{-\sqrt{-\lambda}L}$ (2).
Multiplying (1) by $e^{\sqrt{-\lambda}L}$ and subtracting (2) gives $c_2(e^{\sqrt{-\lambda}L} - e^{-\sqrt{-\lambda}L}) = 0 \Rightarrow c_2 = 0$ and thus $c_1 = 0$ from (1). Thus, $y = 0$ for the cases $\lambda = 0$ and $\lambda < 0$.
(b) $y'' + \lambda y = 0$ has an auxiliary equation $r^2 + \lambda = 0 \Rightarrow r = \pm i\sqrt{\lambda}$ $\Rightarrow y = c_1 \cos\sqrt{\lambda}x + c_2 \sin\sqrt{\lambda}x$ where $y(0) = 0$ and $y(L) = 0$. Thus, $0 = y(0) = c_1$ and $0 = y(L) = c_2 \sin\sqrt{\lambda}L$ since $c_1 = 0$. Since we cannot have a trivial solution, $c_2 \neq 0$ and thus $\sin\sqrt{\lambda}L = 0 \Rightarrow \sqrt{\lambda}L = n\pi$ where n is an integer $\Rightarrow \lambda = n^2\pi^2/L^2$ and $y = c_2 \sin(n\pi x/L)$ where n is an integer.

EXERCISES 15.6

1. $y_c(x) = c_1 e^{3x} + c_2 e^{-2x}$ and try the particular solution $y_p(x) = A\cos 3x + B\sin 3x$. Then we need

$$
\begin{aligned}
-6y_p &= -6A\cos 3x - 6B\sin 3x \\
-y_p' &= -3B\cos 3x + 3A\sin 3x \\
y_p'' &= -9A\cos 3x - 9B\sin 3x \quad \text{or}
\end{aligned}
$$

$\cos 3x = (-15A - 3B)\cos 3x + (3A - 15B)\sin 3x$.

Hence $A = -5/78$, $B = -1/78$ and the general solution is

$y(x) = y_c + y_p = c_1 e^{3x} + c_2 e^{-2x} - (5/78)\cos 3x - (1/78)\sin 3x$.

3. The complementary solution is $y_c(x) = e^{2x}(c_1 x + c_2)$, so try the particular solution $y_p(x) = Ae^{-x}$. Then we need $Ae^{-x} + 4Ae^{-x} + 4Ae^{-x} = e^{-x}$ or $A = 1/9$. Hence the general solution is $y(x) = e^{2x}(c_1 x + c_2) + e^{-x}/9$.

5. The complementary solution is $y_c(x) = c_1\cos 6x + c_2\sin 6x$ and try $y_p(x) = Ax^2 + Bx + C$. Then we need $2A + 36Ax^2 + 36Bx + 36C = 2x^2 - x$. Thus $A = 1/18$, $B = -1/36$, $C = -1/324$ and the general solution is
$y(x) = c_1\cos 6x + c_2\sin 6x + x^2/18 - x/36 - 1/324$.

7. $y_c(x) = c_1 e^{-x/4} + c_2 e^{-x}$ and try $y_p(x) = Ae^x$. Then $(10A)e^x = e^x$ so $A = 1/10$ and the general solution is $y(x) = c_1 e^{-x/4} + c_2 e^{-x} + e^x/10$.

9. Since the roots of $r^2 - 2r + 5 = 0$ are $1 \pm 2i$, $y_c(x) = e^x(c_1\cos 2x + c_2\sin 2x)$. For $y'' - 2y' + 5y = x$ try $y_{p_1}(x) = Ax + B$. Then $5Ax + B - 2A = x$, so $y_{p_1}(x) = (x + 2)/5$. For $y'' - 2y' + 5y = \sin 3x$ try $y_{p_2}(x) = A\cos 3x + B\sin 3x$. Then $-9A\cos 3x - 9B\sin 3x + 6A\sin 3x - 6B\cos 3x + 5A\cos 3x + 5B\sin 3x = \sin 3x$.
Thus $(-9A - 6B + 5A) = 0$ and $(-9B + 6A + 5B) = 1$ so $A = 3/26$ and $B = -1/13$.
Hence the general solution is
$y(x) = e^x(c_1\cos 2x + c_2\sin 2x) + (x + 2)/5 + (3/26)\cos 3x - (1/13)\sin 3x$.
But $1 = y(0) = c_1 + (2/5) + (3/26)$ implies $c_1 = 63/130$ and
$2 = y'(0) = c_1 + 2c_2 + (1/5) - (3/13)$ implies $c_2 = 201/260$. Thus the solution to the initial-value problem is
$y(x) = e^x[(63/130)\cos 2x + (201/260)\sin 2x] + (x + 2)/5 + (3/26)\cos 3x - (1/13)\sin 3x$.

11. $y_c(x) = c_1 e^x + c_2 e^{-x}$. Try $y_p(x) = (Ax + B)e^{3x}$. Then
$e^{3x}(9Ax + 9B + 6A) - e^{3x}(Ax + B) = xe^{3x}$ implies $A = 1/8$, $B = -3/32$ and the general solution is $y(x) = c_1 e^x + c_2 e^{-x} + [(1/8)x - (3/32)]e^{3x}$. But $0 = y(0) = c_1 + c_2 - 3/32$ and $1 = y'(0) = c_1 - c_2 - 9/32 + 1/8$ so the solution to the initial-value problem is

$$y(x) = (5/8)e^x - (17/32)e^{-x} + e^{3x}[(1/8)x - 3/32].$$

13. Since the roots of the auxiliary equation are complex, we need just try
$$y_p(x) = (Ax^4 + Bx^3 + Cx^2 + Dx + E)e^{2x}.$$

15. Since $y_c(x) = e^x(c_1 \cos x + c_2 \sin x)$ we try $y_p(x) = xe^x(A\cos x + B\sin x)$.

Solving Equations (15.46) and (15.48) in the Method of Variation of Parameters gives
$$u_1' = -Gy_2/a(y_1y_2' - y_2y_1') \text{ and } u_2' = Gy_1/a(y_1y_2' - y_2y_1'). \text{ We will use these equations rather than repeatedly resolving the system in each of the remaining exercises.}$$

17. (a) The complementary solution is $y_c(x) = c_1\cos 2x + c_2\sin 2x$. A particular solution is of the form $y_p(x) = Ax + B$. Thus, $4Ax + 4B = x \Rightarrow 4A = 1$ and $4B = 0 \Rightarrow A = \frac{1}{4}$ and $B = 0 \Rightarrow y_p(x) = \frac{1}{4}x$. Thus, the general solution is $y = y_c + y_p$
$= c_1\cos 2x + c_2\sin 2x + \frac{1}{4}x$.

(b) In (a) $y_c(x) = c_1 \cos 2x + c_2 \sin 2x$, so set $y_1 = \cos 2x$, $y_2 = \sin 2x$. Then
$y_1y_2' - y_2y_1' = 2\cos^2 2x + 2\sin^2 2x = 2$ so $u_1' = -x\sin 2x/2$ implies
$u_1(x) = -\frac{1}{2}\int x \sin 2x \, dx = -\frac{1}{4}(-x\cos 2x + \frac{1}{2}\sin 2x)$ by parts
and $u_2' = x\cos 2x/2$ implies $u_2(x) = \frac{1}{2}\int x\cos 2x \, dx = \frac{1}{4}(x\sin 2x + \frac{1}{2}\cos 2x)$ by parts.
Hence $y_p(x) = -\frac{1}{4}(-x\cos 2x + \frac{1}{2}\sin 2x)\cos 2x + \frac{1}{4}(x\sin 2x + \frac{1}{2}\cos 2x)\sin 2x = \frac{1}{4}x$.
Thus, $y(x) = y_c(x) + y_p(x) = c_1\cos 2x + c_2\sin 2x + \frac{1}{4}x$.

19. (a) $r^2 - r = r(r-1) = 0 \Rightarrow r = 0, 1$, so the complementary solution is $y_c(x) = c_1e^x + c_2xe^x$. A particular solution is of the form $y_p(x) = Ae^{2x}$. Thus, $4Ae^{2x} - 4Ae^{2x} + Ae^{2x} = e^{2x} \Rightarrow Ae^{2x} = e^{2x} \Rightarrow A = 1 \Rightarrow y_p(x) = e^{2x}$. Thus, a general solution is $y(x) = y_c(x) + y_p(x)$
$= c_1e^x + c_2xe^x + e^{2x}$.

(b) In (a) $y_c(x) = c_1 e^x + c_2 xe^x$, so set $y_1 = e^x$, $y_2 = xe^x$. Then,
$y_1y_2' - y_2y_1' = e^{2x}(1+x) - xe^{2x} = e^{2x}$ and so $u_1' = -xe^x$ implies
$u_1(x) = -\int xe^x \, dx = -(x-1)e^x$ by parts and $u_2' = e^x$ implies $u_2(x) = \int e^x \, dx = e^x$.
Hence $y_p(x) = (1-x)e^{2x} + xe^{2x} = e^{2x}$ and the general solution is
$y(x) = y_c(x) + y_p(x) = c_1e^x + c_2xe^x + e^{2x}$.

21. As in Example 6, $y_c(x) = c_1\sin x + c_2\cos x$, so set $y_1 = \sin x$, $y_2 = \cos x$. Then
$y_1y_2' - y_2y_1' = -\sin^2 x - \cos^2 x = -1$ so $u_1' = -\sec x\cos x/(-1) = 1$ implies $u_1(x) = x$
and $u_2' = \sec x \sin x/(-1) = -\tan x$ implies $u_2(x) = -\int \tan x \, dx = \ln|\cos x| = \ln \cos x$
on $0 < x < \pi/2$. Hence $y_p(x) = x\sin x + (\cos x)\ln \cos x$ and the general solution is
$y(x) = (c_1 + x)\sin x + (c_2 + \ln \cos x)\cos x$.

23. $y_1 = e^x$, $y_2 = e^{2x}$ and $y_1 y_2' - y_2 y_1' = e^{3x}$. So $u_1' = -e^{2x}/[(1+e^{-x})(e^{3x})]$
$= -e^{-x}/(1+e^{-x})$ and $u_1(x) = \int [-e^{-x}/(1+e^{-x})]dx = \ln(1+e^{-x})$. And
$u_2' = e^x/[(1+e^{-x})(e^{3x})] = e^x/(e^{3x}+e^{2x})$ so $u_2(x) = \int [e^x/(e^{3x}+e^{2x})]dx$
$= \ln[(e^x+1)/e^x] - e^{-x} = \ln(1+e^{-x}) - e^{-x}$.

Hence $y_p(x) = e^x \ln(1+e^{-x}) + e^{2x}[\ln(1+e^{-x}) - e^{-x}]$ and the general solution is
$y(x) = [c_1 + \ln(1+e^{-x})]e^x + [c_2 - e^{-x} + \ln(1+e^{-x})]e^{2x}$.

25. $y_1 = e^{-x}$, $y_2 = e^x$ and $y_1 y_2' - y_2 y_1' = 2$. So $u_1' = -e^x/(2x)$, $u_2' = e^{-x}/(2x)$ and
$y_p(x) = -e^{-x}\int (e^x/(2x))dx + e^x \int (e^{-x}/(2x))dx$. Hence the general solution is
$y(x) = [c_1 - \int (e^x/(2x))dx]e^{-x} + [c_2 + \int (e^{-x}/(2x))dx]e^x$.

EXERCISES 15.7

1. By Hooke's Law $k(0.6) = 20$ so $k = \frac{100}{3}$ is the spring constant and the differential
equation is $3x'' + \frac{100}{3}x = 0$. The general solution is
$x(t) = c_1 \cos(\frac{10}{3}t) + c_2 \sin(\frac{10}{3}t)$.
But $0 = x(0) = c_1$ and $1.2 = x'(0) = \frac{10}{3}c_2$ so the position of the mass after
t seconds is $x(t) = 0.36 \sin(\frac{10}{3}t)$.

3. $k(0.5) = 6$ or $k = 12$ is the spring constant and the initial-value problem is
$2x'' + 14x' + 12x = 0$, $x(0) = 1$, $x'(0) = 0$. The general solution is
$x(t) = c_1 e^{-6t} + c_2 e^{-t}$. But $1 = x(0) = c_1 + c_2$ and $0 = x'(0) = -6c_1 - c_2$.
Thus the position is given by $x(t) = (-1/5)e^{-6t} + (6/5)e^{-t}$.

5. To be critically damped, we need
$c^2 - 4mk = 0$ or $m = c^2/4k = (14)^2/(4)(12) = 49/12$ kg.

7. The differential equation is $mx'' + kx = F_0 \cos \omega_0 t$ and $\omega_0 \neq \omega = \sqrt{k/m}$. Here the
auxiliary equation is $mr^2 + k = 0$ with roots $\pm\sqrt{k/m}\,i = \pm \omega i$
so $x_c(t) = c_1 \cos \omega t + c_2 \sin \omega t$. Since $\omega_0 \neq \omega$, try $x_p(t) = A \cos \omega_0 t + B \sin \omega_0 t$. Then
we need $(m)(-\omega_0^2)(A \cos \omega_0 t + B \sin \omega_0 t) + k(A \cos \omega_0 t + B \sin \omega_0 t) = F_0 \cos \omega_0 t$ or
$A(k - m\omega_0^2) = F_0$ and $B(k - m\omega_0^2) = 0$. Hence $B = 0$ and
$A = F_0/(k - m\omega_0^2) = F_0/m(\omega^2 - \omega_0^2)$ since $\omega^2 = k/m$. Thus the motion of the mass is
given by $x(t) = c_1 \cos \omega t + c_2 \sin \omega t + [F_0/m(\omega^2 - \omega_0^2)]\cos \omega_0 t$.

Section 15.8

9. Here the initial-value problem for the charge is $Q'' + 20Q' + 500Q = 12$, $Q(0) = Q'(0) = 0$. Then $Q_c(t) = e^{-10t}(c_1 \cos 20t + c_2 \sin 20t)$ and try $Q_p(t) = A \Rightarrow 500A = 12$ or $A = 3/125$. Hence the general solution is
$Q(t) = e^{-10t}(c_1 \cos 20t + c_2 \sin 20t) + 3/125$. But $0 = Q(0) = c_1 + 3/125$ and
$Q'(t) = I(t) = e^{-10t}[(-10c_1 + 20c_2) \cos 20t + (-10c_2 - 20c_1) \sin 20t]$ but
$0 = Q'(0) = -10c_1 + 20c_2$. Thus the charge is given by
$Q(t) = (-e^{-10t}/250)(6 \cos 20t + 3 \sin 20t) + 3/125$ and the current is given by
$I(t) = e^{-10t}(3/5) \sin 20t$.

11. As in Exercise 9, $Q_c(t) = e^{-10t}(c_1 \cos 20t + c_2 \sin 20t)$ but $E(t) = 12 \sin 10t$ so try
$Q_p(t) = A \cos 10t + B \sin 10t$. Then we need $-100A \cos 10t - 100B \sin 10t - 200A \sin 10t + 200B \cos 10t + 500A \cos 10t + 500B \sin 10t = 12 \sin 10t$, or $400A + 200B = 0$ and $400B - 200A = 12$. Thus $A = -3/250$, $B = 3/125$ and the general solution is
$Q(t) = e^{-10t}(c_1 \cos 20t + c_2 \sin 20t) - (3/250) \cos 10t + (3/125) \sin 10t$. But
$0 = Q(0) = c_1 - 3/250$ so $c_1 = 3/250$. Also
$Q'(t) = I(t) = (3/25) \sin 10t + (6/25) \cos 10t + e^{-10t}[(-10c_1 + 20c_2) \cos 20t + (-10c_2 - 20c_1) \sin 20t]$ and $0 = Q'(0) = 6/25 - 10c_1 + 20c_2$ so $c_2 = -3/500$. Hence the charge is given by
$Q(t) = e^{-10t}[(3/250) \cos 20t - (3/500) \sin 20t] - (3/250) \cos 10t + (3/125) \sin 10t$.

13. $x(t) = A \cos(\omega t + \delta) \Leftrightarrow x(t) = A[\cos \omega t \cos \delta - \sin \omega t \sin \delta]$
$\Leftrightarrow x(t) = A[\frac{c_1}{A} \cos \omega t + \frac{c_2}{A} \sin \omega t]$ where $\cos \delta = \frac{c_1}{A}$ and $\sin \delta = -\frac{c_2}{A}$
$\Leftrightarrow x(t) = c_1 \cos \omega t + c_2 \sin \omega t$. (Note that $\cos^2 \delta + \sin^2 \delta = 1 \Rightarrow c_1^2 + c_2^2 = A^2$.)

EXERCISES 15.8

1. Assuming $y(x) = \sum_{n=0}^{\infty} a_n x^n$, we have $\sum_{n=0}^{\infty} n a_n x^{n-1} - 6 \sum_{n=0}^{\infty} a_n x^n = 0$ or
$\sum_{n=1}^{\infty} n a_n x^{n-1} - 6 \sum_{n=0}^{\infty} a_n x^n = 0$. Replacing n by $n+1$ in the first series gives
$\sum_{n=0}^{\infty} [(n+1) a_{n+1} - 6 a_n] x^n = 0$. Thus the recurrence relation is
$a_{n+1} = 6 a_n / (n+1)$, $n = 0, 1, 2, \ldots$. Then $a_1 = 6 a_0$, $a_2 = 6 a_1 / 2 = 6^2 a_0 / 2$,
$a_3 = 6 a_2 / 3 = 6^3 a_0 / 2 \cdot 3$, ..., $a_n = 6 a_{n-1} / n = 6^n a_0 / n!$. Thus the solution is
$y(x) = \sum_{n=0}^{\infty} a_n x^n = \sum_{n=0}^{\infty} a_0 (6^n / n!) x^n = \sum_{n=0}^{\infty} [a_0 (6x)^n / n!] = a_0 e^{6x}$.

Section 15.8

3. Assuming $y(x) = \sum_{n=0}^{\infty} a_n x^n$ we have $y'(x) = \sum_{n=1}^{\infty} n a_n x^{n-1} = \sum_{n=0}^{\infty} (n+1) a_{n+1} x^n$ and

$-x^2 y = -\sum_{n=0}^{\infty} a_n x^{n+2} = -\sum_{n=2}^{\infty} a_{n-2} x^n$. Hence the differential equation becomes

$\sum_{n=0}^{\infty} (n+1) a_{n+1} x^n - \sum_{n=2}^{\infty} a_{n-2} x^n = 0$ or $a_1 = 2a_2 x + \sum_{n=2}^{\infty} [(n+1) a_{n+1} - a_{n-2}] x^n = 0$.

Equating coefficients gives $a_1 = a_2 = 0$ and $a_{n+1} = a_{n-2}/(n+1)$ for $n = 2, 3, \ldots$. Since $a_1 = 0$, the recurrence relation implies $a_4 = 0$ which implies $a_7 = 0$, ... so $a_{3n+1} = 0$. Similarly $a_2 = 0$ and the recurrence relation implies $a_{3n+2} = 0$. Finally $a_3 = a_0/3$, $a_6 = a_3/6 = a_0/6 \cdot 3 = a_0/(3^2)(2!)$, $a_9 = a_6/9 = a_0/9 \cdot 6 \cdot 3 = a_0/(3^3)(3!)$, ..., and $a_{3n} = a_0/(3^n)(n!)$. Thus the solution is

$y(x) = \sum_{n=0}^{\infty} a_n x^n = \sum_{n=0}^{\infty} a_{3n} x^{3n} = \sum_{n=0}^{\infty} [a_0 (x^3/3)^n / n!] = a_0 e^{x^3/3}$.

5. Assuming $y(x) = \sum_{n=0}^{\infty} a_n x^n$, $y''(x) = \sum_{n=0}^{\infty} n(n-1) a_n x^{n-2} = \sum_{n=0}^{\infty} (n+2)(n+1) a_{n+2} x^n$,

$3xy'(x) = 3x \sum_{n=0}^{\infty} n a_n x^{n-1} = \sum_{n=0}^{\infty} 3n a_n x^n$ and the differential equation becomes

$\sum_{n=0}^{\infty} (n+2)(n+1) a_{n+2} + \sum_{n=0}^{\infty} 3n a_n x^n + \sum_{n=0}^{\infty} 3a_n x^n = 0$

or $\sum_{n=0}^{\infty} [(n+2)(n+1) a_{n+2} + 3(n+1) a_n] x^n = 0$. Thus the recurrence relation is $a_{n+2} = -3a_n/(n+2)$ for $n = 0, 1, 2, \ldots$. Then assuming a_0 and a_1 are known, $a_2 = -3a_0/2$, $a_4 = -3a_2/4 = (-3)^2 a_0/(2^2)(2!)$, $a_6 = -3a_4/6$
$= (-3)^3 a_0/(2^3)(3!)$, ..., $a_{2n} = (-3)^n a_0/(2^n)(n!) = (-1)^n (3/2)^n a_0/n!$ and
$a_3 = -3a_1/3$, $a_5 = -3a_3/5 = (-3)^2 a_1/5 \cdot 3$, $a_7 = -3a_5/7 = (-3)^3 a_1/7 \cdot 5 \cdot 3$, ...,
$a_{2n+1} = (-3)^n a_1/(2n+1)(2n-1)\cdots(5)(3) = (-3)^n a_1 2^n n!/(2n+1)!$
$= (-1)^n 6^n (n!) a_1/(2n+1)!$ noting $(2n+1)(2n-1)\cdots(5)(3) = (2n+1)!/(2^n)(n!)$. Thus

the solution is $y(x) = \sum_{n=0}^{\infty} a_n x^n = \sum_{n=0}^{\infty} a_{2n} x^{2n} + \sum_{n=0}^{\infty} a_{2n+1} x^{2n+1}$

$= \sum_{n=0}^{\infty} [(-1)^n a_0 (3x^2/2)^n/n!] + \sum_{n=0}^{\infty} [a_1 (-1)^n n! 6^n x^{2n+1}/(2n+1)!]$

$= a_0 e^{-3x^2/2} + a_1 \sum_{n=0}^{\infty} [(-1)^n 6^n (n!) x^{2n+1}/(2n+1)!]$.

7. Assuming $y(x) = \sum_{n=0}^{\infty} a_n x^n$, $(x^2+1) y'' = (x^2+1) \sum_{n=0}^{\infty} n(n-1) a_n x^{n-2}$

$= \sum_{n=0}^{\infty} n(n-1) a_n x^n + \sum_{n=0}^{\infty} (n+2)(n+1) a_{n+2} x^n$, $xy' = \sum_{n=0}^{\infty} n a_n x^n$, and the differential

equation becomes $\sum_{n=0}^{\infty} \{(n+2)(n+1) a_{n+2} + [n(n-1) + n - 1] a_n\} x^n = 0$. Thus the

recurrence relation is $a_{n+2} = -(n-1) a_n/(n+2)$, $n = 0, 1, 2, \ldots$. Then assuming a_0 and

215

a_1 are known, $a_2 = -(-1)a_0/2 = a_0/2$, $a_4 = -a_2/4 = -a_0/(2^2)(2!)$,
$a_6 = -3a_4/6 = (-1)(-3)a_0/(2^3)(3!)$, ..., $a_{2n} = (-1)^{n-1}(1)(3)\cdots(2n-3)a_0/(2^n)(n!)$
$= (-1)^{n-1}(2n-3)!a_0/(2^n)(2^{n-2})(n!)(n-2)!$
$= (-1)^{n-1}(2n-3)!a_0/(2^{2n-2})(n!)(n-2)!$ for $n = 2, 3, \ldots$. And $a_3 = 0 \cdot a_1/3$
$= 0$ with the recurrence relation implies $a_{2n+1} = 0$ for $n = 1, 2, \ldots$. Thus the solution is

$y(x) = \sum_{n=0}^{\infty} a_n x^n$
$= a_0 + a_1 x + a_0 x^2/2 + a_0 \sum_{n=2}^{\infty} [(-1)^{n-1}(2n-3)!x^{2n}/(2^{2n-2})(n!)(n-2)!]$.

9. Assuming $y(x) = \sum_{n=0}^{\infty} a_n x^n$, $y''(x) = \sum_{n=0}^{\infty} (n+2)(n+1)a_{n+2}x^n$, $-xy'(x) = -\sum_{n=0}^{\infty} na_n x^n$
and the differential equation becomes $\sum_{n=0}^{\infty} [(n+2)(n+1)a_{n+2} - (n+1)a_n]x^n = 0$. Thus
the recurrence relation is $a_{n+2} = a_n/(n+2)$ for $n = 0, 1, 2, \ldots$. But $a_0 = y(0) = 1$
so $a_2 = 1/2$, $a_4 = a_2/4 = 1/2\cdot 4$, $a_6 = a_4/6 = 1/2\cdot 4\cdot 6$, ..., $a_{2n} = 1/(2^n)(n!)$. Also
$a_1 = y'(0) = 0$ and so the recurrence formula implies $a_{2n+1} = 0$ for $n = 0, 1, 2, \ldots$.
Thus the solution to the initial-value problem is

$y(x) = \sum_{n=0}^{\infty} a_n x^n = \sum_{n=0}^{\infty} x^{2n}/(2^n)(n!) = \sum_{n=0}^{\infty} (x^2/2)^n/n! = e^{x^2/2}$.

11. Assuming $y(x) = \sum_{n=0}^{\infty} a_n x^n$, $y''(x) = \sum_{n=0}^{\infty} n(n-1)a_n x^{n-2} = \sum_{n=-1}^{\infty} (n+3)(n+2)a_{n+3}x^{n+1}$
$= 2a_2 + \sum_{n=0}^{\infty} (n+3)(n+2)a_{n+3}x^{n+1}$ and the differential equation becomes
$2a_2 + \sum_{n=0}^{\infty} [(n+3)(n+2)a_{n+3} + (n+1)a_n]x^{n+1} = 0$. Then $a_2 = 0$ and the
recurrence relation is $a_{n+3} = -(n+1)a_n/(n+3)(n+2)$, $n = 0, 1, 2, \ldots$.
But $a_0 = y(0) = 0 = a_2$ and the recurrence relation implies $a_{3n} = a_{3n+2} = 0$ for
$n = 0, 1, 2, \ldots$. Also $a_1 = y'(0) = 1$ so $a_4 = -2/(4)(3)$,
$a_7 = -(5)a_4/(7)(6) = (-1)^2(2)(5)/(7)(6)(4)(3) = (-1)^2(2)^2(5)^2/7!$, ...,
$a_{3n+1} = (-1)^n(2)^2(5)^2\cdots(3n-1)^2/(3n+1)!$ Thus the solution is

$y(x) = \sum_{n=0}^{\infty} a_n x^n = x + \sum_{n=0}^{\infty} [(-1)^n(2)^2(5)^2\cdots(3n-1)^2 x^{3n+1}/(3n+1)!]$

Chapter 15 Review

REVIEW EXERCISES FOR CHAPTER 15

1. False. y' cannot be written as a function of $\frac{y}{x}$.

3. True. $\partial(3x^2 + 6xy)/\partial y = 6x$, $\partial(3x^2 + 2y)/\partial x = 6x$, which means it is exact

5. True. $y' + \frac{2}{x}y = \frac{\sin x}{x} \Rightarrow I(x) = e^{\int (2/x)\,dx} = e^{2\ln|x|} = x^2$.

7. True. $\cosh x$ and $\sinh x$ are linearly independent solutions of this linear homogeneous equation.

9. Since exact, there exists f such that $f_x(x,y) = 1 + 2xy^2$ and $f_y(x,y) = 2x^2y$ implies
$f(x,y) = x + x^2y^2 + g(y)$ and $f_y(x,y) = 2x^2y + g'(y)$. Hence the solution is $x + x^2y^2 = c$.

11. Since linear, $I(x) = e^{\int(-2/x)\,dx} = x^{-2}$ and multiplying the differential equation by $I(x)$:
$x^{-2}y' - 2y^{-3} = 1 \Rightarrow (x^{-2}y)' = 1 \Rightarrow y(x) = x^2[\int 1\,dx + C] = x^2[x + C] = Cx^2 + x^3$.

13. $(2y - 3y^2)dy = (x\cos x)dx$ or $y^2 - y^3 = x\sin x + \cos x + C$.

15. $y' = (x^2 + y^2)/(x^2 + xy) = (1 + (y/x)^2)/(1 + (y/x))$. Setting $v = y(x)$ gives
$v' = \{[(1+v^2)/(1+v)] - v\}/x$ or $[(1+v)/(1-v)]dv = (1/x)dx$. Thus
$-v - 2\ln|1-v| = \ln|x| + c_1$ or $-(y/x) - 2\ln|(x-y)/x| = \ln|x| + c_1$ or
$(y/x) + \ln(x-y)^2 = c + \ln|x|$.

17. $y' = (y^2 + 1)/\sqrt{1-x^2}$ so $[1/(y^2+1)]dy = (1/\sqrt{1-x^2})dx$.
Then $\tan^{-1}y = (\sin^{-1}x) + c$ or $y = \tan[(\sin^{-1}x) + c]$.

19. Since linear, $I(x) = e^{\int -dx} = e^{-x}$ and multiplying the differential equation by $I(x)$:
$e^{-x}y' - e^{-x}y = e^x \Rightarrow (e^{-x}y)' = e^x \Rightarrow y(x) = e^x[\int e^x dx + C] = e^{2x} + Ce^x$.

21. The auxiliary equation is $r^2 - 6r + 34 = 0$ with roots $r = 3 \pm 5i$. Thus the solution is
$y(x) = e^{3x}(c_1\cos 5x + c_2\sin 5x)$.

23. The auxiliary equation is $2r^2 + r - 1 = 0$ with roots $r_1 = -1$, $r_2 = 1/2$. Thus
$y(x) = c_1 e^{-x} + c_2 e^{x/2}$.

25. $y_c(x) = e^{-x}(c_1 + c_2 x)$ and try $y_p(x) = A\cos 3x + B\sin 3x$. Then we need
$-9A\cos 3x - 9B\sin 3x - 6A\sin 3x + 6B\cos 3x + A\cos 3x + B\sin 3x$
$= \sin 3x$, or $6B - 8A = 0$ and $-6A - 8B = 1$. Thus $A = -3/50$, $B = -2/25$ and the
general solution is $y(x) = e^{-x}(c_1 + c_2 x) - (3/50)\cos 3x - (2/25)\sin 3x$.

27. $y_c(x) = c_1\cos(3x/2) + c_2\sin(3x/2)$ and try $y_p(x) = Ax^2 + Bx + C$.
Then we need $8A + 9Ax^2 + 9Bx + 9C = 2x^2 - 3$. Thus $A = 2/9$, $B = 0$,
$C = -43/81$, and the solution is $y(x) = c_1\cos(3x/2) + c_2\sin(3x/2) + (2/9)x^2 - (43/81)$.

Chapter 15 Review

29. $y_c(x) = c_1 e^x + c_2 e^{2x}$ so try $y_p(x) = Axe^{2x}$.
Then we need $(4Ax + 4A)e^{2x} + (-6Ax - 3A)e^{2x} + 2Axe^{2x} = e^{2x}$, or $A = 1$. Thus the solution is $y(x) = c_1 e^x + c_2 e^{2x} + xe^{2x}$.

31. Since linear, $I(x) = e^{\int dx} = e^x$ and multiplying the differential equation by $I(x)$:
$e^x y' + e^x y = \sqrt{x} \Rightarrow (e^x y)' = \sqrt{x} \Rightarrow y(x) = e^{-x}[\int (\sqrt{x})dx + c]$
$= e^{-x}[(2/3)x^{3/2} + c]$. But $3 = y(0) = c$, so the solution
to the initial-value problem is $y(x) = e^{-x}[(2/3)x^{3/2} + 3]$.

33. The auxiliary equation is $r^2 + 6r = 0$ and the general solution is
$y(x) = c_1 + c_2 e^{-6x} = k_1 + k_2 e^{-6(x-1)}$. But $3 = y(1) = k_1 + k_2$ and
$12 = y'(1) = -6k_2$. Thus $k_2 = -2$, $k_1 = 5$ and the solution is $y(x) = 5 - 2e^{-6(x-1)}$.

35. The auxiliary equation is $r^2 - 5r + 4 = 0$ and the general solution is $y(x) = c_1 e^x + c_2 e^{4x}$.
But $0 = y(0) = c_1 + c_2$ and $1 = y'(0) = c_1 + 4c_2$, so the solution is
$y(x) = (1/3)(e^{4x} - e^x)$.

37. The curves $kx^2 + y^2 = 1$ form a family of ellipses for $k > 0$, a family of hyperbolas
for $k < 0$, and two parallel lines $y = \pm 1$ for $k = 0$. Differentiating gives $2kx + 2yy' = 0$
or $y' = -kx/y = -(1-y^2)x/yx^2 = (y^2-1)/(xy)$. Thus the orthogonal trajectories
must satisfy $y' = -xy/(y^2-1)$ (for $k \neq 0$) or $(y^2/2) - \ln|y| = (-x^2/2) + c$.
(For $k = 0$, the orthogonal trajectories are given by $x = c_1$ for c_1 an arbitrary
constant.)

39. Assuming $y(x) = \sum_{n=0}^{\infty} a_n x^n$, $y''(x) = \sum_{n=0}^{\infty} n(n-1)a_n x^{n-2} = \sum_{n=0}^{\infty} (n+2)(n+1)a_{n+2} x^n$
and the differential equation becomes $\sum_{n=0}^{\infty} [(n+2)(n+1)a_{n+2} + (n+1)a_n]x^n = 0$.
Thus the recurrence relation is $a_{n+2} = -a_n/(n+2)$ for $n = 0, 1, 2, \ldots$. But
$a_0 = y(0) = 0$ and the recurrence relation implies $a_{2n} = 0$ for $n = 0, 1, 2, \ldots$. Also
$a_1 = y'(0) = 1$, so $a_3 = -1/3$, $a_5 = -(-1)/3 \cdot 5$,
$a_7 = (-1^3/3 \cdot 5 \cdot 7 = (-1)^3(2^3)(3!)/7!, \ldots, a_{2n+1} = (-1)^n(2^n)(n!)/(2n+1)!$
for $n = 0, 1, 2, \ldots$. Thus the solution to the initial-value problem
is $y(x) = \sum_{n=0}^{\infty} a_n x^n = \sum_{n=0}^{\infty} [(-1)^n(2^n)(n!)x^{2n+1}/(2n+1)!]$.

41. Here the initial-value problem is $2Q'' + 40Q' + 400Q = 12$, $Q(0) = 0.01$, $Q'(0) = 0$.
Then $Q_c(t) = e^{-10t}(c_1 \cos 10t + c_2 \sin 10t)$ and try $Q_p(t) = A$. Thus the general
solution is $Q(t) = e^{-10t}(c_1 \cos 10t + c_2 \sin 10t) + (3/100)$. But $0.01 = Q(0) = c_1 + 0.03$
and $0 = Q'(0) = -10c_1 + 10c_2$, so $c_1 = -0.02 = c_2$. Hence the charge is given by
$Q(t) = -0.02e^{-10t}(\cos 10t + \sin 10t) + 0.03$.

43. (a) The differential equation is $\frac{dP}{dt} - kP = -m$, which is linear with integrating factor $I(t) = e^{-\int k\,dt} = e^{-kt}$. So the equation becomes $e^{-kt}\frac{dP}{dt} - ke^{-kt}P = -me^{-kt} \Rightarrow \frac{d}{dt}(e^{-kt}P) = -me^{-kt} \Rightarrow e^{-kt}P(t) = \frac{m}{k}e^{-kt} + C \Rightarrow P(t) = \frac{m}{k} + Ce^{kt}$. But $P(0) = \frac{m}{k} + C = P_0 \Rightarrow C = P_0 - \frac{m}{k}$ so $P(t) = \frac{m}{k} + (P_0 - \frac{m}{k})e^{kt}$.

(b) There will be an exponential expansion $\Leftrightarrow P_0 - \frac{m}{k} > 0 \Leftrightarrow m < kP_0$.

(c) The population will be constant $\Leftrightarrow P_0 - \frac{m}{k} = 0 \Leftrightarrow m = kP_0$. It will decline $\Leftrightarrow P_0 - \frac{m}{k} < 0 \Leftrightarrow m > kP_0$.

APPLICATIONS PLUS (after Chapter 15)

1. (a) Since we are assuming that the earth is a solid sphere of uniform density, we can calculate the density ρ as follows: $\rho = \frac{\text{mass of earth}}{\text{volume of earth}} = \frac{M}{\frac{4}{3}\pi R^3}$. If V_r is the volume of that portion of the earth lying within a distance r of the center, then $V_r = \frac{4}{3}\pi r^3$ and $M_r = \rho V_r = \frac{Mr^3}{R^3}$. Thus $F_r = \frac{-GM_r m}{r^2} = \frac{-GMm}{R^3}r$.

    (b) The particle is acted upon by a varying gravitational force during its motion. By Newton's Second Law of Motion, $m\frac{d^2y}{dt^2} = F_y = \frac{-GMm}{R^3}y$, so $y''(t) = -k^2 y(t)$ where $k^2 = \frac{GM}{R^3}$. At the surface, $-mg = F_R = \frac{-GMm}{R^2}$, so $g = \frac{GM}{R^2}$. Therefore $k^2 = \frac{g}{R}$.

    (c) The differential equation $y'' + k^2 y = 0$ has auxiliary equation $r^2 + k^2 = 0$. (This is the r of Section 15.5, not the r measuring distance from the earth's center.) The roots of the auxiliary equation are ik and $-ik$, so by (15.33), the general solution of our differential equation for y is $y(t) = c_1 \cos kt + c_2 \sin kt$. It follows that $y'(t) = -c_1 k \sin kt + c_2 k \cos kt$. Now $y(0) = R$ and $y'(0) = 0$, so $c_1 = R$ and $c_2 k = 0$. Thus $y(t) = R \cos kt$ and $y'(t) = -kR \sin kt$. This is simple harmonic motion (see Section 15.7) with amplitude R, frequency k, and phase angle 0. The period $T = 2\pi/k$. $R \approx 3960$ mi $= (3960)(5280)$ ft and $g = 32$ ft/s^2, so $k = \sqrt{g/R} \approx 1.24 \times 10^{-3}$ s^{-1} and $T = 2\pi/k \approx 5079$ s ≈ 85 min.

    (d) $y(t) = 0 \Leftrightarrow \cos kt = 0 \Leftrightarrow kt = \frac{\pi}{2} + \pi n$ for some integer n $\Rightarrow y'(t) = -kR \sin\left(\frac{\pi}{2} + \pi n\right) = \pm kR$. Thus the particle passes through the center of the earth with speed $kR \approx 4.899$ mi/s $\approx 17,600$ mi/h.

3. The figure shows the missile as viewed from above, with the x-axis pointing due east and the y-axis pointing due north. The angle θ varies with time. When the missile is fired, it points due west, but the northerly wind pushes the missile in the positive y-direction. As the homing device turns it toward the target, the missile's velocity acquires a component in the negative y-direction. That component counteracts the wind and carries the missile back toward the x-axis and the target.

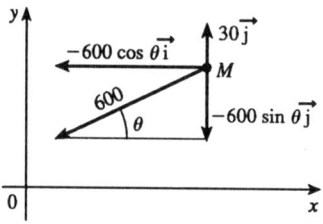

Applications Plus

(a) From the figure, we see that the effective velocity of the missile M, taking the wind into consideration, is $\vec{v} = -600\cos\theta\,\vec{i} + (30 - 600\sin\theta)\vec{j}$. Thus $\frac{dx}{dt} = -600\cos\theta$ and $\frac{dy}{dt} = 30 - 600\sin\theta$.

(b) $\frac{dy}{dx} = \frac{dy/dt}{dx/dt} = \frac{30 - 600\sin\theta}{-600\cos\theta}$. Since $x = r\cos\theta = \sqrt{x^2 + y^2}\cos\theta$, we have $\cos\theta = \frac{x}{\sqrt{x^2+y^2}}$ [for $(x,y) \neq (0,0)$]. Similarly, $\sin\theta = \frac{y}{\sqrt{x^2+y^2}}$. Thus

$$\frac{dy}{dx} = \frac{30 - 600(y/\sqrt{x^2+y^2})}{-600(x/\sqrt{x^2+y^2})} = \frac{30\sqrt{x^2+y^2} - 600y}{-600x} = \frac{\sqrt{x^2+y^2} - 20y}{-20x}.$$

(c) From (b), $\frac{dy}{dx} = \frac{\sqrt{1 + (\frac{y}{x})^2} - 20(\frac{y}{x})}{-20}$, which is a homogeneous differential equation.

Set $v = \frac{y}{x}$, then $y = vx \Rightarrow \frac{dy}{dx} = v + x\frac{dv}{dx}$.

Then the differential equation becomes

$v + x\frac{dv}{dx} = \frac{\sqrt{1+v^2} - 20v}{-20} \Rightarrow \frac{-20\,dv}{\sqrt{1+v^2}} = \frac{dx}{x}$. Using Formula 25 in the Table of Integrals (or substituting $v = \tan t$ and integrating), we get

$-20\ln(v + \sqrt{1+v^2}) = \ln x + C$. When $x = 50$, $y = 0$ and $v = y/x = 0$. Thus $C = -\ln 50$. Now $-20\ln(v + \sqrt{1+v^2}) = \ln x - \ln 50$. But

$v + \sqrt{1+v^2} = \frac{y}{x} + \sqrt{1 + (\frac{y}{x})^2} = \frac{y + \sqrt{x^2+y^2}}{x}$, so

$-20[\ln(y + \sqrt{x^2+y^2}) - \ln x] = \ln x - \ln 50$. That is,

$19\ln x + \ln 50 = 20\ln(y + \sqrt{x^2+y^2})$ or $\ln(50x^{19}) = \ln(y + \sqrt{x^2+y^2})^{20}$. Thus $(y + \sqrt{x^2+y^2})^{20} = 50x^{19}$ and $y + \sqrt{x^2+y^2} = 50^{1/20}x^{19/20}$ is the equation describing the path of the missile.

5. (a) We have $P' = -mP + kC$, where $C(t) = S - P(t)$, and $P(0) = S$. Thus $P' = -mP + k(S - P) = -(m+k)P + kS$, or $P' + (m+k)P = kS$. Following the method of Section 15.2, we multiply both sides by the integrating factor $e^{(m+k)t}$ to get $\frac{d}{dt}\left(e^{(m+k)t}P\right) = kSe^{(m+k)t} \Rightarrow e^{(m+k)t}P = \frac{kS}{m+k}e^{(m+k)t} + C$. When $t = 0$, this says $S = \frac{kS}{m+k} + C$, so $C = S - \frac{kS}{m+k} = \frac{mS}{m+k}$ and so $e^{(m+k)t}P = \frac{kS}{m+k}e^{(m+k)t} + \frac{mS}{m+k}$. Therefore $P(t) = \frac{kS}{m+k} + \frac{mS}{m+k}e^{-(m+k)t}$.

$Q = \lim_{t\to\infty} P(t) = \frac{kS}{m+k}$.

Applications Plus

(b) $C(t) = S - P(t) = S - \dfrac{kS}{m+k} - \dfrac{mS}{m+k}e^{-(m+k)t} = \dfrac{mS}{m+k} - \dfrac{mS}{m+k}e^{-(m+k)t}$
$= \dfrac{mS}{m+k}(1 - e^{-(m+k)t})$, so $\lim\limits_{t \to \infty} C(t) = \dfrac{mS}{m+k}$.

(c) $g(t) = \ln\left(\dfrac{P(t)}{Q} - 1\right) = \ln\left(1 + \dfrac{m}{k}e^{-(m+k)t} - 1\right) = \ln\left(\dfrac{m}{k}e^{-(m+k)t}\right)$
$= -(m+k)t + \ln\left(\dfrac{m}{k}\right)$. If $g(t) = at + b$, then $a = -(m+k)$ and $b = \ln\left(\dfrac{m}{k}\right)$. The second relation says $\dfrac{m}{k} = e^b$, so $m = ke^b$. Now $a = -(m+k) = -ke^b - k$
$= -k(e^b + 1)$. Therefore $k = \dfrac{-a}{e^b + 1}$ and $m = ke^b = -\dfrac{ae^b}{e^b + 1}$.

7. (a) Consider a light ray emanating from the origin with angle of inclination θ.
Suppose that the light ray intersects the graph of f at the point P with coordinates (x, y). Since P lies on the ray and also on the graph, its coordinates satisfy the relations (1) $y = (\tan\theta)x$ and $y = f(x)$. Let ϕ be the angle of inclination of the tangent line to the graph of f at P. Then ϕ is also the angle between the tangent line and the reflected ray (since the reflected ray is parallel to the x-axis).

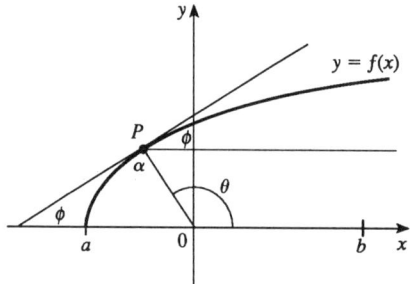

We have (2) $f'(x) = \tan\phi$. Since the angle of incidence equals the angle of reflection at P, we have (3) $\alpha = \phi$. Also, we see from the figure that $\theta = \alpha + \phi$ (4). By (3) and (4), $\theta = 2\phi$, so the coordinates of P satisfy
$\dfrac{f(x)}{x} \stackrel{(1)}{=} \tan\theta = \tan 2\phi = \dfrac{2\tan\phi}{1 - \tan^2\phi} \stackrel{(2)}{=} \dfrac{2f'(x)}{1 - [f'(x)]^2}$. Solving for $f'(x)$, we get
$f(x)[f'(x)]^2 + 2xf'(x) - f(x) = 0$ and
$f'(x) = \dfrac{-2x \pm \sqrt{4x^2 + 4[f(x)]^2}}{2f(x)} = \dfrac{-x \pm \sqrt{x^2 + [f(x)]^2}}{f(x)}$. This condition holds at all points P of the graph of f, so solving for f will determine the nature of the graph.

Writing $y = f(x)$, we have $\dfrac{dy}{dx} = \dfrac{-x \pm \sqrt{x^2 + y^2}}{y}$.

Applications Plus

(b) Rewrite the equation in the form $\dfrac{x}{\sqrt{x^2+y^2}} + \dfrac{y}{\sqrt{x^2+y^2}}\dfrac{dy}{dx} = \pm 1$. This equation is exact by Theorem 15.14. (The theorem applies even when the right–hand side of the equation is a nonzero constant.) Integrating, we get $\sqrt{x^2+y^2} = \pm x + K$. By hypothesis, f is non–negative, so $y(0) \geq 0$. Taking $x = 0$, we get $K = \sqrt{[y(0)]^2} = y(0) \geq 0$. Squaring, we get $x^2 + y^2 = x^2 \pm 2Kx + K^2$ or $y^2 = \pm 2Kx + K^2$. Set $C = \pm K$. Then our result becomes $y^2 = 2Cx + C^2$ where C is an arbitrary constant. When $C > 0$, this is the top half of a parabola with focus at the origin, opening to the right. When $C < 0$, the graph is the top half of a parabola with focus at the origin, opening to the left. In both cases, the vertex is at $\left(-\dfrac{C}{2}, 0\right)$. <u>Another method</u>: The differential equation is also homogeneous, so substitute $v = y/x$.

Appendix D

EXERCISES D

1. Guessing: Calculator results (8 digits, 3 spare digits; don't forget to switch to radians!), followed by microcomputer results (16 digits, base 16).

    The last column shows the values of the sixth-degree Taylor polynomial for $f(x) = \csc^2 x - x^{-2}$ near $x = 0$ (note that the second arrangement of Taylor's polynomial is easier to use with a calculator):

    $$T_6(x) = \tfrac{1}{3} + \tfrac{1}{15}x^2 + \tfrac{2}{189}x^4 + \tfrac{1}{675}x^6 \equiv ((\tfrac{1}{675}\,x^2 + \tfrac{2}{189})\,x^2 + \tfrac{1}{15})\,x^2 + \tfrac{1}{3}$$

x	$f(x)_{\text{calculator}}$	$f(x)_{\text{computer}}$	$f(x)_{\text{Taylor}}$
0.1	0.33400107	.3340010596844678	0.33400106
0.01	0.333341	.3333400001065456	0.33334000
0.001	0.3334	.3333334000781178	0.33333340
0.0001	0.34	.3333333283662796	0.33333333
0.00001	2.0	.3333339691162109	0.33333333
0.000001	100 or 200	.33349609375	0.33333333
0.0000001	10000 or 20000	.328125	0.33333333
0.00000004	-230000	.5	0.33333333
0.00000001	1000000	0	0.33333333

    We see that already at $x = 0.0001$, the results deteriorate, and for smaller x, they become meaningless. The different results "100 or 200" etc. depended on whether we calculated $((\sin x)^2)^{-1}$ or $((\sin x)^{-1})^2$.

    With the microcomputer, the results become meaningless at 4×10^{-8}, when about half of the device's digits are used to store x (compare with the calculator result!) A detailed analysis reveals that the values of the function should be greater than $\tfrac{1}{3}$, but at $x = 0.0001$, the value is less than $\tfrac{1}{3}$.

    The polynomial $T_6(x)$ was obtained by patient simplification of the expression for $f(x)$, starting with $\sin^2(x) = \tfrac{1}{2}(1 - \cos 2x)$, where

    $$\cos 2x = 1 - (2x)^2/2! + (2x)^4/4! - \ldots - (2x)^{10}/10! + R_{12}(x).$$

    Consequently, the exact value of the limit is $T_6(0) = \tfrac{1}{3}$. It can also be obtained by several applications of l'Hospital's Rule to the expression

    $$f(x) = \frac{x^2 - \sin^2 x}{x^2 \sin^2 x}$$

    with intermediate simplifications.

Appendix D

3. From $$f(x) = \frac{x^{25}}{(1.0001)^x} \quad \text{(we may assume } x > 0; \text{ why?)}$$
   we have $$\ln f(x) = 25 \ln x - x \ln (1.0001)$$
   and $$\frac{f'(x)}{f(x)} = \frac{25}{x} - \ln (1.0001).$$

   This derivative, as well as the derivative $f'(x)$ itself, is positive for
   $$0 < x < x_0 = \frac{25}{\ln (1.0001)} \approx 249{,}971.015, \text{ and negative for } x > x_0.$$
   Hence the maximum value of $f(x)$ is $f(x_0) = \frac{x_0^{25}}{(1.0001)^{x_0}}$, a number too large to be calculated directly. Using decimal logarithms,
   $$\log_{10} f(x_0) \approx 124.08987757, \text{ so that } f(x_0) \approx \underline{1.229922 \times 10^{124}}.$$
   The limit $\lim_{x \to \infty} f(x) = 0$; it would be wasteful and inelegant to use l'Hospital's Rule twenty-five times since we can transform $f(x)$ into
   $$f(x) = \left[\frac{x}{(1.0001)^{x/25}}\right]^{25}$$
   and the inside expression needs just one application of l'Hospital's Rule to yield limit 0.

5. For $f(x) = \ln \ln x$ with $x \in [a, b]$, $a = 10^9$, and $b = 10^9 + 1$, we need
   $$f'(x) = \frac{1}{x \ln x}, \quad f''(x) = -\frac{\ln x + 1}{x^2 (\ln x)^2}$$
   In part (a): $f'(b) < D < f'(a)$, where

   $f'(a) \approx 4.82549\,42434 \times 10^{-11}$ $\qquad$ $f'(b) \approx 4.82549\,42383 \times 10^{-11}$

   In part (b): Let us estimate $f'(b) - f'(a) = (b - a)f''(c_1) = f''(c_1)$. Since f'' increases (its absolute value decreases), we have
   $$|f'(b) - f'(a)| < |f''(a)| \approx 5.0583 \times 10^{-20}.$$

7. (a) The 11-digit calculator value of $192 \sin \frac{\pi}{96}$ is $6.28206\,39018$, while the value (on the same device) of p before rationalization is $6.28206\,3885$, less than the trigonometric result by 1.68×10^{-8}.

   (b) The result is
   $$p = \frac{96}{\sqrt{2 + \sqrt{3}} \cdot \sqrt{2 + \sqrt{2 + \sqrt{3}}} \cdot \sqrt{2 + \sqrt{2 + \sqrt{2 + \sqrt{3}}}} \cdot \sqrt{2 + \sqrt{2 + \sqrt{2 + \sqrt{2 + \sqrt{3}}}}}}$$

   but of course we can avoid repetitious calculations by storing intermediate results in a memory:
   $$p_1 = \sqrt{2 + \sqrt{3}}, \quad p_2 = \sqrt{2 + p_1}, \quad p_3 = \sqrt{2 + p_2}, \quad p_4 = \sqrt{2 + p_3}, \quad p = \frac{96}{p_1 p_2 p_3 p_4}.$$
   According to this formula, a calculator gives $p \approx 6.28206\,39016$, in agreement with the

trigonometric result within 2×10^{-10}.

A microcomputer program working with 16 digits gives $p \approx 6.28206\,39017\,81019$ with an error before rationalization about 4.1×10^{-14}, and after rationalization about 8.9×10^{-16}, also a gain of about 2 digits of accuracy.

9. (a) Let $A = \left(\dfrac{27q + \sqrt{729q^2 + 108p^3}}{2}\right)^{\frac{1}{3}}$ and $B = \left(\dfrac{27q - \sqrt{729q^2 + 108p^3}}{2}\right)^{\frac{1}{3}}$.

Then $A^3 + B^3 = 27q$ and $AB = \left(\dfrac{729q^2 - (729q^2 + 108p^3)}{4}\right)^{\frac{1}{3}} = -3p$.

Substitute into the formula $A + B = \dfrac{A^3 + B^3}{A^2 - AB + B^2}$ where we replace B by $-3p/A$:

$$x = \tfrac{1}{3}(A+B) = \dfrac{27q/3}{\left(\dfrac{27q + \sqrt{729q^2 + 108p^3}}{2}\right)^{\frac{2}{3}} + 3p + 9p^2\left(\dfrac{27q + \sqrt{729q^2 + 108p^3}}{2}\right)^{-\frac{2}{3}}}$$

which almost yields the given formula; since replacing q by $-q$ results in replacing x by $-x$, a simple discussion of the cases $q > 0$ and $q < 0$ allows us to replace q by $|q|$ in the denominator expression, so that it involves only positive numbers. The problems mentioned in the introduction to this exercise have disappeared.

(b) A direct attack works best here. To save space, let $\alpha = 2 + \sqrt{5}$, so we can rationalize, using $\alpha^{-1} = -2 + \sqrt{5}$ and $\alpha - \alpha^{-1} = 4$ (check it!):

$$u = \dfrac{4}{\alpha^{2/3} + 1 + \alpha^{-2/3}} \cdot \dfrac{\alpha^{1/3} - \alpha^{-1/3}}{\alpha^{1/3} - \alpha^{-1/3}} = \dfrac{4(\alpha^{1/3} - \alpha^{-1/3})}{\alpha - \alpha^{-1}} = \alpha^{1/3} - \alpha^{-1/3}$$

and we cube the expression for u:

$$u^3 = \alpha - 3\alpha^{1/3} + 3\alpha^{-1/3} - \alpha^{-1} = 4 - 3u,$$

$$u^3 + 3u - 4 = (u-1)(u^2 + u + 4) = 0,$$

so that the only real root is $u = 1$.

A check using the formula from part (a): $p = 3$, $q = -4$, so $729q^2 + 108p^3 = 14580 = 54^2 \times 5$, and

$$x = \dfrac{36}{\left(54 + 27\sqrt{5}\right)^{\frac{2}{3}} + 9 + 81\left(54 + 27\sqrt{5}\right)^{-\frac{2}{3}}},$$

which simplifies to the given form after reduction by 9.

11. Proof that $\lim_{n \to \infty} a_n = 0$:

From $1 \leq e^{1-x} \leq e$ it follows that $x^n \leq e^{1-x}x^n \leq x^n e$, and integration gives

$$\tfrac{1}{n+1} = \int_0^1 x^n\,dx \leq \int_0^1 e^{1-x}x^n\,dx \leq \int_0^1 x^n e\,dx = \tfrac{e}{n+1},$$

Appendix D

that is,
$$\tfrac{1}{n+1} \le a_n \le \tfrac{e}{n+1}$$

and since
$$\lim_{n\to\infty} \tfrac{1}{n+1} = \lim_{n\to\infty} \tfrac{e}{n+1} = 0,$$

it follows from the Squeeze Theorem that $\lim_{n\to\infty} a_n = 0$.

Of course, the expression $\tfrac{1}{n+1}$ on the left side could have been replaced by 0 and the proof would remain valid.

Calculations: An 11-digit scientific pocket calculator gives

n	a_n (red. formula)
0	1.71828 18284
1	0.71828 18284
2	0.43656 36568
3	0.30969 09704
4	0.23876 38816
5	0.19381 94080
6	0.16291 64480
7	0.14041 51360
8	0.12332 10880
9	0.10988 97920
10	0.09889 79200
11	0.08787 71200
12	0.05452 54400
13	$-0.29116\ 92800$
14	$-5.07636\ 992$
15	$-77.14554\ 88$
16	$-1235.32878\ 08$
17	$-21001.58927\ 4$
18	-378029.60693
19	-7182563.5317
20	-143651271.63

It is clear from step 13 that the values calculated from the direct reduction formula will diverge to $-\infty$.

13. We can start with expressing e^x and e^{-x} in terms of $E(x) = (e^x - 1)/x$ ($x \ne 0$), where $E(0) = 1$ to make E continuous at 0 (by L'Hospital's Rule). Namely,
$$e^x = 1 + xE(x), \quad e^{-x} = 1 - xE(-x)$$

and
$$\sinh x = \frac{1 + xE(x) - (1 - xE(-x))}{2} = \tfrac{1}{2}x\bigl[E(x) + E(-x)\bigr]$$

where the addition involves only positive numbers $E(x)$ and $E(-x)$, thus presenting no loss of accuracy due to subtraction.

Another form which calls the function E only once: we write
$$\sinh x = \frac{(e^x)^2 - 1}{2e^x} = \frac{(1 + xE(x))^2 - 1}{2(1 + xE(x))} = \frac{x\bigl[1 + \tfrac{1}{2}|x|\,E(|x|)\bigr]E(|x|)}{1 + |x|\,E(|x|)},$$

taking advantage of the fact that $\tfrac{\sinh x}{x}$ is an even function, hence replacing x by $|x|$ does not change its value.

Appendix F (Complex Numbers)

EXERCISES F

1. $(3+2i)+(7-3i)=(3+7)+(2-3)i=10-i.$

3. $(3-i)(4+i)=12+3i-4i-(-1)=13-i.$

5. $\overline{12+7i}=12-7i.$

7. $\dfrac{2+3i}{1-5i}=\dfrac{2+3i}{1-5i}\cdot\dfrac{1+5i}{1+5i}=\dfrac{2+10i+3i+15(-1)}{1-25(-1)}=\dfrac{-13+13i}{26}=-\tfrac{1}{2}+\tfrac{1}{2}i.$

9. $\dfrac{1}{1+i}=\dfrac{1}{1+i}\cdot\dfrac{1-i}{1-i}=\dfrac{1-i}{1-(-1)}=\dfrac{1-i}{2}=\tfrac{1}{2}-\tfrac{1}{2}i.$

11. $i^3=i^2\cdot i=(-1)i=-i.$

13. $\sqrt{-25}=\sqrt{25}\,i=5i.$

15. $\overline{3+4i}=3-4i,\ |3+4i|=\sqrt{3^2+4^2}=\sqrt{25}=5.$

17. $\overline{-4i}=\overline{0-4i}=0+4i=4i,\ |-4i|=\sqrt{0^2+(-4)^2}=4.$

19. $4x^2+9=0\Leftrightarrow 4x^2=-9\Leftrightarrow x^2=-\tfrac{9}{4}\Leftrightarrow x=\pm\sqrt{-\tfrac{9}{4}}=\pm\sqrt{\tfrac{9}{4}}\,i=\pm\tfrac{3}{2}i.$

21. By the quadratic formula,
$$x^2-8x+17=0\Leftrightarrow x=\dfrac{8\pm\sqrt{8^2-4(1)(17)}}{2(1)}=\dfrac{8\pm\sqrt{-4}}{2}=\dfrac{8\pm 2i}{2}=4\pm i.$$

23. By the quadratic formula,
$$z^2+z+2=0\Leftrightarrow z=\dfrac{-1\pm\sqrt{1-4(1)(2)}}{2(1)}=\dfrac{-1\pm\sqrt{-7}}{2}=-\tfrac{1}{2}\pm\tfrac{\sqrt{7}}{2}i.$$

25. $r=\sqrt{(-3)^2+3^2}=3\sqrt{2},\ \tan\theta=\tfrac{3}{-3}=-1\Rightarrow\theta=\tfrac{3}{4}\pi$ (since the given number is in the second quadrant). Therefore $-3+3i=3\sqrt{2}(\cos\tfrac{3\pi}{4}+i\sin\tfrac{3\pi}{4}).$

27. $r=\sqrt{3^2+4^2}=5,\ \tan\theta=\tfrac{4}{3}\Rightarrow\theta=\tan^{-1}\tfrac{4}{3}$ (since the given number is in the second quadrant). Therefore $3+4i=5[\cos(\tan^{-1}\tfrac{4}{3})+i\sin(\tan^{-1}\tfrac{4}{3})].$

29. For $z=\sqrt{3}+i,\ r=\sqrt{[\sqrt{3}]^2+1^2}=2,$ and $\tan\theta=\tfrac{1}{\sqrt{3}}\Rightarrow\theta=\tfrac{\pi}{6}$ so that $z=2(\cos\tfrac{\pi}{6}+i\sin\tfrac{\pi}{6}).$
For $w=1+\sqrt{3}i,\ r=2,$ and $\tan\theta=\sqrt{3}\Rightarrow\theta=\tfrac{\pi}{3}$ so that $w=2(\cos\tfrac{\pi}{3}+i\sin\tfrac{\pi}{3}).$
Therefore $zw=2\cdot 2[\cos(\tfrac{\pi}{6}+\tfrac{\pi}{3})+i\sin(\tfrac{\pi}{6}+\tfrac{\pi}{3})]=4(\cos\tfrac{\pi}{2}+i\sin\tfrac{\pi}{2}),$
$z/w=\tfrac{2}{2}[\cos(\tfrac{\pi}{6}-\tfrac{\pi}{3})+i\sin(\tfrac{\pi}{6}-\tfrac{\pi}{3})]=\cos(-\tfrac{\pi}{6})+i\sin(-\tfrac{\pi}{6}),$ and
$1=1+0i=\cos 0+i\sin 0\Rightarrow 1/z=\tfrac{1}{2}[\cos(0-\tfrac{\pi}{6})+i\sin(0-\tfrac{\pi}{6})]=\tfrac{1}{2}[\cos(-\tfrac{\pi}{6})+i\sin(-\tfrac{\pi}{6})].$

31. For $z=2\sqrt{3}-2i,\ r=4,\ \tan\theta=\tfrac{-2}{2\sqrt{3}}=-\tfrac{1}{\sqrt{3}}\Rightarrow\theta=-\tfrac{\pi}{6}\Rightarrow z=4[\cos(-\tfrac{\pi}{6})+i\sin(-\tfrac{\pi}{6})].$
For $w=-1+i,\ r=\sqrt{2},\ \tan\theta=\tfrac{1}{-1}=-1\Rightarrow\theta=\tfrac{3\pi}{4}\Rightarrow z=\sqrt{2}(\cos\tfrac{3\pi}{4}+i\sin\tfrac{3\pi}{4}).$
Therefore $zw=4\sqrt{2}[\cos(-\tfrac{\pi}{6}+\tfrac{3\pi}{4})+i\sin(-\tfrac{\pi}{6}+\tfrac{3\pi}{4})]=4\sqrt{2}(\cos\tfrac{7\pi}{12}+i\sin\tfrac{7\pi}{12}),$
$z/w=\tfrac{4}{\sqrt{2}}[\cos(-\tfrac{\pi}{6}-\tfrac{3\pi}{4})+i\sin(-\tfrac{\pi}{6}-\tfrac{3\pi}{4})]=\tfrac{4}{\sqrt{2}}[\cos(-\tfrac{11\pi}{12})+i\sin(-\tfrac{11\pi}{12})]$

$= 2\sqrt{2}(\cos\frac{13\pi}{12} + i\sin\frac{13\pi}{12})$, and $1 = 1 + 0i = \cos 0 + i\sin 0$
$\Rightarrow 1/z = \frac{1}{4}[\cos(0 - (-\frac{\pi}{6})) + i\sin(0 - (-\frac{\pi}{6}))] = \frac{1}{4}(\cos\frac{\pi}{6} + i\sin\frac{\pi}{6})$.

33. For $z = 1 + i$, $r = \sqrt{2}$, $\tan\theta = \frac{1}{1} = 1 \Rightarrow \theta = \frac{\pi}{4} \Rightarrow 1 + i = \sqrt{2}(\cos\frac{\pi}{4} + i\sin\frac{\pi}{4})$.
So by De Moivre's Theorem,
$(1+i)^{20} = [\sqrt{2}(\cos\frac{\pi}{4} + i\sin\frac{\pi}{4})]^{20} = [2^{1/2}]^{20}(\cos\frac{20\pi}{4} + i\sin\frac{20\pi}{4})$
$= 2^{10}(\cos 5\pi + i\sin 5\pi) = 2^{10}(-1 + i(0)) = -2^{10} = -1024$.

35. For $z = 2\sqrt{3} + 2i$, $r = 4$, $\tan\theta = \frac{2}{2\sqrt{3}} = \frac{1}{\sqrt{3}} \Rightarrow \theta = \frac{\pi}{6} \Rightarrow 2\sqrt{3} + 2i = 4(\cos\frac{\pi}{6} + i\sin\frac{\pi}{6})$.
So by De Moivre's Theorem, $(2\sqrt{3} + 2i)^5 = [4(\cos\frac{\pi}{6} + i\sin\frac{\pi}{6})]^5 = 4^5(\cos\frac{5\pi}{6} + i\sin\frac{5\pi}{6})$
$= 4^5(-\frac{\sqrt{3}}{2} + i(0.5)) = -512\sqrt{3} + 512i$.

37. $1 = 1 + 0i = (\cos 0 + i\sin 0)$. Using equation (3) with $r = 1$, $n = 8$, and $\theta = 0$ we have
$w_k = 1^{1/8}\left[\cos\left(\frac{0 + 2k\pi}{8}\right) + i\sin\left(\frac{0 + 2k\pi}{8}\right)\right] = \cos\frac{k\pi}{4} + i\sin\frac{k\pi}{4}$, where $k = 0, 1, 2, \ldots, 7$.
$w_0 = (\cos 0 + i\sin 0) = 1$
$w_1 = (\cos\frac{\pi}{4} + i\sin\frac{\pi}{4}) = \frac{1}{\sqrt{2}} + \frac{1}{\sqrt{2}}i$
$w_2 = (\cos\frac{\pi}{2} + i\sin\frac{\pi}{2}) = i$
$w_3 = (\cos\frac{3\pi}{4} + i\sin\frac{3\pi}{4}) = -\frac{1}{\sqrt{2}} + \frac{1}{\sqrt{2}}i$
$w_4 = (\cos\pi + i\sin\pi) = -1$
$w_5 = (\cos\frac{5\pi}{4} + i\sin\frac{5\pi}{4}) = -\frac{1}{\sqrt{2}} - \frac{1}{\sqrt{2}}i$
$w_6 = (\cos\frac{3\pi}{2} + i\sin\frac{3\pi}{2}) = -i$
$w_7 = (\cos\frac{7\pi}{4} + i\sin\frac{7\pi}{4}) = \frac{1}{\sqrt{2}} - \frac{1}{\sqrt{2}}i$.

39. $0 = 0 + i = \cos\frac{\pi}{2} + i\sin\frac{\pi}{2}$. Using equation (3) with $r = 1$, $n = 3$, $\theta = \frac{\pi}{2}$ we have
$w_k = 1^{1/3}\left[\cos\left(\frac{\frac{\pi}{2} + 2k\pi}{3}\right) + i\sin\left(\frac{\frac{\pi}{2} + 2k\pi}{3}\right)\right]$, where $k = 0, 1, 2$.
$w_0 = (\cos\frac{\pi}{6} + i\sin\frac{\pi}{6}) = \frac{\sqrt{3}}{2} + \frac{1}{2}i$
$w_1 = (\cos\frac{5\pi}{6} + i\sin\frac{5\pi}{6}) = -\frac{\sqrt{3}}{2} + \frac{1}{2}i$
$w_2 = (\cos\frac{9\pi}{6} + i\sin\frac{9\pi}{6}) = -i$.

41. Using Euler's formula (6) with $y = \frac{\pi}{2}$, $e^{i\pi/2} = \cos\frac{\pi}{2} + i\sin\frac{\pi}{2} = i$.

43. Using Euler's formula with $y = \frac{3\pi}{4}$, $e^{i3\pi/4} = \cos\frac{3\pi}{4} + i\sin\frac{3\pi}{4} = -\frac{1}{\sqrt{2}} + \frac{1}{\sqrt{2}}i$.

45. Using Equation 7 with $x = 2$ and $y = \pi$,
$e^{2+i\pi} = e^2 e^{i\pi} = e^2(\cos\pi + i\sin\pi) = e^2(-1 + 0) = -e^2$.

Appendix F (Complex Numbers)

47. Take $r = 1$ and $n = 3$ in De Moivre's Theorem to get
$[1(\cos\theta + i\sin\theta)]^3 = 1^3(\cos 3\theta + i\sin 3\theta) \Rightarrow (\cos\theta + i\sin\theta)^3 = \cos 3\theta + i\sin 3\theta \Rightarrow$
$\cos^3\theta + 3(\cos^2\theta)(i\sin\theta) + 3(\cos\theta)(i\sin\theta)^2 + (i\sin\theta)^3 = \cos 3\theta + i\sin 3\theta$
$\Rightarrow (\cos^3\theta - 3\sin^2\theta\cos\theta) + (3\sin\theta\cos^2\theta - \sin^3\theta)i = \cos 3\theta + i\sin 3\theta$. Equating real and imaginary parts gives $\cos 3\theta = \cos^3\theta - 3\sin^2\theta\cos\theta$ and $\sin 3\theta = 3\sin\theta\cos^2\theta - \sin^3\theta$.

49. $F(x) = e^{rx} = e^{(a+bi)x} = e^{ax+bxi} = e^{ax}(\cos bx + i\sin bx) = e^{ax}\cos bx + i(e^{ax}\sin bx)$
$\Rightarrow F'(x) = (e^{ax}\cos bx)' + i(e^{ax}\sin bx)'$
$= (ae^{ax}\cos bx - be^{ax}\sin bx) + i(ae^{ax}\sin bx + be^{ax}\cos bx)$
$= a[e^{ax}(\cos bx + i\sin bx)] + b[e^{ax}(-\sin bx + i\cos bx)] = ae^{rx} + b[e^{ax}(i^2\sin bx + i\cos bx)]$
$= ae^{rx} + bi[e^{ax}(\cos bx + i\sin bx)] = ae^{rx} + bie^{rx} = (a+bi)e^{rx} = re^{rx}$.